Long Range Transport of Pesticides

David A. Kurtz
Editor

LEWIS PUBLISHERS

Library of Congress Cataloging-in-Publication Data

Long range transport of pesticides / David A. Kurtz, editor.
   p. cm.
Includes bibliographical references.
1. Pesticides—Environmental aspects. 2. Atmospheric diffusion.
3. Ocean circulation. I. Kurtz, David A., 1932- .
TD196.P38L66 1990
628.5′2--dc20   90-40138
ISBN 0-87371-168-8

LEWIS PUBLISHERS, INC.
121 South Main Street, Chelsea, Michigan 48118

PRINTED IN THE UNITED STATES OF AMERICA

This book is dedicated to the memory of

DWIGHT E. GLOTFELTY

who passed away in April 1990 following a long and fruitful career in the study of pesticides: analysis, fate, and transport. He was project leader in the environmental chemistry laboratory of the Natural Resources Institute of the Agricultural Research Service of the U.S. Department of Agriculture Beltsville Agricultural Research Center.

Dr. Glotfelty was instrumental in the use of porous polyurethane foam as a sampling medium for measuring the concentration of pesticide vapors in air. This was a basis for studies of the pesticide movement in the atmosphere. His work on the volatilization of pesticides in the presence of soil moisture advanced the knowledge of transport mechanisms, and his work on the presence of pesticides in fog and aerosol particles developed this new field.

The appearance of triazines and other herbicides in groundwater is another area in which Dr. Glotfelty did early work. His work on the atmospheric transport of triazines and toxaphene from areas in southern U.S. into the Chesapeake Bay is also noteworthy, and is reported in this text.

Dwight Glotfelty will be severely missed for the contributions he would have made, had he lived longer on this, our only earth. For his research and for his encouragement, I want to dedicate this book.

David A. Kurtz is a research pesticide analytical chemist who has spent the bulk of his career at the Pennsylvania State University, College of Agriculture, Department of Entomology. He received his AB degree from Knox College, Galesburg, Illinois, and his MS and PhD degrees from Penn State's Department of Chemistry.

Dr. Kurtz's research specialization has been entirely in the trace residue analysis of pesticide and environmental contaminants in food and environmental samples. He has helped in feeding studies determining the transfer of pesticide residues in feeds to animals, birds, and fish. Dr. Kurtz also studied pesticide exposure in home gardening. Pesticide and contaminant transport in groundwater systems located in karst geological structures has been another research area.

He worked in analytical techniques development, particularly with solid phase adsorption using cellulose triacetate membrane filters, polyurethane foam plugs, and C18 silica filter surfaces. He has been active in the chemometric field of calibration for analysis where the calibration graph and estimates of amount and amount error are made by regression techniques. In this area Dr. Kurtz organized an international symposium on the Role of Chemometrics in Pesticide/Environmental Residue Analytical Determinations, held in Seattle in March 1983. The American Chemical Society text, *Trace Residue Analysis*, was based on this symposium.

He has been especially active in environmental contamination of northern fur seals from the Bering Sea and in pesticide movement in the Pacific Ocean. These interests led to the organization of the international symposium on the Long Range Transport of Pesticides and Other Toxics held in Toronto, June 1988 and this text.

Dr. Kurtz is a long term member of the American Chemical Society with active participation in the Agrochemistry, Environmental Chemistry, and Analytical Divisions. He is also an individual member of the International Union of Pure and Applied Chemistry (IUPAC), Association of Official Analytical Chemists (AOAC), and the Society of Environmental Toxicology and Chemistry (SETAC). He has lectured internationally on chemometric, groundwater, and long range transport topics.

He also has an active interest in international whitewater slalom competitions, having competed on three U.S. teams during 32 years of competition. He is a certified International Canoe Federation Slalom Expert. Kurtz has also advised the Boy Scouts on their flatwater and whitewater canoeing programs.

Acknowledgments

The editor wishes to acknowledge the memory of Donald E. H. Frear, Pennsylvania State University, for the role he played in my academic career. Dr. Frear hired me for my first work in the field of pesticide analysis. Although he made many contributions in the area of disappearance studies for farm-applied pesticides, he may be remembered most for his discovery, with colleague John George, of the presence of DDT in the tissues of Weddell Seals in Antarctica.

In more recent times, I want to thank two men who aided in many of the introductions needed to organize the symposium and this text. Terry Bidleman from the University of South Carolina and James Seiber from the University of California, Davis, alerted me to the names of scientists the world over who are active in the field of long range transport. In addition, Dwight Glotfelty, Agricultural Research Service, U. S. Department of Agriculture, provided encouragement in the development of the general thrust of the symposium text.

Finally, I wish to acknowledge the aid and encouragement of Ralph O. Mumma, Pennsylvania State University, for the release time from other duties in order to work on the text.

Preface

The seed for the symposium upon which this text is based was sewn in my seeking a means of collecting fish samples for northern fur seal research. The sources for the contamination of these marine mammals by PCBs and other chlorinated pesticides were sought by looking at food supplies of the seals. Ships were needed to collect fish that were part of the diet of the northern fur seals along their migratory route. In the process I noticed the NOAA *Oceanographer* was touring the Pacific Ocean in the summer of 1987. It was through this lead that I commenced two studies simultaneously.

The first was the examination of pesticide residues in Pacific Ocean air and water (see Chapter 11). The second was the development of the symposium on the Long Range Transport of Pesticides. I was lucky in the timing, as the forthcoming 195th National Meeting of the American Chemical Society was held jointly with the Third Chemical Congress of North America in Toronto in June 1988. This text, then, is based on the papers presented at that meeting. Further information about the organization of the symposium is found in the Introduction. The Acknowledgments mention some of the researchers who aided in the development and execution of the symposium.

The transport of contaminants throughout the world has become increasingly important as we are discovering the limits of the world to absorb and digest the wastes of human endeavors. In the last century people in the northern hemisphere viewed the eventual dusting effect of the eruption of the volcano Krakatoa. The dust spread thousands of miles—indeed around the world—through the air, darkening the sun from view. In recent times the nuclear power plant explosion at Chernobyl, USSR, sprinkled radioactive dust over Scandinavia and the Arctic, contaminating milk supplies and other food sources. We are accustomed to observe acid rain fallout from power plant source plumes traveling across continents and oceans. We know about the spread of chlorofluorocarbons to the stratosphere and their potential warming effect. These are all examples of the introduction of contaminant dusts or gases into transport media, such as the atmosphere, rivers, groundwater aquifers, and oceans, followed by the movement of these carriers to distant places that can be far removed from the original sources.

Distinct from these "obvious" sources are the semivolatile organic compounds. These compounds are liquids or solids at room temperature and do not generally appear to be volatile. Examples include pesticides, PCBs, polyaromatic hydrocarbons (PAHs), and dioxins. DDT was found in Antarctica in 1965 but never used there. We are now finding DDT, dieldrin, and chlordane compounds in marine life throughout the globe, including the polar regions. Concentrations in polar bears, sea lions, and the like are sufficiently high as to threaten the quality of food eaten by native peoples in these areas. The hexachlorocyclohexanes (HCHs) are ubiquitous throughout the northern hemisphere in concentrations that are easily measured. Environmental groups in Sweden and nearby areas are so concerned by the transport of the organochlorines that they now would like to see general air monitoring for them throughout Europe.

This text brings scientific information about the processes by which pesticides, as a major example of semivolatile organic compounds (SOCs), have been found to enter a transport system, be transported across international boundaries, and be deposited or available in regions far from their starting point of application. We are finding that few compounds are immune from this process. It is doubtful we can ever produce a pristine world. Life's activities are much too complicated and, as well, natural processes self-contaminate, too. We, therefore, look forward to reductions of any amount that may eventually save the earth and its inhabitants from serious problems.

Contents

Introduction

This book is laid out in a mechanistic way; that is, in the same manner as pollution may take place in the environment. The original source of pesticide is where it was first applied. Various transport media shift the compounds to other locations, where they lodge. These secondary locations can then serve as new primary source locations. The result amounts to a continuous process that ceases only when a permanent or long-lasting sink location is found.

Pesticides have initial sources resulting from farm or orchard applications. The first five chapters look at these sources. Two chapters (Spencer and Nash) trace sources from soil into the atmosphere. A third (Jenkins) shows sources from leaf surfaces to the atmosphere. One chapter (Clendening) examines overall movement from application to both the atmosphere and by leaching into groundwater. The final chapter (Woodrow) examines the special case of movement from an aquatic application into the atmosphere.

Each of the next group of four chapters discusses an aspect of atmospheric distribution. In the first chapter (Levy), some aspects of general meteorological distribution patterns are discussed. Gas-particle equilibriums (Pankow) and measuring concentrations of semivolatile organic compounds (SOC), either in the freely dissolved form or the particle-adsorbed form (Johnson) are presented next. The section is brought to a close with a chapter on the potential atmospheric decomposition of organophosphorus compounds, the structure of many common insecticides.

Global transport examples are conveniently measured in ocean areas where the distances are known to be huge. Four chapters discuss concentrations of pesticides as found in global sampling. The first (Tatsukawa) summarizes the concentrations of organochlorine compounds found in western Pacific and Indian Ocean locations over the past two decades. A complete air and water sampling of a circular cruise of the Pacific Ocean gave data presented in the next chapter (Kurtz). The important role of back trajectory calculations is discussed in this group (Atlas) in order to determine the true sources of particular samples. Finally, contaminants found in the south Atlantic and Antarctic Oceans near Brazilian and Argentinian locations are given (Weber).

Aside from the interest in pesticide transport from Californian application areas, the location of highest interest in North America centers in the Great Lakes. A series of chapters discusses transport in this area. Some of the most serious modeling work was developed to explain the present and future situations in the Great Lakes. Three chapters look at air transport from southern United States regions: one looks at the air and rainfall concentrations of herbicides and the insecticide toxaphene in the Chesapeake Bay area (Glotfelty); another shows the model of toxaphene transport from southern states (Voldner), and the third provides deposition data into the Great Lakes for general organochlorine pesticides (Strachan). Not to be overlooked, river transport can be an important transport mode. The research provided here (Baker) explains differences in mass transport between normal and high water river flow. The immobility and movement of the various compounds comprising the chlordane technical mixture are discussed for the Missouri area (Puri). Finally in this section, three chapters discuss the compartment modeling theory for Great Lakes areas: mass balance in the Great Lakes (Strachan), air-soil balance theory (Stiver), and mass balance in Lake Ontario (Mackay).

Contamination of Arctic areas is coming under increased scrutiny. Perhaps because of the refrigerator effect allowing condensable compounds to be sucked up to the cold regions, or because normal air currents sweep up from temperate to Arctic areas and carry these compounds with them, whatever the reason, we have firm evidence of this transport direction. This book shows evidence in northern Canadian wildlife (Muir), arctic atmosphere (Bidleman), and arctic snow (Gregor).

After studying the phenomenon of long range transport, one is left with the feeling that measurable and significant amounts of pesticides and other semivolatile organic compounds can be available in nonapplied areas. Might there be implications for regulatory agencies concerning the sources of the concentrations measured in any field or location? Two chapters discuss this problem. The first (Richards) makes risk assessments on contaminants in drinking water. The second (Rodgers) provides an integrated model for atmospheric and aquatic fate that could be useful in this area.

I asked Bob Risebrough, researcher since the early 1960s in environmental contamination, to dream a little. He suggests an area in which I am also interested: modeling the transport of pesticides in a dynamic way as though the earth's surface were a giant chromatographic column.

Enjoy your reading and studies!

Long Range Transport of Pesticides

## CHAPTER 1

Movement of Pesticides from Soil to the Atmosphere

W. F. Spencer and M. M. Cliath

INTRODUCTION

Volatilization of applied pesticides from soil is a significant source of pesticides for long-range transport. Volatilization can take place during or following application to land surfaces. Volatilization, as used here, is defined as the loss of chemicals from surfaces in the vapor phase; that is, vaporization followed by movement into the atmosphere. Potential volatility of a chemical is related to its inherent vapor pressure, but actual vaporization rates will depend on environmental conditions and all factors that control the chemical at the soil-air-water interface.[1]

This chapter discusses the movement of pesticides into the atmosphere from the standpoints of the magnitude of vapor loss, the mechanism and factors influencing the rate of volatilization, and the progress in developing prediction models. The use of a model for screening pesticides for their environmental behavior, based on relevant physical and chemical properties, and the importance of Henry's law constants in controlling volatilization behavior and distribution of pesticide in the soil are discussed also. Volatilization rates of pesticides are compared with model predictions.

Field-Measured Volatilization Losses

In recent years, several field studies using microclimate techniques have established volatilization as one of the most important pathways for loss of pesticides from treated areas. Volatilization rates from moist soil surfaces or from plants can be very large, with losses approaching 90% within 3 days, even for chemicals with vapor pressures less than 10^{-3} mm Hg.[2]

Table 1, taken from Glotfelty et al.,[3] shows measured volatilization rates from several field studies[4,5,6] along with factors affecting vapor loss rates. Volatilization rates were measured by the aerodynamic method, in which pesticide volatilization losses are calculated from pesticide concentration, wind speed, and temperature gradients obtained simultaneously in the atmosphere

Table 1.   Summary of Pesticide Volatility Studies.

Chemical and Mode of Application	Percentage Lost	Time Elapsed	Data Source
Trifluralin (vapor pressure = 1.1 × 10^{-4} mm Hg at 25°C)			
Soil incorporation to 2.5 cm	22	120 days	White et al.[4]
Soil incorporation to 7.5 cm	3.4	90 days	Taylor[2]
Surface, dry soil	2–25[a]	50 hours	experiment 3[b]
Surface, moist soil	50	3–7.5 hours[c]	experiments 2 and 1
	90	2.5–7 days[c]	experiments 2 and 1
Heptachlor (vapor pressure = 3 × 10^{-4} mm Hg at 25°C)			
Soil incorporation to 7.5 cm	7	167 days	Taylor et al.[5]
Orchard grass	90	7 days	Taylor et al.[6]
Soil surface, dry	14–40[a]	50 hours	experiment 3
Soil surface, moist	50	6 hours[c]	experiment 1
	90	6 days[c]	experiment 1
Lindane (vapor pressure = 6.3 × 10^{-5} mm Hg at 25°C)			
Soil surface, dry	12	50 hours	experiment 3
Soil surface, moist	50	6 hours[c]	experiment 2
	90	6 days[c]	experiment 2
Chlordane (vapor pressure = 1 × 10^{-5} mm Hg at 25°C)			
Soil surface, dry	2	50 hours	experiment 3
Soil surface, moist	50	2.5 days	experiment 1
DCPA (vapor pressure = 2.5 × 10^{-5} mm Hg at 25°C)			
Soil surface, moist	2	34 hours	experiment 1

Source: Glotfelty et al.[3] Used with permission.
[a]Range between integrated flux and soil loss.
[b]Experiment numbers refer to data obtained in (1) 1985 or (2) 1977 (Beltsville site), or (3) 1978 (Salisbury site).
[c]From Equation 1 and Table 3, assuming 12 hours of daylight.

over the field surface.[7] Measured volatilization rates were dependent on the nature of the pesticides and how they were used. Shallow soil incorporation and dry soil surfaces greatly restricted volatilization losses. For example, when trifluralin was applied to a moist soil surface, 50% was lost in 3–7 hr, and 90% in 2.5–7 days. Much smaller amounts were lost when it was soil-incorporated — 22% loss in 120 days when incorporated to 2.5 cm and 3.4% loss in 90 days when incorporated to 7.5 cm. For heptachlor in moist soil, volatilization rates from surface applications were similar to those from foliage — about 90% in 2–7 days. Volatilization rates were much lower when the heptachlor was incorporated to a shallow depth of 7.5 cm — 7% in 167 days.[5] Volatilization of the three organochlorines — lindane, chlordane, and DCPA — were related to their vapor pressures, and volatilization rates were higher from moist than dry soil surfaces.

Similar results have been reported by others. For example, Cliath and co-workers[8] measured volatilization losses of EPTC, a water-soluble herbicide, from an irrigated alfalfa field. The EPTC was applied directly in surface

Table 2.   Amounts of EPTC in Runoff and Volatilized from Water and Wet Soil, During and After a Flood Irrigation Application to Alfalfa.

EPTC	kg/ha	% of Total Applied
Applied in irrigation water (avg. 2.17 ppm)	3.04	100.0
Runoff in tailwater (avg. 1.70 ppm)[a]	0.21	7.0
Volatilized from water[b]	0.86	28.4
Volatilized from wet soil[c]	1.38	45.2
Total volatilized	2.24	73.6
Total lost	2.45	80.6

Source: Cliath et al.[8]
[a]EPTC concentration in runoff varied from 1.44 to 1.97 ppm during volatilization measurements.
[b]Volatilized between 1445 and 2045 hr on May 25.
[c]Volatilized between 2045 hr on May 25 and 1845 hr on May 27.

irrigation water, where the chemical would be expected to move rapidly into the soil. Table 2 shows amounts of EPTC in runoff water and volatilized from the field following its application at 3 kg/ha through a flood irrigation system. The amount of EPTC volatilized from water and wet soil amounted to 73.6% of the total applied — 28.4% volatilized in 6 hr from the irrigation water and 45.2% from wet soil within 46 hr after the water had infiltrated.

Mechanisms and Factors That Influence Volatilization Rates

Volatilization rates of pesticides from nonadsorbing surfaces are directly proportional to their relative vapor pressures. The actual rates of loss, or the proportionality constants relating vapor pressure to volatilization rates, are almost entirely dependent upon external conditions that affect movement away from the evaporating surface, such as surface roughness, wind speed, air turbulence, etc.[1,9]

Volatilization of pesticides from soil is much more complicated and difficult to predict because of the many parameters affecting their adsorption, movement, and persistence. Soil-incorporated pesticides volatilize at a greatly reduced rate dependent not only on the equilibrium distribution between the air, water, and soil matrix as related to vapor pressure, solubility, and adsorption coefficients, but also on their rate of movement to the soil surface. Volatilization from soil involves desorption of the pesticides from the soil, movement to the soil surface, and vaporization into the atmosphere.[1] Mechanisms and factors affecting volatilization of pesticides from soil can be grouped into three categories:

1. those that affect movement away from the evaporating surface
2. those that affect vapor pressure or vapor density of the pesticide within the soil
3. those that affect rate of movement of the pesticide to the evaporating surface

Movement Away from the Evaporating Surface

The rate that a pesticide moves away from the surface is diffusion controlled. Close to a solid surface, there is relatively no vertical movement of air, and vaporized substances are transported from the surface through this stagnant air boundary layer (or laminar flow layer) only by diffusion. The actual rate of mass transfer away from the surface by diffusion will be proportional to the diffusion coefficient and to the vapor density of the pesticide at the evaporating surface.

Outside the surface-air boundary layer, the flow in the overlying atmosphere is turbulent, and dispersal of pesticide vapor into the atmosphere can be described in terms of "eddy diffusion" coefficients. The depth of the turbulent zone is several orders of magnitude greater than that of the laminar flow layer, and the difference between the two diffusion coefficients is also very large, so that dispersal of pesticide vapor in the turbulent zone outside the laminar layer is relatively very rapid. This chapter deals only with processes occurring in the soil or in the so-called "stagnant air" boundary layer below the turbulent zone.

Factors Affecting Vapor Density in Soil

Vapor pressures or vapor densities of pesticides are greatly decreased by their interactions with soil, mainly due to adsorption. The degree of reduction in vapor density in soil due to adsorption depends mainly upon soil water content, the nature of the pesticide and its concentration, and soil properties, particularly soil organic matter content. Partitioning between the soil and the water usually follows the Freundlich equation, and the concentration of the desorbed pesticide in the soil water dictates the vapor density of the pesticide in the soil air in accordance with Henry's law. Thus, soil water adsorption coefficients can be used to calculate vapor densities in the soil atmosphere.

Adsorption effects on vapor density are illustrated in Figure 1, where relative vapor density of dieldrin is plotted against concentration in soil at different soil water contents.[10] Relative vapor density is similar to relative humidity; it is the ratio of the observed vapor density to that of the pure material at the same temperature. In moist soil, the vapor concentration increased rapidly with concentration of dieldrin, but in very dry soil the vapor concentration was almost nil, even at 100 ppm dieldrin. (In this soil, 2.1% water is equivalent to less than 1 molecular layer of water, and 3.9% water where vapor density is high is equivalent to 90 atmospheres matrix suction.) Similar relationships were observed for other relative nonpolar pesticides and for other soils. For example, vapor pressure of lindane,[11] DDT,[12] and trifluralin[13] also decreased to very low values when the water content was decreased below that equivalent to approximately 1 molecular layer. With all of the chemicals, vapor pressures increased to their original values upon rewetting the air-dry soil, indicating that the drying effect is essentially reversible. Such measurements of vapor

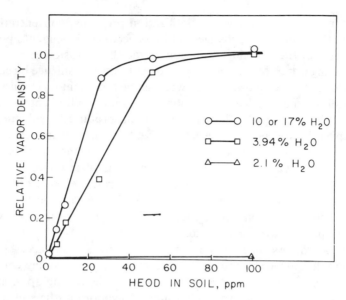

Figure 1. Relative vapor density of dieldrin (HEOD) versus concentration in Gila silt loam as affected by soil water content. *Source*: Spencer et al.[10].

pressure of pesticides in soil at various water contents conclusively demonstrated that the greater vaporization from wet than from dry soils is due mainly to an increased vapor pressure resulting from displacement of the chemical from the soil surface by water.[1]

Temperature influences volatilization rates mainly through its effect on vapor pressure. The change in vapor pressure with temperature for most pesticides is equivalent to an increase in volatility of about 3–4 times for each 10-degree increase in temperature.[10] Temperature may also influence volatility of soil-incorporated pesticides through its effect on movement of the pesticides to the surface by diffusion or mass flow in evaporating water, or through its effect on the soil water adsorption/desorption equilibrium. For all these effects, increases in temperature are usually associated with increases in volatilization rate. However, this may not always be the case since an increase in temperature may be associated with an increase in drying rate at the soil surface, thereby decreasing vapor density and resulting in less volatilization than at the lower temperature. This may occur if the temperature effect on soil drying is greater than its effect on vapor pressure and is probably the only case where an increase in temperature can actually decrease pesticide volatilization rate.

Movement Toward the Soil Surface

The initial volatilization rate of soil-incorporated pesticides will be a function of vapor pressure of the chemical at the soil surface as modified by adsorption. (The small fraction of the pesticide that remains on the soil surface after mixing is readily lost.) As volatilization proceeds, concentration of the chemical at the soil surface changes, and loss becomes dependent upon the rate of movement of the pesticide to the soil surface by diffusion or by convection in evaporating water. Movement of pesticides to the soil surface by convection, or bulk flow, in the evaporating soil water is the dominant mechanism controlling volatilization of pesticides incorporated in moist soil.[14] When water evaporates from the soil surface, the suction gradient produced results in an appreciable upward movement of water to replace that evaporated, and any pesticides in the soil solution will move toward the surface by mass flow with the evaporating water. This accelerates the volatilization of the pesticide. When water is not evaporating, volatilization rate depends on the rate of movement to the soil surface by diffusion only. Usually both mechanisms, diffusion and convection, work together in the field, where the water and pesticide vaporize at the same time.

All the parameters appear to come together at the soil surface. As pesticides move up to the surface by convection in evaporating water, they volatilize into the atmosphere or accumulate at the soil surface, depending upon their physicochemical properties. As water continues to evaporate, the soil dries at the surface and the dry soil further restricts vapor density or volatility. To predict volatilization rates, a model must take into account all these changes at the surface including wetting and drying of the soil and their effects on vapor density. There presently are no models to do this under field conditions.

Predicting Volatilization Rates

Volatilization of pesticides from soil can be estimated from a consideration of the physical and chemical factors controlling their concentrations at the soil surface. Models developed for estimating volatilization rates are based upon equations describing the rate of movement of the chemical to the surface by diffusion and/or by convection, and away from the surface through the air boundary layer by diffusion.[15-17] Additionally, the proportion of a pesticide in soil that will be lost by volatilization depends upon its degradation rate.

In a series of papers, Jury et al.[17-20] described and applied a screening model for assessing relative volatility, mobility, and persistence of pesticides and other trace organics in soil. The model embodies the above principles of volatilization. The model allows the organics to be present in the soil in the adsorbed, solution, and gaseous phases; they are free to move by vapor diffusion, liquid diffusion, and convection with the liquid solution. The soil surface boundary condition consists of a stagnant boundary layer connecting the soil and air through which the organic chemical and water vapor must move to reach the atmosphere. The model assumes that the gas and liquid concentrations are related by Henry's law and that the adsorption isotherms relating liquid and adsorbed concentrations are linear. It also assumes that degradation occurs by first-order rate processes. The model is intended to classify and screen organic chemicals for their environmental behavior based on their physical and chemical properties.

Jury et al.[19] applied the screening model to a set of 20 pesticides for which the benchmark properties of vapor pressure and solubility or Henry's law constant, organic carbon partition coefficient, and degradation rate were obtained from the literature or calculated. The relative magnitude of calculated volatilization rates and their change with time depended upon water evaporation rates and the physicochemical properties of the pesticides. Water evaporation had a considerable effect on volatilization of some chemicals and not on others. Also, volatilization of some pesticides decreased with time, and volatilization of others increased with time when water was evaporating.

The model output indicated that volatilization behavior of a pesticide is controlled mainly by the ratio of its solution to vapor concentration, or Henry's law constant (K_H), which determines the extent to which the air boundary layer restricts volatilization from soil. The extent to which this boundary layer limits the volatilization flux was used as a criterion for classifying organic chemicals into general categories based upon whether control of volatilization is within the soil (Category I), or within the air boundary layer (Category III). Figure 2 shows calculated volatilization rates versus time for three prototype chemicals under conditions of varying water evaporation rates of 0, 2.5, and 5.0 mm/day.[18] A clear distinction is apparent between the behavior of the three categories of chemicals. Category I compounds are those with K_H much greater than 2.65×10^{-5} (dimensionless unit representing the ratio of saturation vapor density to solubility). Their volatility decreases with time under all conditions, whether water is evaporating or not. The flux rate of Category III chemicals increases with time when water is evaporating and decreases slowly with time when water is not evaporating. The Category III chemicals (with K_H

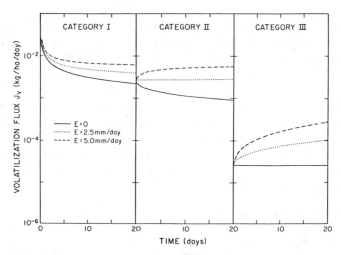

Figure 2. Volatilization flux rates for a prototype chemical from each category of volatilization behavior as affected by water evaporation (E). Source: Jury et al.[18].

much less than 2.65×10^{-5}) move to the surface in evaporating water faster than they can volatilize to the atmosphere through the boundary layer. Consequently, their concentrations increase at the soil surface under evaporative conditions, and their volatilization rates increase with time. Category II chemicals are intermediate in their behavior.

EXPERIMENTAL

Laboratory and field experiments were conducted to test the predictions of the screening model, particularly concerning the importance of K_H in controlling relative volatilization and vapor behavior. Volatilization losses and surface distributions of two pesticides with widely differing K_H were measured, first in the laboratory under controlled conditions, and then in the field under conditions of pesticide use.

Laboratory Experiments

The volatilization apparatus described by Spencer et al.[21] was used to test the model predictions under ideal conditions. The gas flow system includes a rectangular soil chamber from which vapor losses are measured with air at controlled flow rates passing over treated soil. Soil water content is adjusted through porous ceramic tubes installed in the bottom sections; water loss from the soil is controlled by adjusting the relative humidity of the air passing over the surface. Vaporized pesticides are trapped from the moving airstream in polyurethane foam or other trapping agents, extracted, and analyzed. The upper section of the volatilization cell contains an air chamber 0.2 cm deep, and an air flow rate of 1 L/min provides an average wind speed of approximately 1 km/hr and a change of atmosphere in the space over the surface of 2.8 times per second.

Soil containing a mixture of two model compounds—prometon [2,4-bis(isopropylamino)-6-methoxy-s-triazine], a Category III chemical, and lindane (gamma isomer of 1,2,3,4,5,6-hexachloro-cyclohexane), a Category I chemical—was used in the volatilization cell. The relevant chemical properties of the two chemicals are shown in Table 3. The vapor density is much lower and the solubility much higher for prometon than for lindane, resulting in a K_H of 1×10^{-7} for prometon and 1.33×10^{-4} for lindane. The soil used was a San Joaquin sandy loam packed to a bulk density of 1.56. The depth of incorporation was 10 cm, the temperature 25°C, and the atmospheric relative humidity either 42.5% with water evaporating or 100% with no evaporation. For other experimental details see Spencer et al.[22]

Field Experiments

To further test the model under field conditions, a field experiment was established to determine volatilization and movement as related to physical and chemical properties of four herbicides when applied in irrigation water to

Table 3. Physicochemical Properties of Pesticides Used in the Laboratory and Field Experiments.

Pesticide	Category (Jury)	Vapor Pressure (mmHg)	Vapor Density (μg/mL)	Solubility (μg/mL)	Henry's Constant, K_H	Ads. Coeff., K_{oc}	Half-Life (days)
Triallate	I	1.93E−04	3.17E−03	4	7.90E−04	3600	100
Napropamide	III	4.00E−06	5.80E−05	73	7.90E−07	300	70
Prometryn	III	2.13E−06	2.70E−05	48	5.60E−07	610	60
Prometon	III	6.22E−06	7.53E−05	750	1.00E−07	408	100[a]
Lindane	I	6.47E−05	1.00E−03	7.5	1.33E−04	1300	266

[a]Estimated.

a 2.5-ha field. Herbicides were applied in sprinkler irrigation water at the following rates: triallate 4.5, napropamide 2.3, prometryn 13.7, and prometon 11.2 kg/ha. Some physicochemical properties of the four herbicides are shown in Table 3. Triallate, with a relatively high K_H, is a Category I chemical; the others would be classified as Category III chemicals. Volatilization was measured by the aerodynamic method, which requires concurrent gradient measurements of wind speed, air temperature, and vapor concentration above the treated soil surface for extended periods.[7] Soil samples were taken from a depth of 0–1 cm with a 3-cm diameter soil-plug sampler at various times after application in order to follow concentration changes in the surface 1 cm of soil as related to water applications or drying periods. This provided information on pesticide accumulations at the soil surface following upward movement in evaporating water. Volatilization measurements and soil samples were taken during a series of irrigation and drying periods.

RESULTS

Laboratory Experiments

Figure 3 shows measured volatilization of prometon compared with predicted fluxes using the screening model with d = 0.022 cm as the effective boundary layer thickness or stagnant air layer depth.[22] Since predicted volatilization flux is a function of the effective boundary layer thickness, d, the value of d used in the simulations had to represent the boundary layer thickness for the volatilization cells containing the treated soil. We established an "experimental" value for d by measuring volatilization flux with unabsorbed prometon in the volatilization cell in the same geometric configuration as the treated soils. Then d was calculated from the measured vapor flux with the boundary layer transfer equation:

$$d = \frac{D_G (C_s - C_a)}{J_p} \tag{1}$$

where　D_G = diffusion coefficient of prometon in air
C_s = prometon vapor concentration at the surface

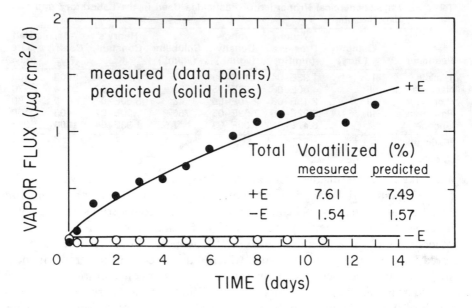

Figure 3.   Volatilization of prometon from San Joaquin sandy loam with (+ E) and without (-E) water evaporating. *Source*: Spencer et al.[22].

C_a = vapor concentration in the outgoing airstream as it passes through the trap

J_p = measured prometon flux

The effective boundary layer thickness was calculated as 0.022 cm at the 1 L/min flow rate. Extremely good agreement was observed between measured and predicted vapor flux values both with and without water evaporating. With water evaporating, the volatilization rate of prometon increased from 0.03 μg/cm^2/day during the first 5 hr, to 1.24 μg/cm^2/day after 14 days with water evaporating at 0.55 cm/day. This is a 41-fold increase in volatilization rate over the 14-day period due to accumulation of prometon at the soil surface as water evaporated. Total prometon volatilized, in percent of applied, was 7.61% with water evaporating, compared with 7.94% predicted, and 1.54% without water evaporating, compared with 1.57% predicted.

Figure 4 shows the measured and predicted volatilization of lindane. Again, good agreement was observed between measured and predicted values. Also, as we have noted many times with other Category I chemicals, volatilization decreased with time with or without water evaporating.

The distribution of the prometon and lindane within the soil columns was determined at the end of the 14-day volatilization period. Prometon accumulated at the soil surface with water evaporating (Figure 5). Good agreement was observed between measured and predicted distributions of prometon in

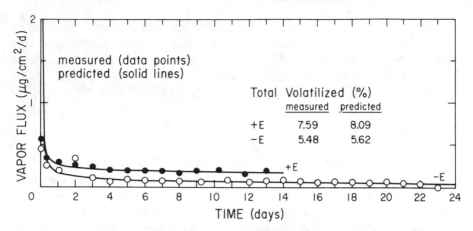

Figure 4.   Volatilization of lindane from San Joaquin sandy loam with (+E) and without (-E) water evaporating. *Source*: Spencer et al.[22].

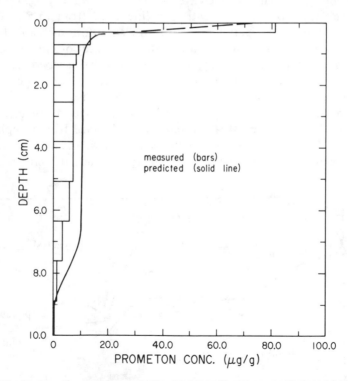

Figure 5.   Distribution of prometon in soil after 14 days with water evaporating at 0.55 cm/day. *Source*: Spencer et al.[22].

the soil column. Concentrations in the surface 0.3 cm increased from 10 to 81 $\mu g/g$. Lindane distribution after 14 days (not shown) with water evaporating indicated a slight decrease in concentration at the surface and good agreement between measured and predicted values throughout the soil column. Additional details of the results from the laboratory experiments are given in Spencer et al.[22]

Field Experiments

All pesticide analyses have not been completed, and consequently volatilization data are not yet available. However, unpublished data on changes in concentration of the four herbicides at a depth of 0–1 cm at various times following application are shown in Figure 6. The irrigation periods are also shown on the graph. Soil drying occurred between the irrigations. With triallate, the herbicide with the highest K_H, concentrations continually decreased following application, whether the soil was drying or being wetted with irrigation water. Prometon, with the lowest K_H, reacted as expected: concentrations in the surface 1 cm markedly increased while the soil was drying. Following the first irrigation, prometon concentrations increased from 15.5 to 61.5 $\mu g/g$; following the second irrigation at day 10, concentrations increased from 5.0 to 27.9 $\mu g/g$; and after the third irrigation, the concentrations increased from 1.3 to 16.0 $\mu g/g$ over the final 13-day drying period. The other two herbicides with intermediate K_H values reacted in an intermediate fashion. Their concentrations increased only slightly during the drying periods.

Distribution of the four herbicides within the 1-x layer of soil after 8 days of drying are shown in Table 4. These data were obtained from samples scraped from the soil surface in 0.2-cm increments, extracted, and analyzed. The Category III compounds are concentrated in the surface 0.2 cm; triallate, the

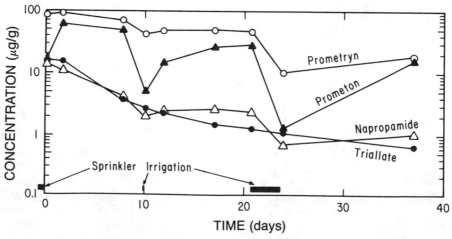

Figure 6.   Concentrations of four herbicides in the 0–1 cm soil depth at various times following application with soil drying between irrigations.

Table 4.   Pesticide Distribution in the Surface 0.8 cm Eight Days After the Application Irrigation.

| Depth, cm | Concentration, μg/g | | | |
	Triallate	Napropamide	Prometryn	Prometon
0.0–0.2	7.1	8.5	270	145
0.2–0.4	3.6	2.5	19	26
0.4–0.6	2.1	0.4	3	9
0.6–0.8	1.6	0.2	2	6

Category I chemical, is more evenly distributed within the 1-cm surface layer. These data indicate that compounds with a sufficiently low K_H will accumulate at the soil surface when water is evaporating as predicted from the screening model. The evidence further indicates that the Category III compounds concentrate right at, or on, the surface as water evaporates. The amount of increase in concentration undoubtedly is related to water evaporation rates, degree of soil drying following irrigation or rainfall, and the initial pesticide concentration. We expect the volatilization rates to reflect these increases in pesticide concentration at the surface, particularly when the soils are rewetted following drying.

The phenomenon of pesticides accumulating at the soil surface following convective movement in evaporating water has several interesting ramifications. First, the mechanism could greatly enhance the volatilization of chemicals with low K_H, formerly considered to be essentially nonvolatile from wet soils. Second, the mechanism could significantly increase the amounts of some chemicals available for photolysis or for runoff into surface water bodies from rainfall or irrigation water.

SUMMARY AND CONCLUSIONS

One important source of pesticides for long-range transport is volatilization of applied pesticides from soil and plant surfaces. Volatilization rates from vegetation or moist soil surfaces can be very large—losses can approach 90% within 3 days—even for chemicals with vapor pressures less than 10^{-3} mm Hg.[2] An example is the herbicide EPTC applied in irrigation water: volatilization losses amounted to 73.6% of the applied EPTC within 52 hr following application.[8]

Studies of mechanisms and factors affecting volatilization rates indicate that volatilization of soil-incorporated pesticides is controlled by their rate of movement away from the surface, their effective vapor pressure within the soil, and their rate of movement through the soil to the vaporizing surface. Movement of pesticides to the soil surface by convection in evaporating water is the dominant mechanism controlling volatilization of pesticides incorporated in moist soil. Vapor pressures of pesticides are greatly decreased by their interaction with soil, mainly due to adsorption. The magnitude of the adsorption effect depends on the nature of the chemical, its concentration, the soil water content, and soil properties such as organic matter content. The soil water content is especially important; volatilization rates of pesticides are essentially zero in dry soil. Consequently, the presence of water in the soil is a major factor in volatilization of pesticides and other organic chemicals. The presence of water increases the vapor density of the pesticide in the soil. Water evaporation from the soil greatly enhances the volatilization rate by carrying pesticides to the soil surface as the water evaporates.

Measurements of volatilization losses and surface accumulation of pesticides with widely differing Henry's law constants agreed with predictions from a screening model, demonstrating that relative volatilization of pesticides from soil is controlled mainly by the ratio of their vapor to solution concentrations. Whether or not the volatilization rate increases with time when water is evaporating depends on the magnitude of the K_H for the chemical. For compounds with high K_H, volatilization rates decrease with time with or without water evaporating; for compounds with very low K_H, volatilization rates increase with time, and their concentrations increase at the soil surface when water is evaporating. Both laboratory and field experiments verified this accumulation of pesticides at the soil surface when water is evaporating.

With respect to long-range transport of pesticides and other toxic chemicals, both the volatilization and redeposition of pesticides will be dependent upon their K_H values or ratio of their volatility to solubility. It is safe to say that most of the compounds that are transported over long distances will be classified as Category I compounds with relatively high K_H values. For example, DDT is classified as a Category I compound with a reported K_H of 2.0×10^{-3}.[19] Most of the organochlorine pesticides fall into this category—many of them are more volatile than the so-called Category III compounds with low K_H. For the organochlorines, volatilization rates are usually greatest soon after application and then, more or less, decrease with time thereafter. For example, atmospheric transport of DDT compounds is well documented.[23] Even in areas where DDT use has been discontinued, the persistence of DDT residues is sufficiently great that they will continue to be redistributed for many years after DDT applications cease. In these areas, volatilization from the soil probably will be the main source of DDT components moving into the atmosphere, which means a shift in the ratio of the various DDT components entering the atmosphere. During years when DDT was used heavily, the various components entered the atmosphere somewhat in the same ratio as the vapor components associated with technical DDT. In subsequent years, the ratio of *p,p'*-DDT to other DDT compounds will decrease in proportion to the degree of degradation of *p,p'*-DDT to other compounds such as *p,p'*-DDE and *p,p'*-TDE, which are more volatile than the parent compound.[12]

ACKNOWLEDGMENTS

This is a contribution of the U.S. Salinity Laboratory, USDA-ARS, and the California Agricultural Experimental Station, Riverside, CA.

REFERENCES

1. Spencer, W. F., W. J. Farmer, and M. M. Cliath. "Pesticide Volatilization," *Residue Reviews* 49:1–47 (1973).
2. Taylor, A. W. "Postapplication Volatilization of Pesticides under Field Conditions," *J. Air Poll. Control Assoc.* 28:922–27 (1978).
3. Glotfelty, D. E., A. W. Taylor, B. C. Turner, and W. H. Zoller. "Volatilization of Surface-Applied Pesticides from Fallow Soil," *J. Agric. Food Chem.* 32:638–43 (1984).
4. White, A. W., Jr., L. A. Harper, R. A. Leonard, and J. W. Turnbull. "Trifluralin Volatilization from a Soybean Field," *J. Environ. Qual.* 6:105–10 (1977).
5. Taylor, A. W., D. E. Glotfelty, B. L. Glass, H. P. Freeman, and W. M. Edwards. "The Volatilization of Dieldrin and Heptachlor from a Maize Field," *J. Agric. Food Chem.* 24:625–30 (1976).
6. Taylor, A. W., D. E. Glotfelty, B. C. Turner, R. E. Silver, H. P. Freeman, and A. Weiss. "Volatilization of Dieldrin and Heptachlor Residues from Field Vegetation," *J. Agric. Food Chem.* 25:542–48 (1977).
7. Parmele, L. H., E. R. Lemon, and A. W. Taylor. "Micrometeorological Measurement of Pesticide Vapor Flux from Bare Soil and Corn under Field Conditions," *Water Air Soil Poll.* 1:433–51 (1972).
8. Cliath, M. M., W. F. Spencer, W. J. Farmer, T. D. Shoup, and R. Grover. "Volatilization of *s*-Ethyl *n,n*-Dipropylthiocarbamate from Water and Wet Soil During and After Flood Irrigation of an Alfalfa Field," *J. Agric. Food Chem.* 28:610–13 (1980).
9. Hartley, G. S. "Evaporation of Pesticides," *Adv. Chem. Series* 86:115–34 (1969).
10. Spencer, W. F., M. M. Cliath, and W. J. Farmer. "Vapor Density of Soil-Applied Dieldrin as Related to Soil Water Content and Dieldrin Concentration," *Soil Sci. Soc. Amer. Proc.* 33:509–11 (1969).
11. Spencer, W. F., and M. M. Cliath. "Desorption of Lindane from Soil as Related to Vapor Density," *Soil Sci. Soc. Amer. Proc.* 34:574–78 (1970).
12. Spencer, W. F., and M. M. Cliath. "Volatility of DDT and Related Compounds," *J. Agric. Food Chem.* 20:645–49 (1972).
13. Spencer, W. F., and M. M. Cliath. "Factors Affecting Vapor Loss of Trifluralin from Soil," *J. Agric. Food Chem.* 22:987–91 (1974).
14. Spencer, W. F., and M. M. Cliath. "Pesticide Volatilization as Related to Water Loss from Soil," *J. Environ. Qual.* 284–89 (1973).
15. Mayer, R., J. Letey, and W. J. Farmer. "Models for Predicting Volatilization of Soil-Incorporated Pesticides," *Soil Sci. Soc. Amer. Proc.* 38:563–68 (1974).
16. Jury, W. A., R. Grover, W. F. Spencer, and W. J. Farmer. "Modeling Vapor Losses of Soil-Incorporated Triallate," *Soil Sci. Soc. Amer. Proc.* 44:445–50 (1980).
17. Jury, W. A., W. F. Spencer, W. J. Farmer. "Behavior Assessment Model for Trace Organics in Soil: I. Model Description," *J. Environ. Qual.* 12:558–64 (1983).
18. Jury, W. A., W. J. Farmer, and W. F. Spencer "Behavior Assessment Model for

Trace Organics in Soil: II. Chemical Classification and Parameter Sensitivity," *J. Environ. Qual.* 13:567–72 (1984).

19. Jury, W. A., W. F. Spencer, and W. J. Farmer. "Behavior Assessment Model for Trace Organics in Soil: III. Application of Screening Model," *J. Environ. Qual.* 13:573–79 (1984).

20. Jury, W. A., W. F. Spencer, and W. J. Farmer. "Behavior Assessment Model for Trace Organics in Soil: IV. Review of Experimental Evidence," *J. Environ. Qual.* 13:580–86 (1984).

21. Spencer, W. F., T. D. Shoup, M. M. Cliath, W. J. Farmer, and R. Haque. "Vapor Pressures and Relative Volatility of Ethyl and Methyl Parathion," *J. Agric. Food Chem.* 27:273–78 (1979).

22. Spencer, W. F., M. M. Cliath, W. A. Jury, and L.-Z. Zhang. "Volatilization of Organic Chemicals from Soil as Related to Their Henry's Law Constants," *J. Environ. Qual.* 17:504–9 (1988).

23. Spencer, W. F. "Movement of DDT and Its Derivatives into the Atmosphere," *Residue Reviews* 59:91–117 (1975).

## CHAPTER 2

Modeling Pesticide Volatilization and Soil Decline Under Controlled Conditions

Ralph G. Nash and Bernard D. Hill

INTRODUCTION

Before long-range aerial transport can occur, pesticides must first enter the air. It is well known that pesticides may enter the air "directly" during agricultural spray applications. However, the subsequent volatilization of the deposited spray may also contribute significantly to the amount of pesticides in air. This chapter will focus on (1) pesticide volatilization and (2) models for pesticide decline and volatilization.

VOLATILIZATION FROM SOIL

Most literature data show that if pesticide dissipation is plotted on a logarithmic scale versus time on a linear scale, a straight line results.[1] Such dissipation is usually said to follow pseudo first-order kinetics (exponential) (Figure 1).[2] However, if enough samples are taken during the period immediately after application (Figure 2), the resulting log-linear plot is often curvilinear rather than linear. Upon replotting the data with two straight lines (one for fast initial dissipation and one for slower later dissipation), a much better fit of the data can be obtained. The data indicate that at least a biphasic dissipation mechanism is present.[3-5] Contributing to this nonlinear relationship, especially during the early part of the dissipation period, is the volatilization of the pesticide. Figure 3 shows the rapid initial dissipation of the herbicide trifluralin from moist soil and the cumulative amount of volatilization contributing to this dissipation.[3]

Trifluralin has a high vapor pressure and is a volatile herbicide unless it is incorporated into the soil or applied in a formulation that will minimize its volatilization. For example, trifluralin volatilization at 20°C 12 hours after application was 0.151 kg/ha/day (Table 1) and demonstrates that a sizeable amount of the 2.5 kg/ha trifluralin treatment was lost under the conditions of this experiment. For the "lower vapor pressure" pesticides (dieldrin, DCPA,

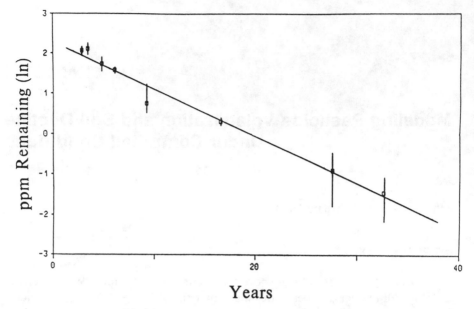

Figure 1. Lindane dissipation from incorporation into a field soil at Beltsville, Maryland, demonstrating exponential decline. *Source:* Nash.[2]

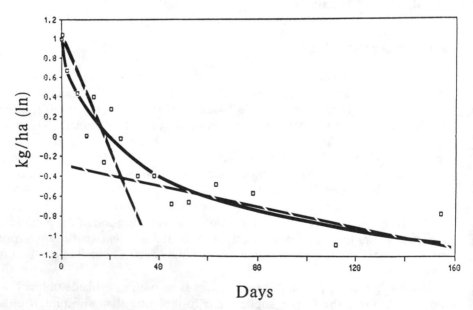

Figure 2. Dieldrin dissipation from a fallow moist soil surface at 35°C in microagroecosystem chambers showing a biphasic pattern. *Source:* Nash and Gish.[3]

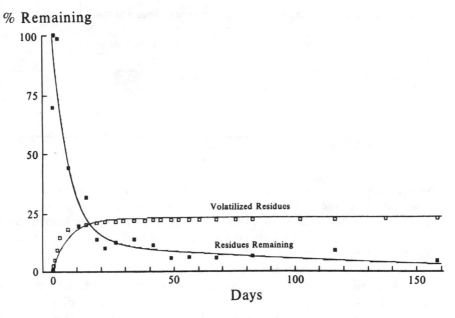

% Remaining

Figure 3. Trifluralin dissipation and cumulative volatilization from a fallow moist soil surface at 20°C in microagroecosystem chambers. *Source:* Nash and Gish.[3]

and atrazine), volatilization can still be considerable on the day of application, especially at 35°C.

Vapor pressure is the partial pressure of a pesticide in equilibrium with its liquid or solid state. In a nonequilibrium matrix, vapor pressure will indicate a potential rate for pesticide volatilization from a nonactive surface but is not necessarily related to the total amount of pesticide volatilized over a given time period. For example, Table 2 shows that over a 154-day period, the total volatilization of dieldrin (a low vapor pressure pesticide) was greater than for trifluralin (a high vapor pressure pesticide). This phenomenon occurred because trifluralin dissipated more rapidly than did dieldrin (Figure 4). The more rapid dissipation of trifluralin resulted in lesser amounts available for

Table 1.   Pesticide Volatilization, Flux (g/ha/day) at 0.5 Day After 2.5 kg/ha Application to Soil Maintained Moist and Fallow.

| Pesticide | Soil Temperature, C° | | | VPa mPa |
	5	20	35	
Trifluralin	21	151	249	32
PCNB	49	195	292	15
Dieldrin	3	23	105	1.3
DCPA	5	35	145	1.3
Atrazine	2	6	29	0.9

Source: Nash and Gish.[3]

aVapor pressure at 30°C.

Table 2.   Pesticide Volatilization (Percent of Applied) After 154 Days from Moist Fallow Soil.

Pesticide	5	20	35
		Soil Temperature, C°	
Trifluralin	17	32	49
PCNB	35	54	70
Dieldrin	28	47	56
DCPA	17	38	62
Atrazine	9	20	25

Source: Nash and Gish.[3]

volatilization. Because dieldrin dissipation is slower (Figure 4), it continued volatilizing from the moist soil at a higher rate for a longer time period (Figure 5). This explains why long-lived pesticides with very low vapor pressures can often volatilize as much (percent of total) as short-lived pesticides with higher vapor pressures.

In contrast to the halogenated pesticides, several acid herbicides volatilized only a fraction as much (Table 3). The acid herbicides were applied either as an amine (dicamba) or low volatile esters (2,4-D, 2,4,5-T, and silvex) or as a potassium salt (picloram). Measurements of samples taken 0.05 and 0.2 day after spraying indicated only the acid herbicide present.[6] Apparently, the herbicides were hydrolyzed almost immediately to their acid or salt, resulting in little volatilization, but increased biodegradation.[6] However, since the samples

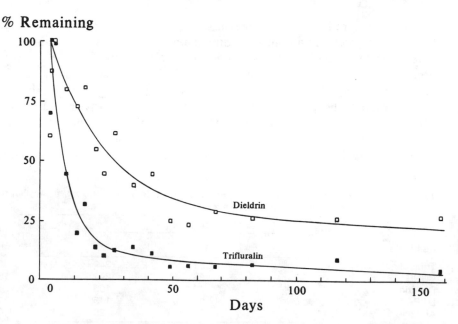

Figure 4.   Trifluralin and dieldrin dissipation from a fallow moist soil surface at 20°C. *Source:* Nash and Gish.[3]

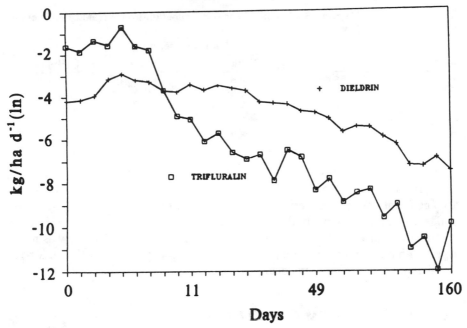

Figure 5.   Trifluralin and dieldrin flux in air after 2.5 kg/ha application to fallow soil surface maintained moist at 25°C. *Source:* Nash and Gish.[3]

were stored in a frozen state for several weeks prior to analysis, hydrolysis could have occurred during storage.

The examples above are an illustration of the relative potential for pesticide volatilization from a field. These results were under conditions that were controlled to maximize volatilization: the pesticides were applied to the soil surface, the soil was maintained moist throughout the experiment, and the soil remained fallow because of the several herbicides applied.[3,6] Although maximum volatilization conditions do exist at some point in almost every field situation where pesticides are applied to the soil surface, they are often short-lived. Presumably, under agricultural conditions, as the field crops grow, not all of the pesticide vaporized from the soil will be free to move beyond the field

Table 3.   Herbicide Volatilization (Percent of Applied) After 154 Days from Moist Fallow Soil.

Herbicide	Soil Temperature		
	5	20	35
Dicamba	0.6	6.3	7.9
2,4–D	0.9	3.6	12.1
Silvex	3.8	9.1	10.4
2,4,5–T	0.6	2.5	5.5
Picloram	0.1[a]	0.2[a]	0.5[a]

Source: Nash.[6]
[a]Values not confirmed.

Table 4.   Insecticides on/in Soybeans: Foliar Contamination from Volatilization Versus Root Sorption.

Pathway	Insecticide, ppm			
	DDT	Dieldrin	Endrin	Heptachlor
Vapor	5.4[a]	8.5	6.7	5.5
Root	0.8	10.8	38.1	20.6
Vapor/root	6.8	0.8	0.2	0.3

Source: Nash and Beall.[7]
[a]Based on ^{14}C-content of soybeans. Soil treatment = 20 ppm.

boundaries. Nash and Beall,[7] for example, showed that when insecticides volatilize from the soil, some of the insecticide vapors condense on the plant foliage (Table 4). This experiment, conducted to determine the magnitude of foliar contamination by both root uptake and vaporization of chlorinated hydrocarbon insecticides, demonstrated that DDT on/in the foliage was more likely to have originated from volatilization from the soil than through root uptake and subsequent translocation.

VOLATILIZATION FROM A WEEDY TURF

Pesticide loss from a field through volatilization occurs from foliage as well as from soil. Under conservation tillage, much of the applied pesticide never reaches the soil surface, but lands on and remains on the crop residue.[8] How does volatilization from crop residue or grass differ from soil volatilization? Dieldrin and trifluralin were applied either to fallow soil or to a weedy turf.[9] (The soil was saturated 3 days before treatment with subsequent alternate wetting and drying.) Initially, there was more volatilization from the weedy turf than from the soil (Table 5). Then by about day 10, volatilization of dieldrin was greater from the fallow soil than from the weedy turf. (The same phenomenon probably would have occurred for trifluralin as well, but by the 10th day, the residues remaining were too low to demonstrate this.) The greater volatilization initially from the weedy turf is probably explained by its larger surface area. Also, air circulation and transfer is probably much greater in the turf than on the soil surface. By the 10th day, most of the pesticide residues probably had been washed from the turf to the soil.[10]

MODELING THE CHANGE IN PESTICIDE CONCENTRATION IN AIR AND ON SOIL WITH TIME AFTER APPLICATION

After application, the concentration of pesticide in air over a treated field is initially high, but the concentration declines rapidly with time. This relationship can usually be expressed mathematically by determining logarithmic pesticide concentration in air versus logarithmic time.[3,11] The decline of pesticide concentration in air appears to follow a series of short-term exponential

Table 5. Volatilization (g/ha/day) of Dieldrin and Trifluralin from Weedy Turf or Fallow Soil After a 2.5 kg/ha Application to Moist Soil and Subsequent Wetting and Drying.

Day	Trifluralin		Dieldrin	
	Turf	Fallow	Turf	Fallow
0.04	381	129	41.2	25.1
0.15	265	113	48.3	17.5
0.27	178	88.9	30.2	13.1
0.44	93.7	81.4	48.5	36.1
0.88	88.7	29.3	27.1	9.22
1.32	27.9	17.1	18.1	6.56
2.01	11.7	8.72	19.7	5.78
3.29	3.62	4.53	9.22	4.61
5.9	11.05	1.92	6.19	3.52
8.08	0.44	0.35	2.39	2.51
10.2	0.09	0.10	0.57	1.17
12.9	0.20	0.23	0.73	1.05
15.2	0.11	0.24	0.59	0.90
17.3	0.12	0.11	0.94	1.12
19.9	0.09	0.04	0.27	0.49
22.1	0.04	0.07	0.47	0.81
24.3	0.06	0.04	0.54	0.45
27.0	0.01	0.01	0.05	0.04
29.2	0.01	0.00	0.18	0.25
30.0	0.01	0.01	0.28	0.27

Source: Nash.[9]

declines because of the changing nature of the remaining pesticide to its matrix (i.e., due to volatilization, sorption, dissipation).

Temperature effects can also be entered into the mathematical description. The concept of using degree-days to remove temperature variations during pesticide decline has been previously reported.[4,12] In one experiment, the soil temperature was set at 5°C, but with hot humid temperatures, we were unable to maintain the 5°C temperature. Dieldrin volatilization increased with the higher temperatures. Dieldrin flux, when plotted logarithmic versus linear time, followed a curve related to the soil temperature. In order to improve the flux-time relationship, degree-days were divided into the flux to give a new smoother curve. Figure 6 shows the original flux values versus time (upper curve), the flux divided by the experimentally measured degree-day values (lower curve) from which a power equation [flux = 3.89 day (exp -1.01)] was developed, and an adjusted flux (middle curve) obtained from the power equation that shows what the flux curve would have been had the temperature been maintained at 5°C throughout the experiment. At 5°C temperature, dieldrin volatilization was almost constant because of very slow dieldrin dissipation from the soil.

The total pesticide dissipation can be modeled by combining data from volatilization and soil dissipation. As stated previously, pesticide dissipation from soil is often biphasic. There is an initial period of fast dissipation followed by a period of slower dissipation. Volatilization contributes to the initial

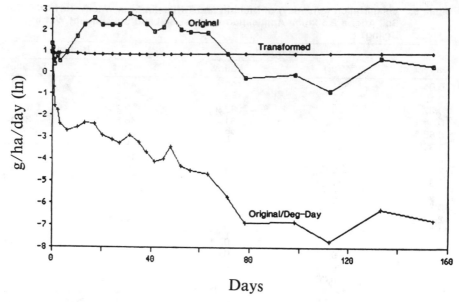

Figure 6.   Dieldrin flux amount in air after 2.5 kg/ha application to fallow soil surface maintained moist near 5°C. The top curve is the original flux data (g/ha/day) reflecting the variable soil temperature. The bottom curve is the original flux divided by degree-day, which reduces the temperature fluctuation. The middle curve is the adjusted flux (g/ha/day), obtained from a power equation describing flux/degree-day, then transformed back to flux day. This curve shows what the flux curve would have been had the soil temperature been maintained at 5°C. *Source:* Nash and Gish.[3]

fast dissipation, but usually does not account for all of the initial pesticide loss.

To model this phenomenon, a "six-compartment" model (6CM) is proposed (Figure 7). The pesticide spray is deposited in the C_1 compartment (i.e., the soil surface) where the residues are initially nonadsorbed, labile, and readily available to several dissipation processes (physical loss, volatilization, photolysis, surface hydrolysis, and other chemical reactions). If the residues were to dissipate from C_1 only, the overall observed effect usually is shown to be an exponential decline of soil residues.[1] However, residue decline on soils is usually biphasic (Figure 2). Thus, a C_2 compartment is proposed in which the residues become adsorbed to the soil, becoming less available for dissipation, and thus concentration on soil declines at a slower rate. Chemical analysis with good exhaustive extraction does not differentiate between C_1 and C_2 residues, thus C_3 is required to represent the total detectable soil residues ($C_1 + C_2$). When data are available that measure the pesticide in the air above the soil surface, a C_4 air compartment may be specified. Otherwise, volatilization would be grouped with the other processes that act on the C_1 residues to produce a pool of "fast" dissipated pesticide represented by C_5. The pesticide lost via the "slow" dissipation of the C_2 residues is represented by C_6. The rate

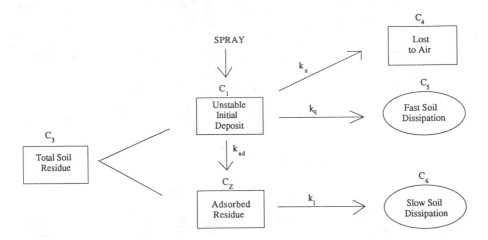

Figure 7.   The six-compartment model (6CM) for pesticide dissipation.

constants (k_{ad}, k_a, k_q, k_i) for the movement of pesticide between compartments are all assumed to be exponential. More specifically, k_{ad} represents the rate of residue adsorption to the soil, k_a represents the rate of volatilization from the soil, k_q represents the fast surface processes (thought to be mainly chemical reactions), and k_i represents the slow dissipation processes (thought to be mainly microbial and enzymic reactions).

A computer program written in BASIC (available from B. D. Hill on request) is used to fit the 6CM simultaneously to the soil residue (C_3) and air (C_4) data from several experiments[3,13] conducted at several temperatures in a microagroecosystem in which the soil was maintained moist.[14] This program allows one to iteratively solve for the best-fit k-values. A solution, with plausible k-values, helps one visualize and predict the movement and dissipation of the pesticide residues using an integrated approach, rather than applying largely empirical equations to the soil residue and air data individually.

The 6CM is applied to the soil residue and air data for dieldrin at 35°C in Figure 8. The soil residue decline (C_3) was biphasic with a fast initial loss for about 30 days, followed by a much slower decline (the percent remaining scale is linear in Figure 8; the biphasic decline is more obvious in the log-linear plot of the same data in Figure 2). There was a significant accumulation of volatilized residue (C_4), which by day 160 was about 50% of the initial soil residue. The 6CM fit the observed C_3 and C_4 data well. It predicted some accumulated "fast" dissipated residues (C_5) during the first 30 days, and very little "slow" dissipated residues (C_6) over the 160 days. Thus, the 6CM predicted significant volatilization, some initial dissipation (presumably physical and chemical losses), and very slow biodegradation, presumably, of dieldrin under the conditions of this experiment.

In contrast, the data indicated and the 6CM predicted a much different type

% Remaining

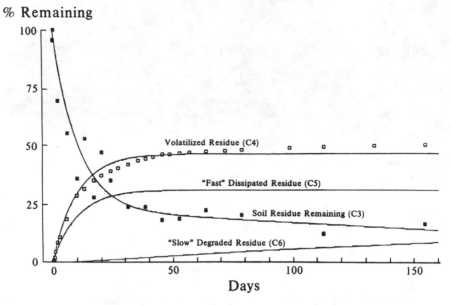

Figure 8.   Dieldrin dissipation using the 6CM model. The rate values (% with time) for air (k_a), fast soil dissipation (k_q), adsorbed (k_{ad}), and slow soil dissipation (k_i) were 0.045, 0.03, 0.022, and 0.00324, respectively. The points are actual data from a 2.5 kg/ha application to fallow soil maintained moist at 35°C. *Source:* Nash and Gish.[3]

of residue decline for silvex at 20°C (Figure 9). The silvex soil residues (C_3) also declined biphasically, but there was very little volatilized residue (C_4) or "slow" dissipated residue (C_6). Soil residue decline was almost entirely due to "fast" dissipation processes (probably surface hydrolysis and subsequent degradation).

We have applied the 6CM to the residue decline of different pesticides at different temperatures. We have found that the 6CM fits the data well and gives plausible k-values in most cases. We hope to see a pattern emerge for predicting the amount of volatilization from soil residues that is dependent on the pesticide and the temperature.

SUMMARY

A portion of many pesticides volatilizes when applied to soil or to crop residues. The magnitude of volatilization depends upon the pesticide's vapor pressure (but not directly so), temperature, soil moisture, and type of surface. Several halogenated pesticides volatilized more rapidly from a weedy turf than from a fallow soil. The amount of volatilization over a period of time also depends upon how fast the pesticides are dissipated. Significant amounts of "low vapor pressure" pesticides can be volatilized on a cumulative basis because they dissipate so slowly. A six-compartment model is proposed to

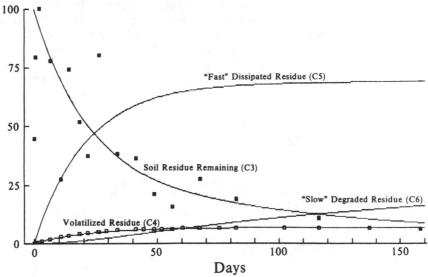

% Remaining

Days

Figure 9. Silvex dissipation using the 6CM model. The rate values (% with time) for air (k_a), fast soil dissipation (k_q), adsorbed (k_{ad}), and slow soil dissipation (k_i) were 0.00317, 0.0321, 0.0117, and 0.0076, respectively. The points are actual data from a 2.5 kg/ha application to fallow soil maintained moist at 20°C. *Source:* Nash.[6]

provide an integrated approach to modeling the movement and dissipation of pesticides from the soil into air. Several examples are given to demonstrate the probable magnitude of pesticide volatilization from fields under different conditions.

REFERENCES

1. Nash, R. G. "Dissipation from Soil," in *Environmental Chemistry of Herbicides,* R. Grover, Ed., Vol. I (Boca Raton, FL: CRC Press, 1988), pp. 131–169.
2. Nash, R. G. "Halogenated Pesticide Dissipation from Moist Soils at Several Temperatures in Chambers," 191st American Chemical Society Meeting, Environmental Chemistry Division, Vol. 26, No. 1 (1986), pp. 5–7.
3. Nash, R. G., and T. J. Gish. "Halogenated Pesticide Volatilization and Dissipation from Soil under Controlled Conditions," *Chemosphere* 18:2353–62 (1989).
4. Hill, B. D., and G. B. Schaalje. "A Two-Compartment Model for the Dissipation of Deltamethrin on Soil," *J. Agric. Food Chem.* 33:1001–6 (1985).
5. Hill, B. D., D. J. Inaba, and W. A. Charnetski. "Dissipation of Deltamethrin Applied to Forage Alfalfa," *J. Agric. Food Chem.* 37:1150–53 (1989).
6. Nash, R. G. "Volatilization and Dissipation of Acidic Herbicides from Soil under Controlled Conditions," *Chemosphere* 18:2363–73 (1989).
7. Nash, R. G., and M. L. Beall, Jr. "Chlorinated Hydrocarbon Insecticides: Root Uptake versus Vapor Contamination of Soybean Foliage," *Science* 168:1109–11 (1970).

8. Nash, R. G., A. R. Isensee, and C. S. Helling. "Rain and Irrigation Washoff of Pesticides from Crop Residue under Conservation Tillage," 197th American Chemical Society Meeting, Agrochemicals Division Paper No. 97, Dallas, TX (1989).

9. Nash, R. G. "Comparative Volatilization of Several Pesticides from Soil or Grass under Controlled Temperature Regimes," 3rd Chem. Congress North America, Agrochemicals Division Paper No. 29, Toronto, Ont. (1988).

10. McDowell, L. L., G. H. Willis, L. M. Southwick, and S. Smith. "Methyl Parathion and EPN Washoff from Cotton Plants by Simulated Rainfall," *Environ. Sci. Technol.* 18:423–27 (1984).

11. Nash, R. G., M. L. Beall, Jr., and W. G. Harris. "Toxaphene and 1,1,1-Trichloro-2,2-bis(*p*-chlorophenyl)ethane (DDT) Losses from Cotton in an Agroecosystem Chamber," *J. Agric. Food Chem.* 25:336–41 (1977).

12. Hill, B. D., W. A. Charnetski, G. B. Schaalje, and B. D. Schaber. "Persistence of Fenvalerate in Alfalfa: Effect of Growth Dilution and Heat Units on Residue Half-Life," *J. Agric. Food Chem.* 30:653–57 (1982).

13. Nash, R. G. "Models for Estimating Pesticide Dissipation from Soil and Vapor Decline in Air," *Chemosphere* 2375–81 (1989).

14. Nash, R. G. "Determining Environmental Fate of Pesticides with Microagroecosystems," *Residue Reviews* 85:199–215 (1983).

## CHAPTER 3

Comparison of Pendimethalin Airborne and Dislodgeable Residues Following Application to Turfgrass

J. J. Jenkins, R. J. Cooper, and A. S. Curtis

INTRODUCTION

The increased use of pesticides on turf in recent years has elevated concern because of the high potential for human exposure. The recent controversy over the reevaluation of the chronic health effects of 2,4-dichlorophenoxyacetic acid derivatives prompted the Massachusetts Pesticide Board to take action to further restrict their use. In addition, regulations have been adopted in Massachusetts that require prior notification of all commercial pesticide application to lawns, and new golf course construction in Massachusetts may require an approved pesticide management plan. Such a plan is designed to minimize environmental hazards and human exposure by requiring the best available pest management practices, which minimize pesticide use.

Risk assessment of increased pesticide use on turfgrass is constrained by the lack of information on the environmental fate of pesticides applied to turf. Evaluation of the potential for human exposure, as well as the degree of exposure, must take into consideration pesticide attenuation and movement in the environment. Following application, potential pathways of pesticide attenuation include sorption to thatch and soil; plant uptake and metabolism; microbial metabolism; and chemical and photochemical degradation.[1] In addition, pesticides may move from the site of application with water movement via runoff[2] or leaching,[3] or as airborne residues resulting from volatilization or wind erosion.[4] Much of the initial pesticide loss from turf may be as airborne residues.[5,6] The characterization of pesticide airborne residues is of interest not only as an attenuation pathway for surface deposits, but also due to the potential for human exposure through inhalation. In addition, pesticide airborne loss is of concern as a factor that may contribute to reduced efficacy.

Field research to date has focused primarily on volatile losses of insecticides and herbicides applied to bare soil or field crops. An in-depth review of field measurements of pesticide volatilization rates has been prepared by Taylor and

Glotfelty.[4] Glotfelty et al.[7] evaluated three insecticides, heptachlor (1,4,5,6,7,8,8-heptachloro-3a,4,7,7a-tetrahydro-4,7-methanoindene), chlordane (1,2,4,5,6,7,8,8-octachloro-2,3,3a,4,7,7a-hexahydro-4,7-methanoindene), and lindane (hexachlorocyclohexane), and two herbicides, trifluralin (a,a,a-trifluoro-2,6-dinitro-N,N-dipropyl-p-toluidene) and dacthal (dimethyl tetra-chloroterephthalate), surface-applied to fallow soil. They reported that except for the herbicide dacthal, volatility reduced surface soil residues by 50% within 3 days. Taylor et al.[5] reported volatile losses of the insecticides dieldrin (1,2,3,4,10,10-hexachloro-6,7-epoxy-1,4,4a,5,6,7,8,8a-octahydro-1,4-endo-5,8-exo-dimethanonaphthalene) and heptachlor totaling 34 and 57%, respectively, within 23 days following application of 5.6 kg/ha to a grass pasture. This magnitude of volatile loss was substantially higher than for application of like rates of soil-incorporated dieldrin and heptachlor, which resulted in losses of 3 and 7%, respectively, over a period of 167 days following application.[8] White et al.[9] used the method first described in Hartley[10] to monitor concentrations of trifluralin in air following application to soil at soybean planting and reported that approximately half of the trifluralin seasonal loss of 25.9% occurred during the first 9 days following application and 90% during the first 35 days. Grover et al.[11] evaluated trifluralin volatility following soil incorporation prior to establishment of wheat and estimated total trifluralin vapor losses of 23.7% over a 67-day period, with 51% of the loss occurring during the first 7 days.

In the research cited above, the predominate method for measuring airborne loss employed an estimate of the vertical flux using the aerodynamic method of Parmele et al.,[12] which requires simultaneous measurements of pesticide concentration, wind speed, and temperature at several heights in the atmosphere over the field surface. This technique works best under conditions of relatively uniform source strength, roughness length, wind field, and turbulence, requiring a uniform fetch on the order of 300 m.[13] More recently the trajectory simulation method of Wilson et al.[13] has been used to estimate the vertical profile of the pesticide horizontal flux from a single wind speed and concentration measurement taken above the center of a circular plot with a radius of ≤ 50 m.[14]

Another approach to understanding pesticide volatility has been through the use of laboratory microcosms. Nash et al.[15] developed a microagroecosystem that has been used in both the laboratory and the greenhouse to examine the effect of a number of variables on pesticide volatility.[16,17] A similar system was employed by McCall et al.[18] to model herbicide volatility and foliar behavior. Sanders and Seiber[19] used a chamber to measure pesticide volatility from model soil and water disposal systems in the laboratory. Subsequently, Sanders et al.[20] used this chamber for measuring volatility from waste disposal sites in the field.

The laboratory setting offers the advantage of allowing for the design of well-controlled experiments to determine the effect of individual variables, which would otherwise be difficult to accomplish with field experiments. Field

Table 1. Selected Physical/Chemical Properties of Pendimethalin.

Molecular Weight	281.31
Vapor Pressure	3×10^{-5} mm Hg at 25°C
Water Solubility	0.3 ppm

experiments, however, have the advantage of permitting the evaluation of pesticide fate under natural conditions, which are often difficult to simulate in the laboratory. The research reported here attempts to benefit from both approaches by taking the laboratory microcosm to the field. This research employs chambers similar in design to that of Sanders et al.,[20] allowing for a replicated small plot design (< 40 m²) to estimate airborne loss of the dinitroaniline herbicide pendimethalin following application to turfgrass. Selected pendimethalin physicochemical properties are given in Table 1.

To estimate surface residues available for volatilization, pendimethalin foliar dislodgeable residues were also determined. Foliar dislodgeable residues are thought to be those residues that are most easily dislodged from the foliar surface (loosely bound) and available for dermal transfer, as opposed to more tightly bound or penetrated residues. A procedure for determining foliar dislodgeable residues was originally developed by Gunther et al.[21] to assess dermal exposure to workers reentering treated fields. In a detailed evaluation of citrus leaf compartmentalization of pesticides, Nigg et al.[22] offered an alternative method for the determination of dislodgeable residues. More recently the method of Gunther et al.[21] has been used to determine safe reentry intervals for treated lawns.[23] Although not the focus of the research cited above, "loosely bound" foliar dislodgeable residues may be characterized as those residues that are most available for airborne loss and movement to other parts of the canopy with infiltrating water or by dry deposition.[24,25]

A detailed description of the airborne residue and foliar dislodgeable residue sampling methodology and analytical procedures is in preparation by the authors.[26] A brief description of this work is given below.

EXPERIMENTAL DESIGN

All field experiments were conducted at the University of Massachusetts Turf Research Farm located in South Deerfield, Massachusetts. Pendimethalin 60 WDG was applied at 3.4 kg a.i./ha. Plot design consisted of seven pendimethalin-treated strips (0.4 m × 3.4 m), separated by walkways with similar dimensions, plus an untreated strip. Each strip was designed to accommodate three chambers. The chambers consisted of 19-L Pyrex bottles with the bottoms removed. Chamber design is shown in Figure 1. Chambers were positioned in slits 2.5 cm deep and sealed by compacting the soil and turf around the outside of the chamber wall to minimize air entering the chamber from this source. Pendimethalin airborne residues were trapped using 25 g (50 mL) of Amberlite XAD-4 polymeric resin. A similar amount of resin was

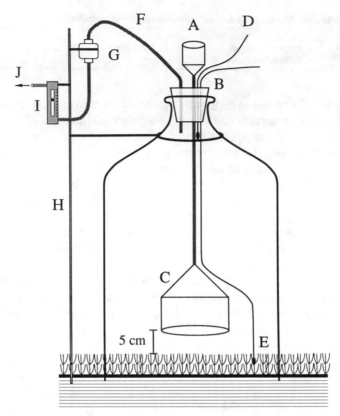

Figure 1.   Diagram of field chamber. (A) intake resin trap, (B) aluminum foil–covered stopper, (C) air dispersion funnel, (D) chamber ambient temperature thermistor, (E) chamber canopy temperature thermistor, (F) Teflon tubing, (G) Teflon resin cartridge, (H) support rod, (I) flow meter, (J) connection to vacuum pump.

placed at the inlet to remove any pendimethalin airborne residues from the incoming air. Air was pulled through each chamber at a rate of 30 L/min. Prior to field studies, the trapping efficiency of the resin cartridges was determined. No measurable breakthrough was determined for concentrations up to 230 μg/m^3, at a limit of detection of 0.3 μg/m^3.[26] Chambers were treated with Sylon-CT silanizing reagent to minimize pendimethalin sorption to the glass.

Air samples and foliar dislodgeable residue samples were collected over a 14-day period following application. Sample collection began at 0700 hr on May 9, 1988, immediately following application. Air samples were collected continuously for the first 48 hr to characterize diurnal variation in pendimethalin airborne flux. For each 24-hr sampling period, 2-hr samples were collected from 0700–1900, followed by a single overnight sample collected from 1900–0700. On successive sampling days (May 12, 13, and 23) 2-hour samples were collected from 0900–1700, with no overnight sampling. To minimize the impact of the sampling method on the "natural environment," for successive diurnal sampling periods the chambers were moved to a new strip, thereby assuring that no treated area was covered by a chamber for more than 2 hr in a given sampling day (except for the overnight samples). In addition to collecting the XAD-4 resin at the end of each sampling period, the chambers were disassembled and the glass bottles and dispersion funnels were rinsed with a total of 200 mL acetone-hexane (1:1) to determine pendimethalin airborne residues that were deposited on those surfaces during sampling.

Foliar dislodgeable samples were collected during each sampling period from an area adjacent to each chamber using a 10.8 cm diameter golf cup cutter, giving a sample area of 91.5 cm². The turfgrass was separated from the thatch and soil and extracted with 80 mL of methanol for 1 min using the method of Nigg et al.[22] All samples were stored in amber glass jars in darkness at 3°C until analysis.

Temperature was monitored at the grass surface both inside and outside the chambers at the beginning, middle, and end of each sampling period using YSI model 401 thermistors (Yellow Springs Instruments). In addition, ambient temperature, relative humidity, rainfall, and total daily solar radiation were recorded using a LI-1200S Data logger (LI-COR, Inc.).

RESIDUE ANALYSIS

Pendimethalin was extracted from XAD-4 resin samples by shaking for 1 hr in 250 mL of acetone-hexane (1:1) using a Wrist-Action shaker. The extracts were then filtered and concentrated by rotary evaporation at 45°C, taken to dryness under nitrogen, and dissolved in toluene for analysis. Chamber/funnel rinsate samples were dried over anhydrous sodium sulfate, concentrated by rotary evaporation at 45°C, taken to dryness with N_2, and dissolved in toluene for analysis.

Distilled water was added to the foliar dislodgeable residue methanol extracts to give a 10% methanol in water solution. To extract and enrich the pendimethalin residues, the methanol-water solution was passed through a 6-mL C18 column at a rate of 20 mL/min., and the eluant discarded. The column was then eluted with 4 mL acetone-hexane (1:1). The acetone-hexane (1:1) eluant was taken to dryness under nitrogen and dissolved in toluene for analysis.

All samples were analyzed by capillary gas liquid chromatography using a Hewlett-Packard 5890A GC equipped with an N-P detector and a 15 m × 0.246 mm DB-17 column (J & W, Inc.). Quantification was by peak area. Recovery efficiency was 94% (SD = 12, n = 13) for chamber/funnel rinsate samples, 93% (SD = 15, n = 11) for XAD-4 resin samples, and 97% (SD = 10, n = 16) for foliar dislodgeable residue extracts. Pendimethalin airborne residues are reported as the sum of the residues in the XAD-4 resin extract and the chamber/funnel rinsate. Pendimethalin airborne residue flux is expressed as $\mu g/m^2$ hr and mg/m^2 day. Hourly flux for each of the three replicate chambers was calculated by dividing the pendimethalin airborne residues by the product of the sampling period and the size of the surface area sampled (0.06 m²/chamber). Daily flux was determined by summing the average pendimethalin airborne residues for each of the 2-hr sampling periods between 0700–1900 hr, and the overnight (1900–0700 hr) sampling period. Foliar dislodgeable residues are expressed as $\mu g/cm^2$ leaf surface area (single side). The leaf area for each sample was determined experimentally from the weight of the turfgrass and a least squares linear regression

$$area = 78.3 \times weight \text{ (in grams)} - 7.3$$

of leaf area to leaf weight, determined experimentally using the method of Goh et al.[23] Limits of detection were 1 μg/chamber for chamber/funnel rinsate samples and 0.3 $\mu g/m^3$ of air sampled for XAD-4 resin samples. For the field experiments, these limits of detection resulted in a limit of detection for airborne flux of approximately 8 $\mu g/m^2$ hr. The limit of detection for the foliar dislodgeable residues was 1 μg/sample. The turf leaf area ranged from approximately 300 to 500 cm²/sample, giving a limit of detection on a leaf area basis of approximately 2.5 ng/cm².

RESULTS AND DISCUSSION

Pendimethalin hourly fluxes, turf canopy surface temperatures inside and outside the chambers, and solar radiation are given in Table 2. For the first 48 hr following application, pendimethalin airborne flux was characterized by continuous sampling as described above. In the initial 24-hr period (day 1), the total flux was 17.1 mg/m², with a flux of 16.0 mg/m² for the 0700–1900 hr sampling period (94% of the total flux for day 1) and 1.1 mg/m² for the 1900–0700 hr overnight sampling period. For the second 24-hr period (day 2), the flux was 3.2 mg/m², with a flux of 1.9 mg/m² from 0700–1900 hr and 1.3 mg/m² from 1900–0700 hr. Airborne loss during the first 48 hr following application represents 6.0% of the pendimethalin applied. Continuous sampling ended at 0700 hr on day 3, and on successive sampling days (4, 5, and 15 days following application) 2-hr samples were collected between 0900–1700 hr only. Figure 2 shows daily airborne loss for the 0900–1700 hr sampling period, along with total solar radiation for each sampling day. Based on this 8-hr sampling period, for days with similar solar radiation, daily airborne loss followed a bilinear pattern. This pattern is characterized by a period of rapid decline in daily flux ending 5–7 days following application, followed by a period of slower decline for the duration of the study. The marked difference in pendimethalin flux on day 2 may be the result of the heavy cloud cover with periods of rainfall totaling 10 mm. This weather resulted in low daily solar radiation, 4.4 MJ/m², as compared to the other sampling days, which aver-

Table 2. Pendimethalin in Airborne Flux from Kentucky Bluegrass Following Application of 3.4 kg a.i./ha 60 WDG

Day and Sampling Period (EDT)	Pendimethalin in Airborne Residues (μg/m² hr)				Chamber Temperature (Degrees C)		Outside Temperature (Degrees C)		Temperature Difference (Inside – Outside)	Solar Radiation (J/m² sec)	
	Rep 1	Rep 2	Rep 3	Ave (S.D.)	Average	Range	Average	Range		Average	Range
Day 1 (May 9)											
0700–0900	515	295	495	435 (122)	24.7	18.0–29.2	17.6	11.0–21.2	7.1	371	200–513
0900–1100	1148	1193	1457	1266 (167)	32.2	29.2–35.2	25.0	21.2–27.7	7.2	749	645–834
1100–1300	1782	1767	2423	1991 (374)	35.0	33.3–36.0	28.5	27.7–29.0	6.5	845	735–874
1300–1500	2177	2595	2295	2356 (216)	36.0	34.9–37.5	26.3	24.0–28.0	9.7	796	735–838
1500–1700	1693	1630	1420	1581 (143)	35.4	37.8–29.6	24.1	26.9–20.0	11.3	504	790–220
1700–1900	400	328	392	373 (39)	22.2	29.6–16.2	16.8	20.0–13.5	5.4	152	220– 47
1900–0700[a]	77	125	83	95 (26)	11.3	13.0– 9.5	10.5	11.5– 9.4	0.9		
Day 2 (May 10)											
0700–0900	105	80	78	88 (15)	14.0	12.3–16.5	12.6	10.9–15.3	1.4	74	20–184
0900–1100	188	150	212	183 (31)	18.0	16.5–19.1	16.1	15.2–17.9	1.9	218	184–255
1100–1300	210	212	238	220 (16)	18.2	20.1–16.4	16.9	17.9–15.4	1.3	167	255–105
1300–1500	203	222	235	220 (16)	17.0	16.6–17.4	14.8	15.8–12.7	2.2	112	128– 98
1500–1700	127	138	158	141 (16)							
1700–1900	115	120	132	122 (9)	13.0	13.4–12.6	11.6	12.0–11.5	1.4		
1900–0700[b]	98	120	97	105 (13)	13.8	12.7–14.9	12.2	11.4–12.9	1.6		
Day 4 (May 12)											
0900–1100	757	692		725 (46)	33.0	30.5–34.7	30.9	27.9–33.3	2.1	749	590–865
1100–1300	1148	1047	1165	1120 (64)	31.7	29.3–34.7	33.3	31.5–36.4	-1.6	927	865–960
1300–1500	1157	1148	1238	1181 (50)	32.9	33.5–32.0	29.1	32.4–25.9	3.8	910	960–820
1500–1700	1263	1263	1658	1395 (228)	35.1	32.9–37.0	28.5	30.2–27.0	6.6	674	820–543
Day 5 (May 13)											
0900–1100	628	510	620	586 (66)	32.0	29.5–33.9	28.1	25.8–30.6	3.9	693	570–806
1100–1300	610	638	757	668 (78)	35.4	33.9–37.0	28.3	30.6–25.9	7.1	882	806–922
1300–1500	1155	757	1088	1000 (213)	37.3	35.0–38.1	27.8	30.1–25.1	9.5	845	915–723
1500–1700	620	655	777	684 (82)	35.4	38.4–31.4	24.6	26.6–22.3	10.8	464	723–310
Day 15 (May 23)											
0900–1100	107	92	100	99 (8)	30.4	23.6–35.6	32.3	22.4–42.1	-1.9	643	290–827
1100–1300	143	93	98	112 (28)	34.8	33.3–36.0	41.7	39.8–43.0	-6.9	732	918–450
1300–1500	170	170	173	171 (2)	35.2	34.6–36.1	36.2	42.0–29.6	-1.0	513	360–729
1500–1700	165	180	193	179 (14)	35.1	36.0–34.7	30.7	32.9–29.5	4.4	466	729–205

[a]May 10.
[b]May 11.

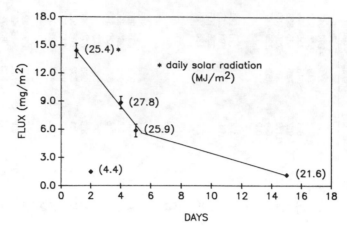

Figure 2. Decline in daily airborne loss (0900–1700 hr) following application of pendimethalin 60 WDG at 3.4 kg a.i./ha to turfgrass.

aged 25.2 ± 2.6 MJ/m². The lack of solar radiation, which controls the temperature at the evaporating surface, is thought to be responsible for low airborne flux on day 2.

Diurnal flux of pendimethalin airborne residues during the daylight hours for the first 48 hr is shown in Figure 3. The curves show the relationship between hourly flux and the diurnal variation in solar radiation shown in Figure 4, with peak flux occurring between 1300–1500 hr. For the purpose of estimating flux during daylight hours on successive sampling days, described

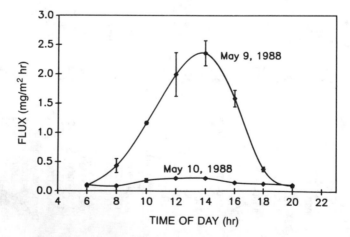

Figure 3. Diurnal flux of airborne residues following application of pendimethalin 60 WDG to turfgrass. Airborne loss was monitored continuously for the first 48 hr beginning at 0700 hr on the day of application (May 9, 1988). For the purpose of estimating flux during daylight hours on successive sampling days, the average hourly flux for the overnight sampling period was used for the values shown at 0600 and 2000 hr.

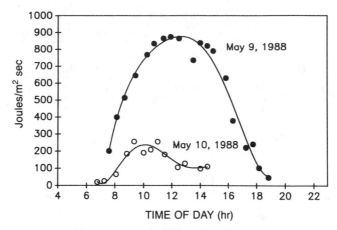

Figure 4.   Solar radiation (J/m² sec) for the daytime air sampling period (0700–1900 hr) for the first 48 hr following pendimethalin 60 WDG.

below, the average hourly flux for the overnight sampling period was used for the values at 0600 and 2000 hr shown in Figure 3.

Method for Estimating Daily Airborne Flux from Foliar Surfaces

An alternate approach to estimating daily airborne loss from foliar surfaces is to assume that

1. peak flux occurs during the 2-hr sampling period between 1300–1500 hr
2. airborne loss occurs only during daylight hours (0600–2000 hr in Figure 3)
3. the loss rate varies linearly between the endpoints and the peak

Integration of the area under the triangle formed gives a daily airborne loss of seven times the peak hourly flux measured between 1300–1500 hr. The advantage to this approach is that only the peak hourly flux must be measured to estimate the daily flux. Although sampling did not begin until 0700 hr on day 1, for the purpose of estimating flux on successive days "daylight hours" was determined to be 0600–2000 hr. Using the overnight average hourly flux to estimate the flux from 0600–0700 hr and 1900–2000 hr, the flux for the 0600–2000 hr period on day 1 is 16.2 mg/m² . Integration of the area under the triangle formed for this period gives a value of 16.5 mg/m², in close agreement with the measured value. Performing a similar integration using the peak hourly flux for day 2 gives a value of 1.5 mg/m², 70% of the measured value of 2.2 mg/m². This underestimate of the measured value suggests that weather conditions of minimal or nonuniform diurnal variation in solar radiation, such as would result from heavy or intermittent cloud cover, are likely to result in a flux pattern that is not well represented by this simple integration technique. Using this technique and the peak hourly flux values for the remaining sample days, the daily flux was estimated to be 8.3, 7.0, and 1.2 mg/m² for days 4, 5,

Table 3.   Measured vs Estimated Pendimethalin Daily Airborne Loss.

| Sampling Interval | Peak Flux (mg/m^2 hr) (1300–1500) | Daily Airborne Loss (mg/m^2) | | |
		Sampling Period	Measured	Estimated (0600–2000)
Day 1	2.356	(0700–1900)	16.0	16.5
Day 2	0.220	(0700–1900)	1.9	1.5
Day 4	1.181	(0900–1700)	8.8	8.3
Day 5	1.000	(0900–1700)	5.9	7.0
Day 15	0.171	(0900–1700)	1.1	1.2

and 15, respectively. Measured and estimated pendimethalin daily airborne loss is shown in Table 3. The bilinear pattern of residue decline determined from the measured airborne loss during the 0900–1700 hr sampling period is also indicated by the estimated daily airborne loss values (0600–2000 hr) given in Table 3.

Effect of Temperature and Solar Radiation on Airborne Flux Measurements

As has been observed by other researchers,[5-7] variation of pendimethalin hourly airborne flux was greatest on the day of application and followed the diurnal variation in solar radiation. However, in the present experiment, hourly flux was more strongly correlated to the chamber canopy temperature than solar radiation. The stronger correlation of airborne flux to chamber canopy temperature, as compared to solar radiation, is thought to be primarily due to the "greenhouse effect" inside the chambers, resulting in higher afternoon temperatures inside the chambers as compared to ambient. The temperature differential was most pronounced between 1300 and 1700 hr on clear days, and averaged 10.5°C on day 1, 5.2°C on day 4, and 10.2°C on day 5. Intermittent cloud cover on day 15 was thought to be responsible for the 1.7°C average temperature differential observed during the 1300–1700 hr sampling period.

Contribution of Airborne Flux to the Attenuation of Foliar Residues

To characterize the relative contribution of airborne loss to the overall attenuation of pendimethalin from turf, other processes acting on the residue deposits were also considered. These processes include chemical, photochemical, and microbial degradation, as well as plant uptake and metabolism, and redistribution within the turf canopy. While the observed attenuation of pesticide foliar residues is the sum of these processes, their individual contribution is generally not known. However, in the present study, attenuation of pendimethalin foliar dislodgeable residues, by redistribution or other mechanisms described above, is thought to be primarily responsible for the decrease in the observed daily airborne flux. This relationship is indicated by the similarity between the pattern of decline of daily airborne flux (Figure 2) and the pattern observed for the attenuation of the foliar dislodgeable residues shown in Figure 5 (average pendimethalin foliar dislodgeable residues for the midday sampling period, 1100–1300 hr). The decline in foliar dislodgeable residues can also be characterized as bilinear, with an initial period of rapid decline ending 5–7 days following application, followed by a period of slower decline for the duration of the study. The natural break in both sets of data between days 5 and 7 supports the observation that dislodgeable residues are most available for airborne loss. Conversely, airborne loss may contribute more to pendimethalin attenuation during the initial rapid phase than during the slow phase.

Conceptual Model for Airborne Loss of Foliar Residues

To aid in the understanding of the relationship between the overall attenuation of foliar residues and airborne loss, a simple two-compartment model may be employed in which pesticide foliar residues are conceptualized as either residing on or below the leaf surface. Subsurface residues may be embedded in the epicuticular waxes or penetrated into the plant tissues. Surface residues are thought to be available for airborne loss and movement into the canopy with infiltrating water or by dry deposition, while subsurface or penetrated residues are thought to be unavailable for airborne loss or movement.[27]

Use of the two-compartment model requires the ability to selectively remove surface residues. For this reason the foliar dislodgeable residue method was used to follow the attenuation of dislodgeable surface residues. Although not originally developed for this purpose, by selectively removing only the loosely bound or dislodgeable surface residues, this method should provide an estimate of those residues that are most available for airborne loss.

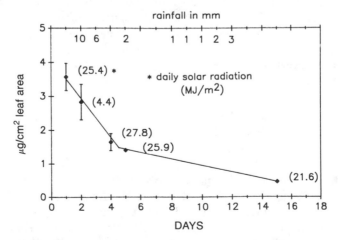

Figure 5. Foliar dislodgeable residues (μg/cm^2 single-sided leaf area) following application of pendimethalin 60 WDG at 3.4 kg a.i./ha to turfgrass.

Table 4.   Daily Airborne Loss as a Percentage of Foliar Dislodgeable Residues Following Application of Pendimethalin 60 WDG at 340 mg a.i./m²

| | Foliar Dislodgeable Residues | | Airborne Loss | |
Sampling Interval	Leaf Area[a] (μg/cm²)	Surface Area (mg/m²)	(0900–1700 hr) (mg/m²)	Percent Dislodgeable Residues Volatilized
Day 1	3.6	178	14.4	8.1
Day 2	2.8	138	1.5	1.1
Day 4	1.7	71	8.8	12.4
Day 5	1.4	53	5.9	11.1
Day 15	0.5	22	1.1	5.0

[a]Leaf area was determined experimentally from sample weight based upon the linear regression of area to weight (area = 78.3 × [weight] − 7.3), using the method of Goh et al.[23]

The data in Table 4 suggest that the two-compartment model may not be adequate to describe the relationship between foliar residues and airborne loss. These data indicate that during the rapid phase of attenuation, for days with similar solar radiation (days 1, 4, and 5), airborne loss accounts for approximately 10% (8–12%) of the daily foliar dislodgeable residue attenuation. By the end of the study (day 15), this percentage has dropped to 5%. The observation that airborne loss accounts for less of the overall attenuation of pendimethalin foliar dislodgeable residues on day 15 as opposed to earlier in the study suggests that the dislodgeable residues remaining after 15 days of attenuation are more tightly bound and less available for airborne loss than the dislodgeable residues present earlier in the study. This finding would suggest that a three-compartment model, composed of loosely bound, dislodgeable, and subsurface or penetrated residues, might better represent the foliar residue dynamics of pendimethalin. Using this model, the dislodgeable residue method would remove both the loosely bound and dislodgeable residues. The loosely bound residues would be more susceptible to airborne loss or redistribution to other parts of the canopy and would contribute less to the dislodgeable residue values determined later in the study. An alternative model has been suggested by McCall et al.[18] They propose that although penetration into the plant tissues further reduces the pesticide available to volatilize, this may be an equilibrium process, and as surface residues are depleted, penetrated residues may diffuse back to the surface following the concentration gradient.

The use of compartmental models to represent foliar residues available for airborne loss is complemented by the predominant theories used to characterize pesticide airborne loss at the foliar surface. Airborne loss of surface residues, as it pertains to atmospheric diffusion and transport of pesticide vapors, is thought to be controlled by two processes: molecular diffusion and eddy dispersion. This implies that laminar flow exists near the foliar surface, with turbulent flow being predominant elsewhere.[28] Assuming that a sufficient pesticide gradient exists and diffusion through the stagnant air layer is rate limiting, for an evenly distributed layer of pesticide on a nonabsorbing surface at constant wind speed, airborne loss of the surface deposit will be a function of temperature and the inherent vapor pressure of the pesticide.[10] This would suggest that for conditions where there is complete coverage of the foliar surface (i.e., pesticide is layered on the surface and is in effect volatilizing from itself), pesticide airborne flux is controlled primarily by the temperature at the foliar surface, which determines the rate of vapor formation and diffusion through the stagnant air layer closely surrounding the surface.[6,29,30] However, as the pesticide surface deposits become increasingly smaller, airborne flux is no longer strictly a function of temperature-dependent molecular diffusion, but is also a function of the degree of sorption to the foliar surface.[6] Sorption and/or penetration into the cracks and fissures of the epicuticular wax will therefore result in an effective vapor pressure lower than the inherent vapor pressure, resulting in reduced volatilization.

Effect of Rainfall on Airborne Flux from Turf

Rainfall that occurred during the study must also be considered. The plots received 16 mm of rainfall between day 1 and day 4, and 10 mm of rainfall between day 5 and 15 (see Figure 2 for daily rainfall), which most likely resulted in redistribution of foliar dislodgeable residues to sites within the canopy and underlying soil.[24,25] Foliar residues that are redistributed into the canopy with infiltrating water may be less available for airborne loss. Lower temperatures and sorption of pendimethalin residues to the turf thatch (underlying detritus) and soil would be expected to result in further reduction in the effective vapor pressure. Average soil temperatures recorded at a depth of 10 cm ranged from 18 to 20°C during the study; the soil was a heat sink relative to surface temperatures. Heavy rainfall on unsaturated soils may also move initially displaced residues to lower soil depths, rendering these residues less available for airborne loss.[30,31] However, infiltrating water may also enhance airborne loss if the soil moisture is low prior to the rainfall event, as the increased soil moisture may displace sorbed pesticide, increasing its effective vapor pressure at the soil surface.[32-34] For pesticides applied to turfgrass, rainfall may result in redistribution of dislodgeable residues from the foliar surface to the thatch only, with some movement to the soil, depending on the thickness of the thatch layer.[35,36] Lower temperatures and increased sorption to organic matter should result in a lower rate of airborne loss from the thatch layer. As pendimethalin foliar surface deposits become depleted, and as residues are redistributed to a variety of sites with differing sorption properties and microclimates, there is no longer a simple relationship between solar radiation (as it affects surface temperature) and airborne loss. This is indicated by the strong correlation between the diurnal variation in chamber canopy temperature and flux observed early in the study (r = 0.94 for days 1 and 2) as opposed to later sampling dates (r = 0.48 for day 4, 0.83 for day 5, and 0.57 for day 15).

CONCLUSIONS

The sampling methodology used in this study allowed the design of a small plot field experiment ($<$ 40 m^2) with replication. Presumably, the large plot size required for the conventional methodology used in measuring pesticide airborne flux in the field has discouraged replication. Another advantage to the air sampling methodology reported here is that pesticide airborne residue measurements are made in a closed system and need not rely heavily upon obtaining simultaneous micrometeorological and pesticide airborne residue measurements, which are required to parameterize micrometeorological models (i.e., aerodynamic or trajectory-simulation models) used for the estimation of pesticide airborne flux. Such models must also rely on a number of simplifying assumptions to estimate airborne flux from the field data. Depending on which model is chosen, results may vary by 2–3 times.[14]

A disadvantage of the chamber method is the modification of the natural environment during sampling, particularly with regard to temperature and wind speed. To minimize the effect of the chambers on the natural environment, no area was sampled for more than 2 hr in any sampling day (except for overnight samples taken during the first 48 hr following application). For all but the overnight sampling locations, this sampling procedure resulted in modification of the turf environment for a total of 10 hr over the course of the 360-hr study. One reason for choosing pendimethalin for this study is that it is applied once in the spring for preemergent weed control, when ambient temperatures in New England are moderate compared to the summer months. Although temperatures inside the chambers were as much as 11°C greater than outside, chamber temperatures never exceeded 38°C. Another consideration in using the chamber technique is the effect of the modified environment during sampling on the flux values measured. Higher canopy temperatures inside the chambers should result in increased volatilization. A number of researchers have reported a three-to fourfold increase in pesticide volatilization for each 10°C increase in temperature.[18,30,37] These data suggest that the chamber method will *overestimate* flux by three-to fourfold when the temperature inside the chamber is 10°C greater than the outside temperature.

In addition, the low wind speed inside the chambers during sampling (0.03 km/hr) must also be considered. Assuming that diffusion across the stagnant air layer is rate limiting, wind speed will affect airborne flux as it affects the depth of the stagnant air layer.[38] Increased wind speed will reduce the depth of the stagnant air layer resulting in increased flux.[29] However, the relationship between wind speed and pesticide airborne flux has rarely been demonstrated. Figure 6 shows a log-linear relationship between wind speed and dieldrin flux using laboratory data reported by Spencer and Cliath.[30] Data was derived from experiments in which dieldrin was applied to the "nonadsorbing" surfaces of glass[29] or quartz sand.[39] For comparison purposes, where different temperatures were used, flux was adjusted to 30°C. These data suggest that between wind speeds of 0.0033 and 4.8 km/hr, there is an exponential relationship

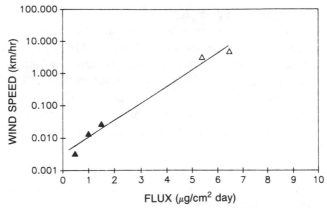

Figure 6. Comparison of dieldrin flux and wind speed. Open triangles: dieldrin flux from quartz sand at 30°C following treatment at 1000 μg/g, reported in Farmer and Letey.[39] Closed triangles: dieldrin flux from glass treated at a rate of 4–9 μg/cm², data of Phillips[29] adjusted by Spencer and Cliath[30] for differences in vapor pressure between 30°C and the temperatures used by Phillips.

between wind speed and flux. In a more recent study, McCall et al.[18] used the model

$$dC_{ar}/dt = k_v C_{sf}$$

where C_{ar} = amount of chemical in the air
         C_{sf} = amount of chemical on the plant surface
         k_v = rate constant for volatility

to examine the foliar behavior of tridiphane. Using data obtained from experiments run in a laboratory chamber, they determined k_v to be 0.15 ± 0.03, 0.15 ± 0.03, and 0.19 ± 0.03 at wind speeds of 0.4, 0.8, and 1.6 km/hr, respectively, and reported no significant difference in these values at the 95% confidence interval. However, given flux is proportional to k_v and using the relationship shown in Figure 6, the narrow range of wind speeds chosen would result in only a slight change in k_v: about a 29% increase between 0.4 and 1.6 km/hr. Interestingly, the k_v of 0.19 determined at 1.6 km/hr is 27% greater than the k_v of 0.15 determined at 0.4 km/hr, which is in close agreement with the relationship shown in Figure 6.

In the present study the wind speed in the chambers during sampling of 0.03 km/hr is well below what might be found in the field under all but "still air" conditions. Using the relationship shown in Figure 6, extrapolating from 0.03 km/hr to 4.8 km/hr (3 mi/hr) would result in approximately a threefold increase in flux; a 25-km/hr wind would be required for a fourfold increase. Whether the relationship shown in Figure 6 holds for wind speeds above 4.8 km/hr is not known. Phillips[29] concluded that the effect of increasing wind speed on increasing flux must ultimately reach a limiting value. Based on boundary layer theory, this limiting value is reached when increasing the wind speed is no longer capable of reducing the depth of the stagnant air layer. Therefore, extending the relationship shown in Figure 6 beyond 4.8 km/hr will eventually result in an overestimation of flux. Taking this into consideration, the chamber method will *underestimate* flux by three-to fourfold compared to "natural conditions" for a range of wind speeds of 3 to 25 km/hr.

The ability to reproducibly measure pendimethalin airborne loss in the field using the chamber technique suggests that this technique may be useful for determining airborne loss of other pesticides from turfgrass. The advantages of this method are

1. greater control over experimental conditions (most notably wind speed)
2. small plot size (< 40 m^2)
3. replicated design

The major disadvantage of this technique is that air sampling is conducted under artificial conditions. However, it appears that for pendimethalin, under the chamber conditions that prevailed during this study, the effects of the chamber environment on two of the critical parameters that control airborne loss from foliar surfaces—wind speed and temperature—were offsetting. Increased temperature in the chambers of approximately 10°C over the natural environment would result in a three-to fourfold increase in airborne flux, whereas the low wind speed in the chambers (compared to wind speeds of 3–25 km/hr, typical of field conditions) would result in a three-to fourfold decrease in flux.

The relationship shown between pendimethalin airborne loss and foliar dislodgeable residue attenuation demonstrates the utility of the dislodgeable residue method. Although this method was not originally intended for this purpose, it has proved to be a useful tool for investigating the redistribution of pesticide foliar residues to the atmosphere and other parts of the environment.

The meteorological conditions that prevailed during the course of the experiment afforded the opportunity to observe the direct relationship between solar radiation and pendimethalin airborne loss, as demonstrated by the marked deviation in daily flux observed on the day following application compared to the other sampling days. In addition, as is the case with most field research, the results presented are unique to the weather patterns that developed, both diurnally and over the two-week course of the study. Additional field research, under a variety of environmental conditions, is necessary to develop a better understanding of the relationship between the various environmental factors that influence pendimethalin airborne loss.

REFERENCES

1. Helling, C. S. *J. Environ. Qual.* 5(1):1–15 (1976).
2. Leonard, R. A., G. W. Bailey, and R. R. Swank, Jr. In *Land Application of Waste Materials* (Ankeny, IA: Soil Conservation Society of America, 1976).
3. Garner, W. J., R. C. Honeycutt, and H. N. Nigg, Eds. *Evaluation of Pesticides in Groundwater,* American Chemical Society Symposium Series 315 (Washington, DC: American Chemical Society, 1986).
4. Taylor, A. W., and D. E. Glotfelty. In *Environmental Chemistry of Herbicides,* R. Grover, Ed., Vol. 1 (Boca Raton, FL: CRC Press, 1988), pp. 89–129.
5. Taylor, A. W., D. E. Glotfelty, B. C. Turner, R. E. Silver, H. P. Freeman, and A. Weiss. *J. Agric. Food Chem.* 25:542–48 (1977).
6. Taylor, A. W. *J. Air Pollut. Control Assoc.* 28:922–27 (1978).
7. Glotfelty, D. E., A. W. Taylor, B. C. Turner, and W. H. Zoller. *J. Agric. Food Chem.* 32:638–43 (1984).
8. Taylor, A. W., D. E. Glotfelty, B. L. Glass, H. P. Freeman, and W. M. Edwards. *J. Agric. Food Chem.* 24:625–31 (1976).
9. White, A. W., Jr., L. A. Harper, R. A. Leonard, and J. W. Turnbull. *J. Environ. Qual.* 6:105–10 (1977).
10. Hartley, G. S. In *Pesticide Formulations Research: Physical and Colloidal Chemical Aspects,* Advanced Chemistry Series 86 (Washington, DC: American Chemical Society, 1969).
11. Grover, R., A. E. Smith, S. R. Shewchuk, A. J. Cessna, and J. H. Hunter. *J. Environ. Qual.* 17:543–50 (1988).
12. Parmele, L. H., E. R. Lemon, and A. W. Taylor. *Water Air Soil Poll.* 1:433–51 (1972).
13. Wilson, J. D., G. W. Thurtell, G. E. Kidd, E. G. Beauchamp. *Atmos. Environ.* 16(8):1861–67 (1982).
14. Majewski, M. S., D. E. Glotfelty, M. M. McChesney, and J. N. Seiber. Paper presented at the 194th National Meeting of the American Chemical Society, New Orleans, LA, Division of Agrochemicals, Abstract 121 (1987).
15. Nash, R. G., M. L. Beall, Jr., and W. G. Harris. *J. Agric. Food Chem.* 25(2):336–41 (1977).
16. Nash, R. G. *J. Agric. Food Chem.* 31:210 (1983).
17. Nash, R. G. *Chemosphere* 18:2363–73 (1988).
18. McCall, P. J., L. E. Stafford, and P. D. Gavit. *J. Agric. Food Chem.* 34:229–34 (1986).
19. Sanders, P. F., and J. N. Seiber. *Chemosphere* 12(7):999–1012 (1983).
20. Sanders, P. F., M. M. McChesney, and J. N. Seiber. *Bull. Environ. Contam. Toxicol.* 35:569–75 (1985).
21. Gunther, F. A., W. E. Westlake, J. H. Barkley, W. Winterlin, and L. Langbehn. *Bull. Environ. Contam. Toxicol.* 9:243–49 (1973).
22. Nigg, H. N., L. G. Albrigo, H. E. Nordby, and J. H. Stamper. *J. Agric. Food Chem.* 29(4):750–56 (1981).
23. Goh, K. S., S. Edmiston, K. T. Maddy, and S. Margetich. *Bull. Environ. Contam. Toxicol.* 37:33–40 (1986).
24. Jenkins, J. J., M. J. Zabik, R. Kon, and E. D. Goodman. *Arch. Environ. Contam. Toxicol.* 12:99–110 (1983).

25. Goodman, E. D., J. J. Jenkins, and M. J. Zabik. *Arch. Environ. Contam. Toxicol.* 12:111–19 (1983).
26. Jenkins, J. J., R. J. Cooper, and A. S. Curtis. Unpublished data (1988).
27. Jenkins, J. J. Ph.D. Dissertation, Michigan State University (1981).
28. Glotfelty, D. E., A. W. Taylor, and W. H. Zoller. *Science* 219:843–45 (1983).
29. Phillips, F. T. *Pestic. Sci.* 2:255–66 (1971).
30. Spencer, W. F., and M. M. Cliath. In *Fate of Pollutants in the Air and Water Environments,* I. H. Suffet, Ed., Part I (New York, NY: John Wiley and Sons, 1977), pp. 107–26.
31. Helling, C. C., P. C. Kearney, and M. Alexander. *Adv. Agron.* 23:147–241 (1971).
32. Spencer, W. F., and M. M. Cliath. *J. Environ. Qual.* 2:284–89 (1973).
33. Spencer, W. F., and M. M. Cliath. *J. Agric. Food. Chem.* 22(6):987–91 (1974).
34. Jury, W. A., R. Grover, W. F. Spencer, W. J. Farmer. *Soil Sci. Soc. Am. J.* 44:445–50 (1980).
35. Sears, M. K., and R. A. Chapman. *J. Econ. Entomol.* 72:272–74 (1979).
36. Niemczyk, H. D., and H. R. Krueger. *J. Econ. Entomol.* 80:950–52 (1987).
37. Burkhard, N., and J. A. Guth. *Pestic. Sci.* 12:37–44 (1981).
38. Scorer, R. S. In *Environmental Aerodynamics* (New York, NY: John Wiley and Sons, 1978), pp. 50–78.
39. Farmer W. J., and J. Letey. In "Volatilization Losses of Pesticides from Soils," Office of Research and Development, EPA-660/2-74-054 (1974).

## CHAPTER 4

A Field Mass Balance Study of Pesticide Volatilization, Leaching, and Persistence

L. D. Clendening, W. A. Jury, and F. F. Ernst

INTRODUCTION

There has been a considerable research effort directed over the years towards characterizing the movement and transformations of pesticides in soil after they are applied in farm operations. Awareness of the pollution potential of pesticides has been heightened recently by the discovery of their widespread appearance in groundwater,[1] and also by the appearance of concentrated pesticide residues in atmospheric aerosols.[2] As a result, California state legislation that is aimed at controlling pesticide contamination and restricting the use of compounds with adverse pollution potential has been passed,[3] and it seems likely that national regulations will soon follow.

If a pesticide regulatory program is to be successful, it must have at its foundation a quantitative means of estimating the potential for environmental pollution of a pesticide before it works its way to groundwater or contaminates the atmosphere. In principle, such an assessment is best made by conducting realistic experiments in natural field soils; in practice, however, there is no feasible way to assess the large number of registered and proposed pesticides in this manner.

Instead it probably will be necessary to evaluate most of the compounds using a model that requires only the limited experimental information readily available for the pesticides in use today. However, at the present time there is no one transport model that has been proven to describe chemical movement to the atmosphere and groundwater accurately under natural conditions.[4]

The majority of pesticide research studies have been conducted in the laboratory under controlled conditions that do not adequately represent natural field conditions. As a result, there is still a deficiency of information on the principal mechanisms controlling the fate of pesticides in field soils. However, within the last ten years, a few field-scale experiments have been conducted that characterize either pesticide transport properties in the field[5-10] or pesticide volatilization losses under field conditions.[11-14]

The field studies of pesticide transport that have been conducted to date have revealed some disturbing results. Rao et al.[15] found that the predicted movement of picloram using its laboratory-derived adsorption coefficient did not match the observed transport, and that picloram was recovered down to depths of 143 cm when the predicted maximum depth of penetration was only 42 cm. Elabd[9] conducted a field experiment on a structureless soil using a moderately adsorbed herbicide, napropamide, which had a predicted retardation factor with respect to a water tracer of approximately 9.5. His results showed that about 20% of the applied chemical exhibited extreme mobility, reaching depths far greater than those predicted from equilibrium adsorption models. Clendening[10] conducted two field experiments on the same field that Elabd used, and also reported higher than predicted mobility for both napropamide and prometryn, an even more strongly adsorbed herbicide. Up to 45% of the applied chemical was randomly located between the surface zone (where it was expected to reside) and depths reached by the bromide water tracer.

Rapid volatilization losses have also been recorded for certain compounds. Glotfelty[14] reported that 50% of heptachlor, chlordane, lindane, and trifluralin volatilized in less than 3 days from a moist soil; Taylor[13] reported volatilization losses of 90% in 2–3 days for dieldrin, trifluralin, and heptachlor.

There are virtually no simulation models in existence that simultaneously model transport, reactions, and volatilization of pesticides. As an alternative, Jury et al.[16-19] proposed using an analytical environmental fate screening model that considers pesticide volatilization, degradation, and mobility in a simplified transport scenario, which allows pesticides to be grouped according to their transport properties. The screening model requires only the input of basic environmental fate properties for each compound and can be used to estimate the behavior of one chemical relative to another under a standard set of conditions. The results of these screening tests can be used both to identify compounds with unusual behavior and also to sort large numbers of different compounds into smaller groups with similar behavior.

As discussed above, experimental field studies of pesticide transport and fate in soil, which are essential in evaluating the accuracy of modeling approaches like the screening model, are very limited, both in size and in scope. Even those few studies that have been conducted have focused either on leaching or on volatilization. The experimental field research described in this chapter is unique in that both processes were monitored, producing a data set that can be used both for model validation and for direct characterization of pesticide fate.

The field study consisted of the application and subsequent monitoring of five herbicides over a month with alternate periods of irrigation and evaporation. One of the objectives of this study was to evaluate the environmental fate of five different pesticides subjected to identical environmental conditions. A second objective was to compare the actual field measured values of the loss by volatilization, leaching, and degradation with screening model predictions using the behavior assessment model published earlier by Jury et al.[16]

Table 1. Environmental Fate Properties for the Five Pesticides Used in the Field Study.

Pesticide	Organic C Partition K_{oc} (cm^3/g)	Henry's Constant K_h	Half-Life T 1/2 (days)	Retardation Factor R^a
Atrazine	160	2.5×10^{-7}	71	5
Bromacil	72	3.7×10^{-9}	350	2
EPTC	280	5.9×10^{-4}	30	9
Prometon	408	1.0×10^{-7}	100	13
Triallate	3600	7.9×10^{-4}	100	100

Source: adapted from Jury et al.[18]
$^a R = 1 + \rho_b f_{oc} K_{oc}/\theta$

MATERIALS AND METHODS

Chemical Properties

The experiments were conducted at a 2-ha field site in southern California whose surface matter is predominantly a sandy loam (Tujunga loamy sand; mixed, thermic, typic Xeropsamments) classified as visually structureless, well drained, with low organic carbon content. A mixture of five different pesticides (EPTC, prometon, atrazine, triallate, and bromacil), ranging in volatility, mobility, and persistence over a substantial range, were selected for the mass balance study. Their principal environmental fate properties are given in Table 1. The organic carbon partition coefficient K_{oc} ranges among the compounds from a low of 70 to a high of 3600 cm^3/g, producing a predicted retardation with respect to a water tracer (such as chloride) ranging between a factor of 2 for the mobile compound bromacil to a factor of 100 for the virtually immobile triallate. The Henry's constant, an index of volatility, ranges from relatively high values for EPTC and triallate to extremely low values for bromacil.

These numbers indicate the type of information that is generally available for chemical fate characterization. All of the numbers are taken from the literature and might be expected to vary substantially in different sources, particularly the half-lives and the organic carbon partition coefficients. Improvements in and standardization of laboratory protocol have made variations in Henry's constants less prevalent.

The screening model of Jury et al.[16] has been modified by dividing the volatilization tendencies of organic compounds into two categories instead of three, based on values of their Henry's constant. Category I compounds with $K_H > > 2 \times 10^{-5}$, such as EPTC and triallate, are volatile and tend to have a volatilization flux that decreases with time. Their concentration remains low at the surface even when water is evaporating. Category III compounds with $K_H < < 2 \times 10^{-5}$, such as bromacil, atrazine, and prometon, are predicted to have insignificant volatilization without water evaporation. However, the surface concentration builds up under water evaporation and volatilization may become significant over time.[17]

Table 2.   Irrigation Amounts and Soil Coring Schedule.

Date	Activity
	Plot 1
10/27/86	1.83 cm irrigation
10/28/86	soil coring
11/03/86	1.75 cm irrigation
11/04/86	soil coring
11/10/86	1.90 cm irrigation
11/11/86	soil coring
	Total water applied 5.48 cm
	Plot 2
11/05/86	1.75 cm irrigation
11/06/86	soil coring
11/12/86	1.83 cm irrigation
11/13/86	soil coring
11/17/86	1.58 cm rain
11/20/86	soil coring
	Total water applied 5.16 cm
	Plot 3
11/10/86	1.68 cm irrigation
11/11/86	soil coring
11/17/86	1.55 cm irrigation
11/17/86	1.58 cm rain
11/18/86	soil coring
11/24/86	2.74 cm irrigation
11/25/86	soil coring
	Total water applied 7.55 cm

Application and Monitoring

Technical-grade samples of the five pesticides were predissolved in water to produce application solution concentrations of 408, 4, 32, 375, and 370 mg/L for bromacil, triallate, atrazine, prometon, and EPTC, respectively. Atrazine and triallate were initially dissolved with acetone to increase their solubility. The acetone:water ratio was 5 mL acetone per 9 L of solution formed. In addition, the solution contained 10 meq/L of KBr so that bromide could be used as a water tracer. Eight liters of the solution were sprayed on the three 4 m × 4 m plots to initiate the experiment. All three plots had been preirrigated extensively and were allowed to dry for 2 days prior to pesticide spraying. Three irrigations were applied to each plot during the experiment at weekly intervals, with the first irrigation applied 2 days after the plots were sprayed. Three soil samplings of four cores each per plot were taken within 24 hr after each irrigation. The complete schedule of irrigations and corings is given in Table 2.

Pesticide volatilization flux from each plot was monitored with portable flux chambers. Two frames were permanently positioned on each plot to provide an airtight seal between the surface and the chamber. The flux chambers used are a modification of a design developed for the U.S. EPA by the Radian Corporation.[20] They consist of clear acrylic skylights that have been treated with Teflon spray and have openings for temperature, relative humidity, and pressure flow rate measurements and for sampling.

The pesticides were trapped on porous polyurethane foam (PUF), a highly efficient and inexpensive trapping media for pesticide vapors. The pesticides can be recovered from PUFs by standard extraction techniques and then analyzed by gas chromatography.

Immediately following the spraying of a plot, one flux chamber was placed on a frame, and samples were taken for a period of 1 hr. The chamber was then removed for a 2-hr period to minimize disturbance of the soil surface underneath the chamber. During the first 36 hr, the chambers were monitored for 1 hr out of every 3 to give a continuous record of the early volatilization. After the initial 36-hr time period, the chambers were operated for 4 hr daily for four weeks.

Each plot was sampled by taking four cores during each sampling event at one-week intervals, for a total of 12 cores per plot. Soil coring was accomplished with a truck-mounted Central Mine equipment model 55 drilling rig. Segments of the soil profile were collected in 50-cm increments with an 80-cm hollow core sampler that unscrewed at each end. The core sampler contained aluminum cylinders (50 cm long by 7.6 cm diameter) to enclose the samples and also a 30-cm blank aluminum cylinder above the 50-cm cylinder to decrease the chance of compaction in case the drilling went deeper than 50 cm. Coring and soil sampling occurred simultaneously, and the hollow core did not rotate during the sampling but was hydraulically pressed into the soil. After pressing the tube to the 50 cm depth, the sampler was withdrawn, and the few centimeters of soil that were contained in the cutting tip were pressed into the aluminum sampling tube. To determine if any compaction had occurred, the depth of the hole and also the amount of soil that was contained in the sampler were measured. A few centimeters of soil from the upper portion of the 50-cm cylinder were discarded as a precaution against possible contamination. The cutting tip of the hollow core sampler was rinsed with acetone after each 50 cm of sampling. Soil samples remained encased in the aluminum cylinders when removed from the hollow core sampler. After removal, the cylinder ends were sealed with aluminum foil and stored on ice until they could be processed in the laboratory later in the day. Three soil segments were taken to 150 cm, or until the cutting tip reached a rocky lens that the truck could not push through. After the soil cores were taken, the holes were refilled with similar soil obtained from offsite cores.

Once the stainless steel cylinders were back at the laboratory, the soil was removed from the cylinders and divided into 10-cm increments. Only the center portion of each 10-cm interval was used for pesticide extraction, which commenced immediately after the core was divided, and the outer edges were used for water content determination and bromide analysis. The soil cores were completely extracted within 5 days of sampling.

THEORETICAL ANALYSIS

The behavior assessment model of Jury et al.[16] was used to predict the mass balance components of the pesticide chemicals qualitatively during the experiment. The behavior assessment model assumes that each chemical is incorporated at time zero into a finite layer of soil at uniform concentration and is subsequently subjected to volatilization, biodegradation, and leaching under uniform water flow, which may be upward, downward, or zero. Volatilization occurs through a stagnant air boundary layer at the soil surface, which was assumed to be 0.5 cm as recommended by Jury et al.[16] Predictions of the model are presented as percentages of the initial mass which either volatilize, degrade, or remain in the soil at the end of the experimental period. The purpose of the model is to evaluate the relative potential of different compounds to partition into different loss pathways under identical soil conditions. Although the conditions assumed in the model are uniform and, therefore, differ greatly from the dynamic soil environment encountered in this experiment, the relative predictions may still be tested as an indication of the degree to which the model can classify pesticides for relative behavior under real conditions.

RESULTS

Volatilization

Figure 1 shows the volatilization flux collected from the flux chambers for the first 17 days of the experiment for atrazine, a Category III pesticide, which is expected to have insignificant volatilization without water evaporation and volatilization that will increase with time when evaporation is present. The volatilization flux peaks occurred at approximately noon, and volatilization tended to decrease rapidly in the evening and increase again the following noon. This rapid decrease in midafternoon might have been caused by surface drying. The first irrigation occurred 48 hr after chemical application, causing the third peak to be broader and volatilization to stay near maximum for a longer time. The second irrigation occurred at 220 hr, which was followed by an increase, and the third irrigation occurred around 350 hr, followed by another increase in flux. This suggests that rewetting the surface decreased the adsorption in this zone and remobilized some of the pesticide.

The flux-time graphs for prometon are very similar to those for atrazine, showing the same kind of response over time and a similar magnitude. In contrast, the other Category III compound, bromacil, which has a very low Henry's constant and vapor pressure, was never detected in the flux chambers.

The volatilization flux for EPTC, shown in Figure 2 for 17 days, behaved qualitatively as expected for a Category I compound: the flux decreased con-

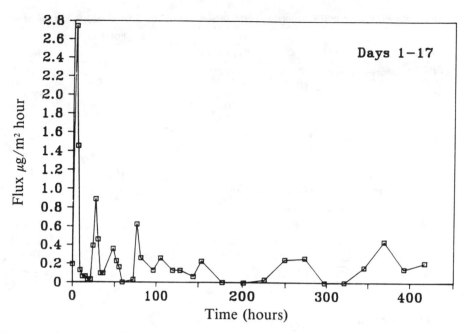

Figure 1.   Atrazine volatilization flux measured over a 17-day period.

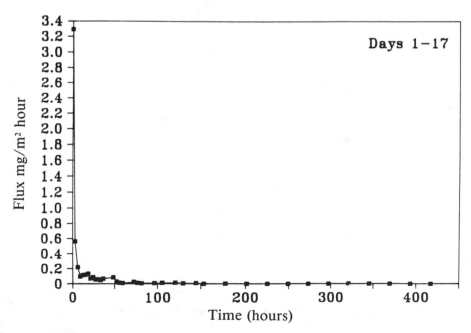

Figure 2.   EPTC volatilization flux measured over a 17-day period.

Table 3.   Predicted and Measured Volatilization over First 3 Days of the Experiment.

	Percent volatilized	
Pesticide	Predicted[a]	Measured
EPTC	26.80	32.00
Triallate	10.50	4.40
Atrazine	0.03	0.16
Prometon	0.0	0.13
Bromacil	0.0	0.0

[a]Using the screening model of Jury et al.[16]

tinuously with time. The irrigations did not appear to affect volatilization by any substantial amount. However, low concentrations of EPTC were detected up to 30 days following application.

We found very good agreement between the predicted relative volatility using the screening model of Jury et al. and the measured volatility (Table 3) — EPTC had the largest percentage of mass volatilized, followed by triallate, atrazine, prometon, and bromacil. The screening model simulations were carried out over a 3-day period using average estimated soil properties.

Soil Concentrations

Figure 3 shows the soil profile concentrations for the five pesticides during the three drillings taken at one-week intervals during the experiment. Although plot 3 received an extra input of water due to an unexpected rainfall, its final concentrations were not significantly different from those of plot 1 or 2 and, therefore, were averaged into the final concentration profiles. Several features

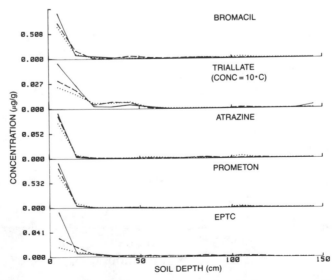

Figure 3.   Average pesticide concentrations (N = 12) recovered at each soil coring (solid curve = first coring; dashed curve = second coring; dotted curve = third coring).

are notable in Figure 3. First of all, the majority of the pesticide in each experiment resided within the top 20 cm during all three drillings, and the concentrations in these two surface layers decreased with time. Second, there is a component of unexpected deep leaching in each of the profiles for all chemicals during the experiment, to depths in excess of 20 cm. This effect is shown in magnified form in Figure 4, which replots Figure 3 with the first layer neglected so that the concentrations of the deep leaching components are magnified considerably. As shown in Figure 4, each compound had a portion of its mass migrate below the surface 20 cm during the first irrigation, and residual concentrations of the same magnitude were picked up at both subsequent samplings. Furthermore, this deep component did not migrate in subsequent samplings after the first drilling except in the case of bromacil, which has extremely high mobility. Since substantial precautions were taken to avoid contamination during sampling, we believe the deep concentrations are real and probably indicate a deep movement during the first irrigation after application of the compounds. This phenomenon has been observed in our laboratory before[9,10] and seems characteristic of the soil in which the experiments were conducted. Notable in Figure 4 is the accumulation of compound at about the 1-m depth, which corresponds to the depth of the rooting zone of the crops that had been grown on this soil during previous experimental years. Therefore, it seems likely that this deep movement is a manifestation of preferential flow through structural voids in the soil in a manner not consistent with the normal retardation experienced by chemicals that adsorb to soil solids. In all cases, this migration represents only a small fraction of the applied compound, the majority of which remained in the top 20 cm, where it was expected to reside.

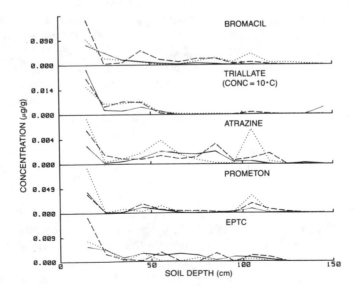

Figure 4. Replot of Figure 3 from 20 cm downward.

Table 4.   Percent of Applied Mass Recovery ± One Standard Deviation in Soil and Volatilization Samples After Each Drilling and Percent Degradation Calculated by Difference.

	% in Soil			% Volatilized			% Degraded		
				EPTC					
Drill 1	33.4	±	11.5	21.0	±	9.1	45.6	±	20.3
Drill 2	20.3	±	7.9	23.2	±	9.2	56.5	±	14.1
Drill 3	11.3	±	5.9	23.5	±	9.3	65.2	±	10.0
				Prometon					
Drill 1	99.8	±	1.4	0.11	±	0.05	0.8	±	0.6
Drill 2	91.4	±	5.7	0.26	±	0.11	8.3	±	4.1
Drill 3	81.6	±	3.5	0.31	±	0.08	18.1	±	3.5
				Atrazine					
Drill 1	99.7	±	3.2	0.2	±	0.1	1.4	±	2.0
Drill 2	98.9	±	3.1	0.4	±	0.1	1.7	±	1.7
Drill 3	89.2	±	1.5	0.6	±	0.2	10.3	±	1.5
				Triallate					
Drill 1	90.0	±	4.1	1.8	±	1.8	2.3	±	2.3
Drill 2	76.5	±	3.3	2.6	±	2.5	21.0	±	3.0
Drill 3	66.5	±	2.4	2.7	±	2.5	30.9	±	0.1
				Bromacil					
Drill 1	91.9	±	2.0	—			8.1	±	2.0
Drill 2	88.1	±	4.0	—			11.9	±	4.0
Drill 3	76.3	±	2.5	—			23.7	±	2.5

Mass Balance Estimates

Table 4 summarizes measured mass balances for each chemical taken immediately after the soil core drilling events. The values in Table 4 refer to the percentages relative to the initial amount applied and the standard deviations estimated from the replicates (i.e., 12 soil cores per drilling event and 3 volatilization estimates).

With the exception of EPTC, the mass balance summaries shown in Table 4 are calculated relative to the amount actually sprayed on the surfaces. For EPTC, however, it became apparent that a significant amount of the chemical had volatilized into the air without reaching the surface. A modified estimate of the amount actually reaching the surface was calculated by calibration with the volatilization model during the first hour of measured volatilization. Using this procedure, it was estimated that 61% of the chemical did not reach the surface but volatilized into the air as an aerosol.

Some results in Table 4 are worth noting. First of all, the standard deviations of mass recovery averaged over the 12 cores per drilling event are relatively small except in the case of EPTC, where the added uncertainty about the mass applied to the surface increased the likelihood of variability between individual samples. Second, the three replicates of volatilization obtained on each plot are correspondingly larger, indicating that individual differences in the water content, temperature, or other variables in the plots may have con-

Table 5. Estimated Degradation Half-Lives Calculated from Mass Balance and Published Half-Lives of the Five Pesticides Studied in the Experiment.

Compound	Measured Half-Life (days)		Literature Value (days)
EPTC	23 ±	8	30
Prometon	57 ±	21	100
Atrazine (7–21 days)	94 ±	18	71
Triallate	31 ±	11	100
Bromacil	152 ±	245	350
Bromacil (7–14 days)	269 ±	330	
Bromacil (14–21 days)	34 ±	4	

Note: Standard deviation estimated from distanced between sampling times and plots (N = 6).

tributed to uncertainty in the measurements. Consequently, since degradation is calculated by the difference between the amount applied and the amount remaining or volatilized, the uncertainty in the degradation estimates is quite high for all of the compounds.

Calculated degradation half-lives from the mass balance estimates are summarized in Table 5. Except where noted, the estimates were obtained for each plot individually at two time periods — between the first and second coring intervals, and between the second and third — for a total of six replicates. Also shown in Table 5 is the average literature value summarized in Jury et al.[18] The bromacil estimates differed significantly between the first and second coring, and between the second and third coring, as shown on Table 5. We postulate that this was caused by leaching of a small amount of bromide below the 150-cm depth of sampling. Despite the errors involved in the degradation estimates, agreement between literature values and measured estimates are reasonable, except in the case of triallate, which was significantly less persistent on this field than observed elsewhere. Because the coefficient of variation is small among the six replicates, we believe this phenomenon is real in the field studied.

Leaching

As shown in Figures 3 and 4, the leaching observed on the field was excessive for a small portion of the compound applied. According to the retardation factors presented in Table 1 and the estimates of the screening model, none of the compounds should have migrated below 20 cm with the small amount of water they were given during the study. Therefore, the deep leaching is in violation of the equilibrium adsorption hypothesis used in the model. However, the movement below 20 cm was not excessive in the study and represented a small fraction of the mass applied. Table 6 summarizes the percent of mass found in the soil at each sampling averaged over all plots and replicates. Two summaries in the table are given: (1) as a percent of mass found in the soil at the time of sampling and (2) compared to the percent found at the first sampling at 7 days. As shown, the amounts recovered below 20 cm relative to the

Table 6.   Percent of Mass Found in Soil Below 20 cm at Each Sampling.

Compound	Drill No.	% of Mass Present in Soil	% of Mass Present in Soil at 7 days
EPTC	1	20	20
	2	23	14
	3	31	10
Prometon	1	4	4
	2	7	7
	3	8	6
Atrazine	1	9	9
	2	12	12
	3	15	14
Triallate	1	16	16
	2	29	23
	3	36	25
Bromacil	1	9	9
	2	15	14
	3	18	15

initial amount do not increase substantially during the experiment, indicating that the principal hypothesis of confinement within the upper 20 cm is validated for the majority of the mass applied. Further, of the compounds studied, only triallate shows a substantial fraction of the mass found below the initial depth — in clear contrast to the behavior expected for the large retardation factor of the compound — and perhaps was rendered more mobile than average during its initial application by the small formulation with acetone at the time of dissolution. Figure 3 shows that the initial displacement of triallate was greater than that of any other chemical even though its retardation would have suggested the opposite. Therefore, we think it likely that the predissolution with acetone produced a stable complex that persisted through the first irrigation.

Volatilization Comparisons with the Screening Model

The volatilization predictions agreed qualitatively with measurements of volatilization loss during the first 3 days of the experiment (Table 3). In addition, the volatilization classification scheme proposed by Jury et al.[16] appears to provide useful guidance about the time behavior of volatilization as well. EPTC, a Category I pesticide, is predicted to decrease its volatilization loss continuously over time, which is confirmed by the volatilization flux time graph shown in Figure 2. In contrast, atrazine, a Category III pesticide, is predicted to regenerate its volatilization as a result of upward water flow, which is consistent with the pattern shown in Figure 1. Thus, although the volatilization model uses assumptions that differ from the dynamic conditions of the field study, both qualitative and quantitative features are reproduced sufficiently well to indicate that it is a useful guide for behavior assessment predictions.

CONCLUSIONS

The field mass balance study using five pesticides of contrasting environmental properties illustrates many of the difficulties associated with modeling movement and fate under field conditions. Mobility of the portion of the compounds undergoing deep migration could not be predicted, even in a relative sense, from the standard equations describing equilibrium adsorption of dissolved compounds. This implies that preferential pathways, such as structural voids or biological channels, may be moving a portion of the compounds rapidly to depths far in excess of those that could be reached under slow movement through the entire soil matrix. Consequently, at least in the near-surface regime, adsorption indices do not appear to be good indicators of relative mobility under field conditions.

Volatilization behavior of the five compounds was more successfully described using conventional modeling techniques than was leaching. Compounds with higher Henry's constants were more volatile and displayed behavior over time consistent with predictions of the screening model of Jury et al. That model was also able to predict relative volatility in the same order as observed.

Since this experiment was confined to the top 1.5 m during soil sampling, it is not clear that the excessive mobility demonstrated in the soil coring will persist to depths sufficient to move compounds below the biologically active zone. However, until experiments are performed over longer time periods and samples analyzed to greater depths, that conclusion cannot be drawn with certainty.

ACKNOWLEDGMENTS

The authors would like to thank the California Department of Health Services, the Southern California Edison Company, and the University of California, Riverside, Toxic Substances Program for financial assistance on this research project.

REFERENCES

1. Cohen, D. B., and G. W. Bowes. "Water Quality and Pesticides: A California Risk Assessment Program," California State Water Research Control Board Report, Sacramento, CA (1984).
2. Glotfelty, D. E., J. N. Seiber, and L. A. Liljedahl. "Pesticides in Fog," *Nature* 325:602–5 (1987).

3. "The Organic Chemical Contamination Prevention Act," California State Assembly, Bill 2021, Sacramento, CA (1986).
4. Jury, W. A., and M. Ghodrati. "Overview of Organic Chemical Environmental Fate and Transport Modeling Approaches," Soil Science Society of America, Special Publication (1988).
5. Biggar, J. W., and D. R. Nielsen. "Miscible Displacement and Leaching Phenomenon," *Agron. Mon.* 11:254–74 (1976).
6. Wild, A., and I. A. Babiker. "The Asymmetric Leaching Pattern of NOs and Cl in a Loamy Sand under Field Conditions," *J. Soil Sci.* 27:460–66 (1976).
7. Van de Pol, R. M., P. J. Wierenga, and D. R. Nielsen. "Solute Movement in Field Soil," *Soil Sci. Soc. Am. J.* 41:10–13 (1977).
8. Jury, W. A., L. H. Stolzy, and P. Shouse. "A Field Test of the Transfer Function Model for Predicting Solute Transport," *Water Resour. Res.* 18:369–75 (1982).
9. Elabd, H. S. "Spatial Variability of the Pesticide Distribution Coefficient," PhD dissertation, University of California, Riverside, CA (1984).
10. Clendening, L. D. "Evaluation of Mobility Screening Parameters under Field Conditions," Master's thesis, University of California, Riverside, CA (1985).
11. Parmele, L. H., E. R. Lemon, and A. W. Taylor. "Micrometeorological Measurement of Pesticide Vapor Flux from Bare Soil and Corn under Field Conditions," *Water Air Soil Poll.* 1:433 (1972).
12. Willis, G. H., and B. R. Carroll. "Volatilization of Dieldrin from Fallow Soils as Affected by Different Soil Water Regimes," *J. Environ. Qual.* 1:193 (1972).
13. Taylor, A. W. "Post-Application Volatilization of Pesticides under Field Conditions," *J. Air Poll. Control Assoc.* 28(9):922–27 (1978).
14. Glotfelty, D. E., A. W. Taylor, B. C. Turner, and W. H. Zoller. "Volatilization of Surface-Applied Pesticides from Fallow Soil," *J. Agron. Food Chem.* 32(3) (1984).
15. Rao, P. S. C., R. E. Green, V. Balusubramanian, and Y. Kanehiro. "Field Study of Solute Movement in a Highly Aggregated Oxisol with Intermittent Flooding: I. Picloram," *J. Environ. Qual.* 3:197–202 (1974).
16. Jury, W. A., W. F. Spencer, and W. J. Farmer. "Behavior Assessment Model for Trace Organics in Soil: I. Model Description," *J. Environ. Qual.* 12:558–63 (1983).
17. Jury, W. A., W. F. Spencer, and W. J. Farmer. "Behavior Assessment Model for Trace Organics in Soil: II. Chemical Classification and Parameter Sensitivity," *J. Environ. Qual.* 13:567–72 (1984).
18. Jury, W. A., W. F. Spencer, and W. J. Farmer. "Behavior Assessment Model for Trade Organics in Soil: III. Application of Screening Model," *J. Environ. Qual.* 13:573–79 (1984).
19. Jury, W. A., W. F. Spencer, and W. J. Farmer. "Behavior Assessment Model for Trace Organics in Soil: IV. Review of Experimental Evidence," *J. Environ. Qual.* 13:580–86 (1984).
20. Dupont, R. R., and J. A. Reineman. "Evaluation of Volatilization of Hazardous Constituents at Hazardous Wasteland Treatment Sites," U.S. EPA Report 600/2–86/071 (1986).

## CHAPTER 5

Modeling the Volatilization of Pesticides and Their Distribution in the Atmosphere

James E. Woodrow, Michael M. McChesney, and James N. Seiber

INTRODUCTION

The atmospheric movement of pesticide residues may lead to unintentional exposures for humans, animals, and plants near treatment sites. Drift during application and volatilization or wind erosion of deposited residues are the most common routes of entry to the atmosphere for pesticides.[1-3] For relatively volatile and stable pesticides applied to surfaces, the most significant dissipation process in residue decline is often volatilization.[1,4] Entry to the atmosphere is dependent on such factors as the method of application, type of formulation, pesticide physicochemical properties, and meteorological conditions at the application site.[2,3,5,6] Some of these factors are amenable to mathematical modeling and laboratory simulation, but few field measurements have been made in aid of model design and validation.[7,8]

Even less has been done to quantify ambient distribution and persistence of airborne pesticides, particularly over broad geographic areas of heavy use. At best, only general descriptions of temporal or geographic trends in airborne residue concentrations have been made because of the relatively small number of samples and/or sampling sites and the nonpoint nature of the source.[7,9-14] The amount and quality of data are limited by the expense of collecting and analyzing large sets of air samples.

In recent years, however, there has been renewed interest in California in airborne toxics, especially pesticides. This interest has been reflected in the passage of the Toxic Air Contaminant Act (AB 1807), which includes a requirement for determining which pesticides are toxic air contaminants and the appropriate degree of control measures needed.[15] To fulfill this requirement of AB 1807, state regulatory agencies must conduct sampling of candidate pesticides in geographic areas of significant use to determine source emissions, atmospheric concentrations, and distribution patterns in relation to potential human exposure, and then conduct risk assessments based on these results and toxicological data.

The design and conduct of a program capable of fulfilling the demands of AB 1807 will necessarily include field measurements, laboratory simulation, and computer modeling. In this chapter, we describe the determination of volatilization rate (flux) for three rice herbicides—MCPA, molinate, and thiobencarb—and one insecticide—methyl parathion—from flooded rice fields using three techniques: (1) field measurement of flux over commercial rice fields, (2) laboratory chamber measurement, and (3) computer model prediction using the EXAMS (Exposure Analysis Modeling System) aquatic fate model.[16] We also describe the measurement of ambient airborne residue levels of methyl parathion; its oxon conversion product, methyl paraoxon; molinate; and thiobencarb at small town and rural locations in the general area of use. Downwind distances from possible sources along with observed residue levels at those distances were correlated with predictions of the ALOHA (Areal Location of Hazardous Atmospheres) Gaussian plume dispersion computer model.[17] An important goal in these studies was to evaluate computer models and laboratory simulations as tools for predicting emissions and dispersion of pesticide residues.

EXPERIMENTAL

Volatilization Rate (Flux)

Evaporative loss from water was studied in the laboratory using a chamber described elsewhere[18,19] and shown in Figure 1. The sample dish was filled to the brim with separate aqueous solutions of MCPA (95 mg/L; pH = 3.5), thiobencarb (16.3 mg/L), molinate (25.1 mg/L), and methyl parathion (14 mg/L).[20,21] Humidified air (85% RH) was passed over the water surface at a flow rate corresponding to a wind velocity of 2.2 m/sec, and the chamber exit air passed through a vapor trap containing XAD-4 cross-linked polystyrene macroreticular resin. The vapor trap was extracted and analyzed to determine chemical loss due to volatilization, and the unvolatilized chemical remaining in

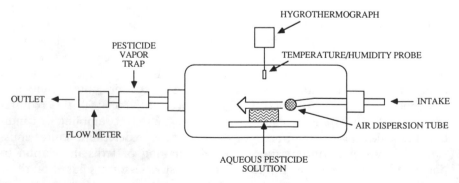

Figure 1.   Laboratory environmental chamber.

solution was also analyzed to assess mass balance. Measured flux was compared with flux calculated using the EXAMS aquatic fate model, with chemical properties and chamber conditions as input. EXAMS is a FORTRAN program that uses differential equations to calculate the amount of chemical lost from water by volatilization, hydrolysis, microbial degradation, etc. Variables include the physicochemical properties of the chemical, properties of the environment (pH, wind speed, temperature, microbial activity, etc.), and parameters describing the size of the water, soil, air, and biotic compartments.

Flux from water was determined in the field using two separate rice fields, of approximately 37 ha and 41 ha, treated by fixed-wing aircraft.[21,22] The herbicides were applied in the granular form at a rate of about 4.5 kg/ha each to control broadleaf and graminaceous weeds. Methyl parathion was applied to the 41-ha field as the emulsified aqueous suspension at a typical rate of about 0.75 kg/ha to control freshwater shrimp. The fields were flooded prior to treatment, and the average water depth was maintained at 15–26 cm throughout the study.

Each field contained a wooden pier located about midway along the length of a rice check (Figure 2). The piers were about 7 m x 0.3 m, and air sampling and meteorological equipment was placed on masts located near the open end

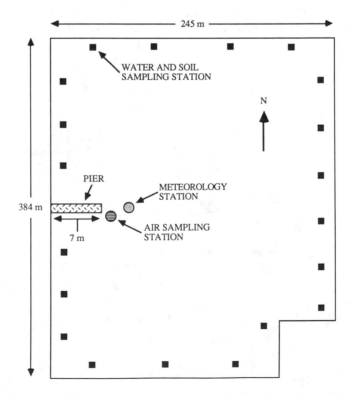

Figure 2. Typical rice-field layout for environmental sampling (not drawn to scale).

of the piers. Airborne chemical residues above the fields were sampled in two separate tests, using high-volume and low-volume air samplers. In one test, high-volume samplers were charged with 100–200 mL of a cross-linked polystyrene adsorbent (XAD-2 or 4) and operated at two heights above the field at flow rates of about 1000 L/min for 1–2 hours per sampling period. In the other test, high-volume samplers were used to pull air through six cartridges containing 30 mL adsorbent each, at flow rates of 50 L/min for 1–2 hours per sampling period. The cartridges were located at six heights above the field along a logarithmic progression (Figure 3). For both tests, wind speed was measured with anemometers located at heights corresponding to the air sampler heights. Air temperature was also measured at two heights, in the second test using 10-junction copper constantin thermopiles, to define the flux plane.

Water samples were taken at points around the perimeter of the rice paddy (Figure 2); three aliquots per point combined to constitute one sample. The composite was filtered to remove suspended particulate matter and then pumped through C_{18} reverse-phase disposable cartridges to extract the chemicals. The extracted samples were placed in dry ice-filled chests for transport to the laboratory where they were analyzed.

Two or three composite soil samples were taken each day of sampling. The composited samples consisted of one soil core (10 cm × 10 cm) taken at each water sampling point around the perimeter of the rice paddy. The composite was mixed in a tub, and a 500-g subsample was removed and filtered by

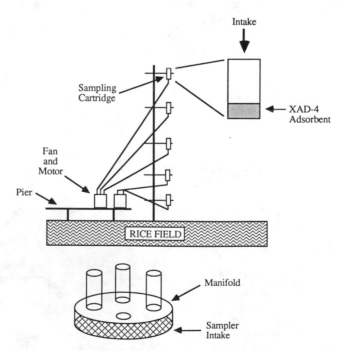

Figure 3. Air sampler configuration for pesticide flux measurement.

suction to remove excess water. The filtered subsamples were sealed in glass jars and stored in dry ice-filled chests for transport to the laboratory where they were analyzed.

To assess the possible pathways available to molinate for loss, about 2.5 cm of sediment was added to 20 separate 0.95-L glass jars, to which was also added about 13 cm of field water containing 1–2 mg/L of added molinate. The jars, 10 sealed and 10 uncovered, were immersed in the water and mud in a line along a check of the rice paddy. From each set of 10 jars, a composite sample was taken by withdrawing 2 mL of water from each jar of the set and mixing the resulting 20 mL in another sample jar, which was then sealed and frozen for storage until analysis. Sampling was done periodically in this manner over a 222-hr period.

Airborne residue levels and wind speeds at two heights were used in the aerodynamic expression to calculate flux.[23,24] Along with the aerodynamic results, flux was calculated using the EXAMS model, with the chemical properties for molinate and methyl parathion and field environmental conditions as input.

Ambient Residue Levels

Ambient residue levels of methyl parathion, molinate, and thiobencarb were monitored at four locations in the Sacramento Valley.[25] High-volume air samplers were used to pull air through two and three adsorbent-filled (XAD-4) cartridges at three sites and one site, respectively. The samplers were placed on rooftops away from obstructions in or near residential/business areas, and the sampling interval was about 24 hr at a flow rate of about 50 L/min per cartridge (Figure 4). Spent sealed cartridges were transported to the laboratory for analysis.

Field-measured flux and ambient residue levels were used as input for the ALOHA model to calculate approximate distances downwind from a source for a given flux-ambient level combination. The ALOHA model plots the distribution of a pollutant gas or vapor resulting from a defined source as a series of Gaussian distributions. As the pollutant moves downwind from the source, the Gaussian bell shape of the distribution spreads in a crosswind direction and becomes continually wider and flatter. In other words, the chemical plume grows in both the crosswind and vertical directions. The model assumes that the source is at, or near, ground level and calculates the concentration distribution near the ground in the human breathing zone.

RESULTS

Volatilization Rate (Flux)

Molinate and thiobencarb were applied to rice fields as the dry granular formulations, and MCPA and methyl parathion were applied as emulsified

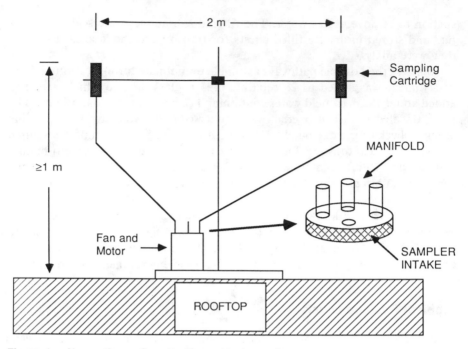

Figure 4.   Air sampler configuration for ambient sampling.

aqueous suspensions. The emulsified formulations remained largely in the rice water after application, followed perhaps by slow partitioning (into bottom sediment and plants) and breakdown. However, the granular formulations sank to the bottom sediment, where the herbicides slowly dissolved. Although thiobencarb reached the maximum water concentration of about 0.58 mg/L 4–6 days after application,[22] molinate required only about 17 hr to reach the maximum observed water concentration of 2.45 mg/L.[24] This behavior reflects the order of magnitude difference in solubility between molinate and thiobencarb (Table 1). Because of this, most of the thiobencarb remained associated with the bottom sediment, while molinate was found primarily in the water.

From the properties of the test chemicals (Table 1), their relative volatilities would give the following order of decreasing volatility, when all other conditions are the same:

molinate > thiobencarb > methyl parathion > > MCPA > > MCPA (DMA salt)

This is the order that was actually observed (Table 2), where volatilization of MCPA in the laboratory was of the undissociated acid (pK ≈ 3).[21] The same order of volatility was predicted by the EXAMS aquatic fate model, and, overall, the model flux values compared well with the observed values (Table 2). One exception was the EXAMS-estimated volatilization rate for thiobencarb. The model rate was about 1/5 of those rates observed in the laboratory and field. The Henry's law constant for molinate is about 5–6 times that for

Table 1. Physical Properties of Rice Chemicals

Chemical	Molec. Wt.	Melt. Pt. (°C)	Water Sol. (mg/L)	Vap. Press. (torr)	H (atm m³/mole)
Molinate	187.3	—	800 (20°C)	3.1×10^{-3}	9.6×10^{-7}
Thiobencarb	257.8	—	30 (20°C)	1.5×10^{-5} (20°C)	1.7×10^{-7}
MCPA (acid)	200.6	118–119	1,500	5.9×10^{-6}	1.0×10^{-9}
MCPA (DMA salt)	244.7	—	>300,000	1.0×10^{-7}	$<10^{-13}$
4-Chloro-o-cresol	142.6	51	4,000	2.4×10^{-3}	1.1×10^{-6}
Methyl parathion	263.2	35–36	37.7 (22°C)	1.1×10^{-5}	1.0×10^{-7}
Methyl paraoxon	247.2	—	~4,000[a]	$<1.1 \times 10^{-5}$	$<9 \times 10^{-10}$

Note: 25°C, unless otherwise specified.

[a]Estimated from the general observation that oxon solubilities are about two orders of magnitude greater than thion solubilities (Bowman, B. T., and W. W. Sans, "The Aqueous Solubility of Twenty-Seven Insecticides and Related Compounds," J. Environ. Sci. Health B14(6):625–34 [1979].

Table 2.   Summary of Normalized Flux Values

| Chemical | H (atm m^3/mole) | Flux (ng/cm^2/hr/ppm)[a] | | Rice Field | |
		EXAMS[b]	Lab Chamber	Air[c]	Water[d]
MCPA (acid)	1.0×10^{-9}	8.1×10^{-3} (pH 3.5)	4.1×10^{-3} (pH 3.5)	1.9	29
MCPA-DMA (salt)	$< 10^{-13}$	0.0000	—	—	—
4-Chloro-o-cresol[e]	1.1×10^{-6}	—	—	243	8
Thiobencarb	1.7×10^{-7}	4.5	23.8	23	58
Molinate	9.6×10^{-7}	51.5	62.8	47	43
		57.8	—	40	100
Methyl parathion	1.0×10^{-7}	2.4	5.6	3.6	161[f]

Source: Seiber et al.[21]
[a]Flux (ng/cm^2/hr) normalized to water concentration (ppm = mg/L).
[b]Computer calculations were based on field conditions for the last two entries; laboratory conditions were used for all other entries.
[c]Average of three days. Derived from air samples and aerodynamic flux measurement method.
[d]Average of three days. Derived from analysis of water samples and assumes that all observed loss was due to volatilization.
[e]Photoproduct of MCPA.
[f]Losses due primarily to chemical and microbial degradation.

thiobencarb (Table 1). However, in the field, half-lives for both compounds in rice water were approximately the same[22,24] for normalized flux values that differed by only a factor of about 2–3 (Table 2). These results imply that the actual Henry's law constant for thiobencarb may be somewhat greater than the value reported in Table 1 because of possibly incorrect vapor pressure and/or solubility data for thiobencarb.

The utility of the EXAMS model is illustrated in Figure 5, where observed variations in field flux for molinate and methyl parathion are compared with the model prediction. The model was able to reproduce the variations in the flux for both compounds when field conditions were used as input. However, for molinate at the highest observed flux, the model "overpredicted", and for methyl parathion, the model tended to "underpredict" at the highest observed flux. For the latter, air concentrations (and flux) were much lower than those for molinate, and consequently the measurement error was greater; thus, the "underprediction" was perhaps due to faulty measured values, resulting, for example, from air sample contamination. The ability of EXAMS to reproduce the diurnal variability of molinate flux was due in part to our assumption that molinate water solubility varied inversely with the temperature. Rummens and Louman[26] reported that many thiocarbamates (e.g., EPTC) exhibit an inverse solubility behavior because the neutral, less polar N-C=O is favored over the more polar $^+$N=C-O$^-$ resonance structure as temperature is increased. Because molinate belongs to this same chemical class, we used EPTC as a model for adjusting the literature solubility for molinate (800 mg/L at 20°C) so that its water solubility would decrease as the temperature increased.[24] The temperature dependence of solubility is critically important to flux because Henry's law constant—a prime determinant of flux from water—has a direct

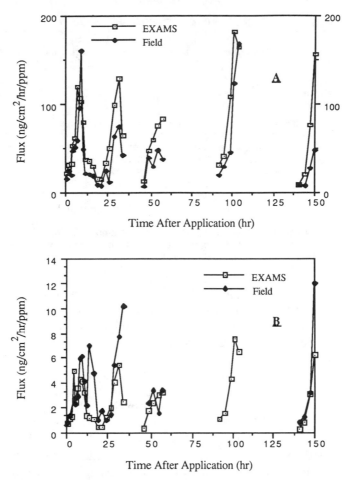

Figure 5. Diurnal volatilization profiles for (A) molinate and (B) methyl parathion dissolved in rice-field water.

dependence on the ratio of vapor pressure to water solubility. Thus, field-measured wind speed and water temperature, along with the vapor pressure and water solubility corresponding to the field temperature, were used as input to EXAMS to calculate the volatilization rates for molinate under the constantly varying conditions that existed in the field.

Dissipation (loss) rates for molinate and thiobencarb, based on analysis of water samples taken from the field and the jar experiment (molinate), were about 2–3 times the rates derived from the aerodynamic method (Table 3). However, it was assumed that loss was primarily by volatilization and that loss by other routes was insignificant in the time frame of this study; the slow loss of molinate from the closed jars ($t_{1/2} > 200$ hr) confirmed that volatilization was the major loss route for molinate under field conditions. The utility of the aerodynamic method depends upon the equilibration of the chemical residues

Table 3.   Approximate Half-Lives (hrs) of Pesticides in Water

Chemical	Dissipation from Water (all routes)[a]	Loss by Volatilization Only[b]	Other Processes
Thiobencarb	104	260	>2000; 95[c]
Molinate	100	250	>8000; 219[c]
Methyl parathion	44	2900	52; 192; 259[d]
4-Chloro-o-cresol	>1000	42	>100[e]

[a]Derived from analysis of water samples and reflects losses from all routes.
[b]Derived from air samples and aerodynamic method.
[c]Photolysis in pure water and water plus oxidant, respectively. From Draper, W. M., and D. G. Crosby, "Solar Photooxidation of Pesticides in Dilute Hydrogen Peroxide," *J. Agric. Food Chem.* 32:231–37 (1984).
[d]Biodegradation, photolysis, and hydrolysis, respectively. From Smith, J. H., W. R. Mabey, N. Bohomes, B. R. Holt, S. S. Lee, T. -W. Chou, D. C. Bomberger, and T. Mill, "Environmental Pathways of Selected Chemicals in Freshwater Systems—Part II: Laboratory Studies," U.S. EPA Report-600/7-78-074 (1978).
[e]Photolysis. From Crosby, D. G., and J. B. Bowers, "Composition and Photochemical Reactions of a Dimethlyamine Salt Formulation of (4-Chloro-2-methylphenoxy) Acetic Acid (MCPA)," *J. Agric. Food Chem.* 33:569–73 (1985).

above the field. Equilibration occurs when the upwind distance from the sampler is at least 100 times the height of the sampler above the field, a condition more than satisfied in this study. These results, and the results from other studies, have called into question some of the assumptions made in the derivation of the aerodynamic expression, casting some doubt on the accuracy of this method for measuring pesticide volatilization under any circumstances, not just from rice-field water.[27,28]

Loss of MCPA and methyl parathion from rice-field water would be primarily due to photochemical (MCPA, methyl parathion) and microbial (methyl parathion) degradation of these chemicals (Tables 2 and 3). Therefore, volatilization of MCPA from treated rice-field water would be expected to be negligible, but in fact the observed flux of MCPA was almost the same as that for methyl parathion. Based on the results of earlier work,[7,29] it can be concluded that postapplication airborne residues of MCPA were due primarily to volatilization of the free acid formed on the dry soil of checks surrounding the paddies and plant surfaces within the paddies from the breakdown of the dimethylamine salt. However, water analysis indicated relatively rapid loss of MCPA due primarily to conversion to the major breakdown product, 4-chloro-o-cresol, which subsequently volatilized from water at the highest rate measured in the field (Table 2). The dramatic difference between water analysis-based and flux-based dissipation half-lives for 4-chloro-o-cresol was due to its formation in water from breakdown of the parent MCPA.

Ambient Residue Levels

The loss of a pesticide through volatilization from a treated site, or drift during application, will result in airborne residues occurring in nontarget areas. Of course, distribution of airborne residues will depend on the physicochemical properties of the chemical and the meteorological conditions. Dissipation processes open to chemical residues in air include precipitation (dry deposition, rainout) and chemical conversion (e.g., direct and indirect photolysis). Residues that occur in nontarget areas will consist of the parent compounds as well as any possible conversion products. California's Sacramento Valley, where most of the rice in the state is grown, is bounded on three sides by ranges of mountains and hills and is open only to the south. Prevailing winds during the rice-growing season move from the coast through the delta and partly move northward up into the Sacramento Valley, resulting in a cyclonic motion of the air in the Valley (Figure 6). It is expected that this situation would lead to distribution of chemical residues throughout much of the air basin. This possibility was strongly suggested by the occurrence of propanil up to 88 km from target areas in the Valley[30] and the results of a study that measured the ambient levels of MCPA in the Valley during the application season. In this study, two high-volume air sampling stations, separated by about 18 km, were operated 24 hr/day for approximately two months.[29] Ambient residues of MCPA remained elevated because of applications in the area and dropped to near or below the detection limit (~1 ng/m³) only at the close of the spraying season (Figure 7). However, ambient residues of the cresol conversion product remained elevated sometime after the MCPA residues declined, indicating continuing breakdown of already deposited MCPA residues. It is also possible that some of the cresol may have been formed photochemically from precursors derived from agriculturally unrelated sources.[31]

Summaries of the average ambient airborne concentrations for methyl parathion, molinate, and thiobencarb are presented in Table 4.[25] The highest averages for methyl parathion and molinate were observed at the same site

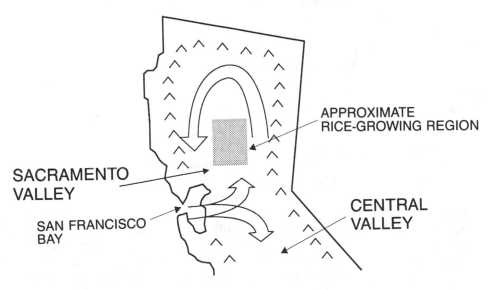

Figure 6. Map of Sacramento Valley, California, showing typical summertime wind patterns (open arrows).

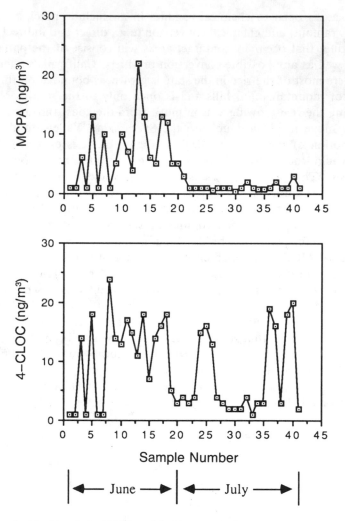

Figure 7.   Ambient levels for MCPA and 4-chloro-*o*-cresol.

(Maxwell), a small town that was almost completely surrounded by sections of land under rice cultivation; the highest average for thiobencarb was observed at a different site (Trowbridge), which was also close to rice cultivation. The Davis site was used to collect background samples because of its remote location in relation to rice-growing areas. Comparing airborne residue levels with usage for methyl parathion during the period of the study indicated that the two profiles were essentially the same (Figure 8a), suggesting that the air concentrations occurred as a result of drift during application; however, the profiles of air concentration and usage for molinate did not correlate as well (Figure 8b), suggesting that volatilization from treated rice fields followed by vapor drift was primarily responsible for the air concentrations at the sampling

Table 4.   Summary of Ambient Air Concentrations for Three Rice Chemicals During the Period May12–June 12, 1986

Location	Average (Highest) Concentration (ng/m^3)			
	Methyl Parathion	Methyl Paraoxon	Molinate	Thiobencarb
Maxwell	6.2 (25.7)	0.8 (3.1)	650 (1700)	29.0 (64.5)
Williams	3.1 (21.8)	0.6 (1.1)	200 (420)	12.9 (23.3)
Trowbridge	0.32 (1.1)	<0.5	150 (280)	67.8 (250)
Robbins	0.27 (0.72)	<0.5	60 (140)	14.2 (40.8)
Davis	<0.5	<0.5	<20	<2.0

Note: Ambient air concentrations ranged from the highest listed to less than the minimum detection limit. Three replicates were taken at the Maxwell site, and two each at the other sites for each sampling period, giving a total of about 220 samples. Total number of sampling days for methyl parathion, methyl paraoxon, and molinate: 20; Total number of sampling days for thiobencarb: 19.

sites. Postapplication volatilization from treated rice fields followed by vapor drift for methyl parathion is less likely because of its low vaporization, as indicated by its Henry's law constant (Table 1), and its relatively rapid degradation by hydrolysis in water. Furthermore, methyl parathion is applied as the emulsified aqueous spray; this condition produces aerosols of a range of sizes, many of which may remain suspended in air above the treated field and produce downwind drift as a result of rapid loss of water from the aerosols under rice-field conditions. Molinate, on the other hand, is applied as the dry, pelletized material, minimizing drift, and the compound is relatively volatile (Table 1) and stable in water under rice field conditions. The much higher ambient concentrations of molinate, when compared with methyl parathion, are due not only to the compound's higher volatility but also to its higher total use and greater application rate in the Sacramento Valley.

The site with some of the highest observed air concentrations for molinate and thiobencarb (Maxwell) was surrounded by rice fields—the closest within about 1 km of the sampling site. Using measured molinate flux from rice fields as the source strength (127 ng/cm^2/hr) and the measured daily ambient air concentrations for molinate (0.14–1.7 μg/m^3) with a typical wind speed (4.3 m/sec), the ALOHA Gaussian plume dispersion model calculated the distances from the source to the sampling sites to be about 0.58 km (1.7 μg/m^3) and 2.09 km (0.14 μg/m^3). Since the uncertainty in the model is one-half to twice the calculated value, the model results suggest that the range of observed ambient concentrations would be due to sources about 0.29–1.16 km (1.7 μg/ m^3) and 1.04–4.18 km (0.14 μg/m^3) away, a reasonable result considering the physical layout of the study. Similar results for thiobencarb at the Maxwell site were 0.38–1.52 km (64.5 ng/m^3) and 1.20–4.78 km (7.1 ng/m^3), for a source strength of about 9 ng/cm^2/hr. Comparing these results gave molinate:thiobencarb ratios of downwind air concentrations that were $\geq$20, due in part to their difference in volatilization rates from water. Field-measured ratios at the Maxwell site fell in the range from 4 to 67 with an average of 32, reflecting the fact that rice fields are not treated concurrently with both compounds and that

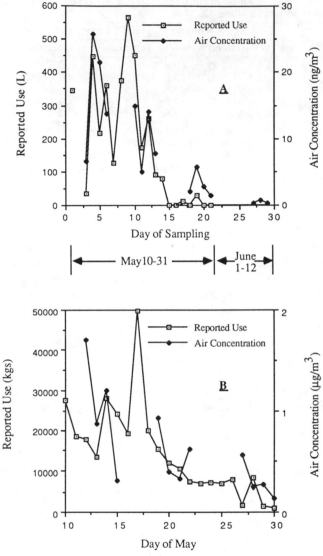

Figure 8. Reported use on rice vs air concentration profiles for (A) methyl parathion and (B) molinate.

more total molinate than thiobencarb is used on rice (in 1986, molinate use was about three times that for thiobencarb).[25] Treating the methyl parathion data in the same way gave 0.13–0.51 km (25.7 ng/m³) and 0.56–2.24 km (1.4 ng/m³), for a source strength of about 0.4 ng/cm²/hr. Average field-measured molinate:methyl parathion and thiobencarb:methyl parathion ratios were 93 (22–267) and 7 (0.5–23), respectively. However, the ALOHA-calculated ratios were about 1200 and 46 for molinate and thiobencarb, respectively. These results suggest that volatilization of methyl parathion from treated rice fields alone could not account for the measured ambient levels and, therefore, give further support to the conclusion that the ambient residues were more likely due to residue drift during application, and also possibly to volatilization from relatively dry surfaces (e.g., foliage).

Methyl parathion can undergo light-catalyzed oxidation to the corresponding oxon. In fact, the main reaction for thiophosphates in air is oxidation to the oxon, but half-lives can range from 4 to 630 min, depending upon the chemical's structure.[32,33] Assuming that the ambient levels of the oxon measured in three-hour accumulative samples at the Maxwell site, for example, were due only to the vapor-phase photooxidation of methyl parathion residues drifting during application and correlating the ambient residues with known applications in the area,[25] approximate half-lives for the conversion in air were less than 30 min. The oxon would not be readily lost from flooded rice fields through volatilization because of its water solubility (about 100 times that of the thion) and hydrolytic instability. While it was assumed that trapping efficiency for the parent compound and its conversion product were the same and that no conversion of trapped methyl parathion took place in the air sampler, the observed conversion in field air compared reasonably well with the results of earlier studies that examined the vapor-phase photooxidation of organophosphates.[34]

Risk Assessment

Results from field measurements and laboratory and computer models can be used to estimate human exposures, which then may serve as input to risk assessment protocols (Figure 9). These data, along with toxicological data, can be used to determine relative risk to human populations located at differing distances from sources of toxic chemicals. The risk assessment results can then be used as input into risk management protocols, which would then take appropriate action, if needed, to reduce exposure by minimizing source emissions and drift or by increasing spray buffer zones. The primary goal of measuring ambient air concentrations of methyl parathion and its oxon was to provide data for risk assessment and risk management. The major initial observable effect in mammals is cholinesterase inhibition (30% inhibition is the threshold), caused by the oxon; a margin of safety (no observed effect level [NOEL] divided by the exposure estimate) of 10 is considered to be minimally adequate. Threshold effects (i.e., noncancer effects) can also be addressed by dividing the NOEL for methyl parathion and methyl paraoxon by uncertainty factors (in increments of 10) to obtain a "safe" exposure or, equivalently, a "safe" distance from the source.[35] Extrapolating from NOEL data for ethyl parathion[36] and the measured ambient level range of 1.4 to 25.7 ng/m³ for methyl parathion at the Maxwell site, estimated margins of safety were about 80–1400 for the human adult and 40–700 for the human child, assuming chronic exposure (24 hr/day, 7 days/week). Of course, the margin of safety

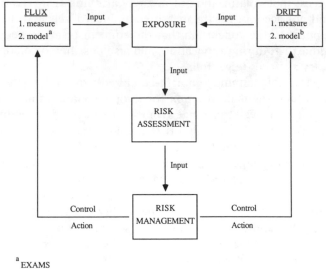

Figure 9. Risk assessment flow diagram.

will vary depending upon the proximity of the person to the source. In agricultural areas it is common for pesticide applications to take place immediately adjacent to residences and thoroughfares. This has given rise to suggestions that "set back" or buffer zones be established to allow drifting residues to be diluted by fresh air before reaching frequented areas. Risk assessments based on data taken at varying distances from pesticide sources would be essential for establishing the sizes of the buffer zones.

CONCLUSIONS

All chemicals applied intentionally to rice fields dissipated measurably from rice-field water. However, only molinate and, to a lesser extent, thiobencarb showed significant loss through volatilization; MCPA and methyl parathion underwent significant chemical/microbial degradation. MCPA, in particular, chemically degraded into a product (4-chloro-*o*-cresol) whose volatilization rate from water was about an order of magnitude greater than the rates for the chemicals applied to rice. The laboratory chamber and the EXAMS computer model were able to predict the volatilization of the applied chemicals well — except for MCPA, which mainly volatilized as the free acid from dry deposits on soil and plant surfaces (a situation not amenable to EXAMS modeling). In addition, the EXAMS model did not predict well the volatilization of thiobencarb, possibly because of inaccurate physicochemical data. Overall, however, EXAMS appeared to be promising as a predictive tool for estimating volatilization, when the appropriate chemical properties and environmental conditions were used as input data. It may also be useful for estimating loss by other dissipation routes as well, and for estimating overall dissipation by all routes — a potentially useful capability for calculating rice-water holding periods and perhaps other information needed to ensure safe use of pesticides in rice culture.

The occurrence of methyl parathion, molinate, and thiobencarb residues in air in residential/business areas was due to drift during application (methyl parathion) and drift of residues volatilized from treated rice water (molinate and thiobencarb). The chemical/microbial instability of methyl parathion, combined with its low Henry's law constant, precluded any appreciable loss from field water. On the other hand, the application of the dry granular formulations and the relative stability of molinate and thiobencarb in water led to measurable downwind residues of the volatilized deposited materials. Volatilization rate, normalized to water concentration, for molinate from rice water was about 2–3 times that for thiobencarb under the same field conditions and for the same application rate. Using the ALOHA computer model with measured downwind air concentrations, the relative volatilization rates for the two herbicides translated into downwind molinate:thiobencarb concentration ratios of greater than 20 at the same downwind distance, assuming a single source for both compounds. The predicted results compared well with observed results, although the observed results were also influenced by differing application schedules for molinate and thiobencarb and different total amounts used (molinate was about three times thiobencarb). On the other hand, measured ratios of methyl parathion with molinate and thiobencarb were considerably less than the model-predicted ratios, suggesting that there were sources for methyl parathion more substantial than volatilization from water (i.e., drift during application and/or volatilization from surfaces resulting from uses of methyl parathion other than on rice in the same general vicinity). The ALOHA model is relatively recent and, therefore, has not been fully evaluated, but its utility is the ability to predict order of magnitude distance-concentration profiles for toxics if chemical source strength and pertinent meteorological data are known. In this way, it would be possible to estimate downwind concentrations expected from nearby applications given the source strength, wind conditions, and topography. From this, exposure can be estimated and thus the potential health impacts as well.

It is clear that good risk assessment of human exposure to pesticides requires reliable exposure data, which would include both field measurements and model (laboratory and computer) evaluations of pesticide source strength and drift. Determining exposure would necessarily involve recommending a "safe" distance from the source. Involved in the process of assessing exposure should be, however, a concern not only for human health, but for animal and plant health as well. As a consequence, risk assessment procedures should expand their scope beyond human health to include the effects of toxic chemicals on total ecosystems. The scope quickly becomes complex — it would be necessary to determine the relative effects of various toxic compounds and their conversion products and to determine the relative distribution of toxics among the various environmental compartments. Laboratory and computer models would go far in describing the extent of potential problems with relatively little investment of time and resources. For example, earlier work with model ecosystems determined the relative distribution of DDT and related chemicals among the soil, water, plants, and animals that made up the system.[37] Computer models, in particular, have the potential of providing exposure scenarios relatively rapidly and economically compared to actual field measurements. When field data are needed, models can aid in experimental design to decide where, and how many, samples should be taken.

SUMMARY

Volatilization rate (flux) was determined for three rice herbicides (MCPA, molinate, and thiobencarb) and an insecticide (methyl parathion) from flooded rice fields by (1) field measurement of flux over commercial rice fields, (2) laboratory chamber measurement, and (3) computer model prediction using an aquatic fate model (EXAMS). Under field conditions, molinate and thiobencarb showed significant loss through volatilization, but loss of MCPA and methyl parathion was due primarily to chemical/microbial degradation. Volatilization predicted by the laboratory chamber and the EXAMS computer model compared well with the field results, except for MCPA which showed higher than expected residues in air due to volatilization of the free acid from dry deposits on soil and plant surfaces. In general, EXAMS is a promising predictive tool for estimating volatilization from water.

Ambient airborne residue levels of methyl parathion; its oxon conversion product, methyl paraoxon; molinate; and thiobencarb were measured at small town and rural locations in the general area of use. These residue levels and their downwind distances from possible sources were correlated with the predictions of a Gaussian plume dispersion computer model (ALOHA). Measurable air concentrations of methyl parathion, methyl paraoxon, molinate, and thiobencarb in residential/business areas were due to drift during application (methyl parathion), to photooxidation of drifting residues (methyl paraoxon), and to drift of residues that volatilized from treated water (molinate and thiobencarb). These conclusions were derived partly from correlations of air concentrations with chemical use and were further supported by the predictions of the ALOHA computer model, using ambient residue levels and flux from rice-field water as input. This model is able to predict downwind concentrations of toxic chemicals if source strength, wind conditions, and topography are known. From this, exposure and potential hazard to health can be estimated.

Taken together, measurements of flux and the ambient residue levels of drifting chemicals and data from laboratory and computer models form the basis for determining exposure, from which relative risk, in terms of margins of safety, can be estimated. Because these margins of safety will vary depending upon the proximity of the chemical source, risk assessment determinations based on data taken at varying distances from chemical sources are essential so that "safe" distances can be established.

REFERENCES

1. Spencer, W. F., W. T. Farmer, and M. M. Cliath. "Pesticide Volatilization," *Residue Reviews* 49:1–47 (1973).
2. Lewis, R. G., and R. E. Lee, Jr. "Air Pollution from Pesticides: Sources, Occurrence, and Dispersion," in *Air Pollution from Pesticides and Agricultural Processes,* R. E. Lee, Jr., Ed. (Cleveland, OH: CRC Press, 1976).
3. Seiber, J. N., G. A. Ferreira, B. Hermann, and J. E. Woodrow. "Analysis of Pesticidal Residues in the Air near Agricultural Treatment Sites," in *Pesticide Analytical Methodology,* J. Harvey, Jr., and G. Zweig, Eds., ACS Symp. Series #136 (Washington, DC: ACS, 1980).
4. Seiber, J. N., S. C. Madden, M. M. McChesney, and W. L. Winterlin. "Toxaphene Dissipation from Treated Cotton Field Environments: Component Residual Behavior on Leaves and in Air, Soil, and Sediments Determined by Capillary Gas Chromatography," *J. Agric. Food Chem.* 27:284–91 (1979).
5. Taylor, A. W. "Post-Application Volatilization of Pesticides under Field Conditions," *J. Air Pollut. Control Assoc.* 28:922 (1978).
6. Taylor, A. W., and D. E. Glotfelty. "Evaporation from Soils and Crops," in *Environmental Chemistry of Herbicides,* R. Grover, Ed., Vol. I (Boca Raton, FL: CRC Press, 1988).
7. Seiber, J. N., and J. E. Woodrow. "Airborne Residues and Human Exposure," in *Determination and Assessment of Pesticide Exposure,* M. Siewierski, Ed. (New York, NY: Elsevier, 1984).
8. Spencer, W. F. "Volatilization of Pesticide Residues," in *Fate of Pesticides in the Environment (Proceedings of a Technical Seminar),* J. W. Biggar and J. N. Seiber, Eds., Agricultural Experiment Station, Division of Agriculture and Natural Resources, University of California, publication #3320 (1987).
9. Que Hee, S. S., R. G. Sutherlund, and M. Vetter. "Glc Analysis of 2,4-D Concentrations in Air Samples from Central Saskatchewan in 1972," *Environ. Sci. Technol.* 9:62–66 (1975).
10. Arthur, R. D., J. D. Cain, and B. F. Barrentine. "Atmospheric Levels of Pesticides in the Mississippi Delta," *Bull. Environ. Contam. Toxicol.* 15:129–34 (1976).
11. Kutz, F. W., A. R. Yobs, and H. S. C. Yang. "National Pesticide Monitoring Networks," in *Air Pollution from Pesticides and Agricultural Processes,* R. E. Lee, Jr, Ed. (Cleveland, OH: CRC Press, 1976).
12. Baunok, I. "Analysis of Airborne Pesticides at Selected Sites in South Africa," *Suid-Afrik. Tydskrif vir Wetenskap.* 80:277 (1984).
13. Kilgore, W., C. Fischer, J. Rivers, N. Akesson, J. Wicks, W. Winters, and W. Winterlin. "Human Exposure to DEF/Merphos," *Residue Reviews* 91:71–102 (1984).
14. Glotfelty, D. E. "Pathways of Pesticide Dispersion in the Environment," in *Agri-*

cultural Chemicals of the Future, J. L. Hilton, Ed., Beltsville Symposia in Agricultural Research, Vol. 8 (Totowa, NJ: Rowan and Allanheld, 1985).

15. "Air Pollution: Toxic Air Contaminants, AB 1807," Legislative Counsel's Digest, Chapter 1047 California Assembly Record, Sacramento, CA (March 27, 1983).
16. "Exposure Analysis Modeling System (EXAMS)," EPA, Environmental Systems Branch, Environmental Research Laboratory, Office of Research and Development, Athens, GA (1980).
17. "Areal Locations of Hazardous Atmospheres (ALOHA, v. 4.0)," National Oceanic and Atmospheric Administration, Hazardous Materials Response Branch, Modeling and Simulation Studies, Seattle, WA (December, 1987).
18. Sanders, P. F., and J. N. Seiber. "A Chamber for Measuring Volatilization of Pesticides from Model Soil and Water Disposal Systems," *Chemosphere* 12:999–1012 (1983).
19. Sanders, P. F., J. N. Seiber. "Organophosphorus Pesticide Volatilization: Model Soil Pits and Evaporation Ponds," in *Treatment and Disposal of Pesticide Wastes,* R. F. Krueger and J. N. Seiber, Eds., ACS Symposium Series, No. 259 (1984).
20. Sanders, P. F. "Pesticide Volatilization from Model Pits and Evaporation Ponds," PhD Dissertation, University of California, Davis, CA (1984).
21. Seiber, J. N., M. M. McChesney, P. F. Sanders, and J. E. Woodrow. "Models for Assessing the Volatilization of Herbicides Applied to Flooded Rice Fields," *Chemosphere* 15:127–38 (1986).
22. Ross, L. J., and R. J. Sava. "Fate of Thiobencarb and Molinate in Rice Fields," *J. Environ. Qual.* 15:220–25 (1986).
23. Taylor, A. W., D. E. Glotfelty, B. L. Glass, H. P. Freeman, and W. M. Edwards. "Volatilization of Dieldrin and Heptachlor from a Maize Field," *J. Agric. Food Chem.* 24:625–31 (1976).
24. Seiber, J.N., and M. M. McChesney. "Measurement and Computer Model Simulation of the Volatilization Flux of Molinate and Methyl Parathion from a Flooded Rice Field," Final Report to the Department of Food and Agriculture, Sacramento, CA (April, 1987).
25. Seiber, J. N., M. M. McChesney, and J. E. Woodrow. "Airborne Residues Resulting from Use of Methyl Parathion, Molinate, and Thiobencarb on Rice in the Sacramento Valley, California," *Environ. Toxicol. Chem.* 8:577–88 (1989).
26. Rummens, F. H. A., and F. J. A. Louman. "Proton Magnetic Resonance Spectroscopy of Thiocarbamate Herbicides," *J. Agric. Food Chem.* 18:1161–64 (1970).
27. Glotfelty, D. E. "Measuring Pesticide Volatilization in the Field," paper presented at the 192nd National Meeting (AGRO 98), American Chemical Society, Anaheim, CA, September 12, 1986.
28. Majewski, M. S., D. E. Glotfelty, M. M. McChesney, and J. N. Seiber. "An Alternative to the Aerodynamic Method for Determining Pesticide Evaporative Flux from Treated Surfaces," paper presented at the 192nd National Meeting (AGRO 102), American Chemical Society, Anaheim, CA, September 12, 1986.
29. Crosby, D. G., M.-Y. Li, J. N. Seiber, and W. L. Winterlin. "Environmental Monitoring of MCPA in Relation to Orchard Contamination," Final Report to the Department of Food and Agriculture, Sacramento, CA (March 16, 1981).
30. Akesson, N. B., D. E. Bayer, C. L. Elmore, S. Wilce, and W. Winterlin. "Air Transport of Herbicide Chemicals and Symptoms on Sensitive Crops," Report to the California Rice Growers Association (1970).
31. Rippen, G., E. Zietz, R. Frank, T. Knacker, and W. Klöpffer. "Do Airborne

Nitrophenols Contribute to Forest Decline?" *Environmental Technology Letters* 8:475–82 (1987).

32. Woodrow, J. E., D. G. Crosby, T. Mast, K. W. Moilanen, and J. N. Seiber. "Rates of Transformation of Trifluralin and Parathion Vapors in Air," *J. Agric. Food Chem.* 26:1312–16 (1978).

33. Klisenko, M. A., and M. V. Pis'mennaya. "Photochemical Conversion of Organophosphorus Pesticides in the Air," *Gig. Tr. Prof. Zabol.* (Russian) 56 (1979), through *Chem. Abst.* 94:126596h.

34. Woodrow, J. E., D. G. Crosby, and J. N. Seiber. "Vapor-Phase Photochemistry of Pesticides," *Residue Reviews* 85:111–25 (1983).

35. Dourson, M. L., and J. F. Stara. "Regulatory History and Experimental Support of Uncertainty (Safety) Factors," *Regulatory Toxicology and Pharmacology* 3:224–38 (1983).

36. Oudiz, D., and A. K. Klein. "Evaluation of Ethyl Parathion as a Toxic Air Contaminant," California Department of Food and Agriculture, Report No. EH-88-5 (June, 1988).

37. Metcalf, R. L., L. K. Cole, S. G. Wood, D. J. Mandel, and M. L. Milbrath. "Design and Evaluation of a Terrestrial Model Ecosystem for Evaluation of Substitute Pesticide Chemicals," Terrestrial Ecology Branch, Corvallis (OR) Environmental Research Laboratory, U.S. EPA Report-600/3–79–004 (1979).

## CHAPTER 6

Regional and Global Transport and Distribution of Trace Species Released at the Earth's Surface

Hiram Levy II

INTRODUCTION

Most atmospheric trace constituents, including pesticides and other toxics, are transported as mere passengers and do not influence the motions of the winds. Their time-dependent behavior is influenced by surface or near-surface emissions, the atmospheric winds, physical removal processes, and chemical destruction and production. This is summarized in the following descriptive equation:

$$\frac{\partial \text{Tracer}}{\partial t} = \text{Source} + \text{Transport} - \text{Sink} - \text{Chemical Destruction} + \text{Chemical Production}$$

where SOURCE is some form of surface or near-surface emission, TRANSPORT is the result of three-dimensional winds acting on gradients in the TRACER mixing ratio (mole fraction), SINK is a physical removal process that is a function of TRACER solubility and surface reactivity, CHEMICAL DESTRUCTION can be either direct chemical reaction or photodissociation, and CHEMICAL PRODUCTION converts an emitted compound into a toxic. If the trace gas or particle is uniformly mixed throughout the globe (i.e., there are no gradients in its mixing ratio or mole fraction), TRANSPORT will be 0. However, this is not the case for modern pesticides. The long-lived ones have only been released for a few decades and have not had time to become well mixed, and the short-lived ones, although they have reached a balance between emission and destruction, have large spatial gradients in their distribution. In all cases, the atmospheric winds play a key role in the global distribution of pesticides and other toxics.

SOURCE

We will first consider a number of observations and model simulations that demonstrate the impact of source nonhomogeneity on tracer transport and distribution.

In the first simulation, a tracer with no sinks (i.e., an infinite lifetime) was uniformly released for 10 days from a 10° midlatitude belt stretching around the globe in the Northern Hemisphere boundary layer. The simulated time evolution of the resulting distribution, averaged around latitude circles, is presented in Figure 1 as a function of latitude on the horizontal axis, and pressure (mb) on the left or height (km) on the right vertical axis. After 10 days of integration, the tracer remains in the northern midlatitudes, though it has started to mix into the free troposphere. After 3 months, it is relatively well mixed throughout the Northern Hemisphere, but very little has been carried across the equator. In 1 year, the tracer has mixed into both hemispheres, though there is still an interhemispheric gradient, which should disappear in approximately 10 years.

When the emissions of very long-lived chemicals are continuous, though not uniform, around the globe (both nitrous oxide and chlorofluorocarbons are good examples), we still find gradients and are able to observe transport events. Chlorofluorocarbons, with their very long lifetimes (≈ 100 yrs), have distributions that have not yet equilibrated in the nearly 40 years of their release. They still show an excess ($\approx 5\%$) in the Northern Hemisphere.[1] Nitrous oxide, which has been released naturally for millions of years, has a very small interhemispheric gradient ($\approx 0.3\%$) that appears to be the residual result of a Northern Hemisphere bias in the source distribution. Both trace gases show considerable variability in the neighborhood of strong local sources,[2] as the simulated time series in Figure 2 demonstrate. Even with the very well-mixed background of nitrous oxide, one can simulate transport events in the vicinity of strong local sources, such as Panama, where the surface mixing ratio varies between 290 ppbv (parts per billion by volume) and 315 ppbv. However, remote regions with no sources, such as Samoa, show almost no variability in the simulated mixing ratio.

TRANSPORT

In this chapter, long-range transport refers to transport from continental source regions to the oceans and to other continents, as well as transport between the hemispheres.

Winds

Not only are atmospheric motions three-dimensional, but they occur on spatial scales ranging from meters for local turbulence, to kilometers for cloud

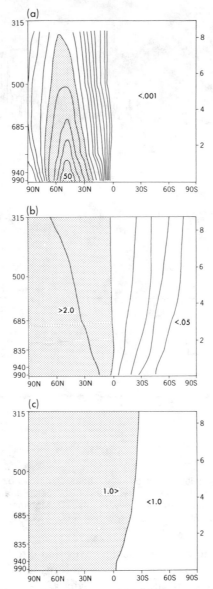

Figure 1. Latitude-altitude plots of zonal-average tracer mixing ratios (in dimensionless units) are shown for (a) 10 days after release, (b) 3 months after release, and (c) 1 year after release. The contours are in logarithmic intervals (1–2–5) with the dotted area > 1.0.

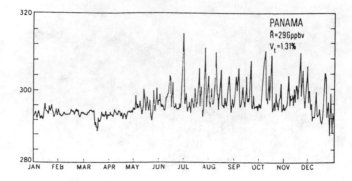

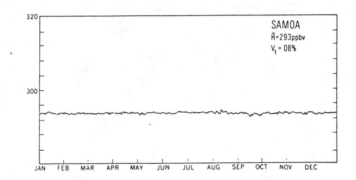

Figure 2.   One-year simulated surface time series of nitrous oxide (ppbv) for two tropical locations, one that has nearby emission sources (Panama) and one that is near no sources (Samoa). *Source*: Levy, H., II, J. D. Mahlman, and W. J. Moxim. "Tropospheric N_2O Variability," *J. Geophys. Res.* 87:3061–80 (1982).

convection, and up to planetary-scale waves that circle the globe. To simplify these time-dependent three-dimensional motions and identify basic transport processes, we average the atmospheric winds over space and time. Having identified potential transport paths from this highly averaged picture, we will see that the average winds do not quantitatively reproduce the average transport and, in some cases, may be quite misleading.

The simplest picture of the winds, generated by averaging over many years and over longitude (i.e., around latitude circles), does identify some important features of atmospheric transport. The yearly averaged E-W and N-S wind velocities are shown in Figures 3a and 3b as a function of latitude and pressure. Pressure is given in db, 1/10 of the surface atmospheric pressure; thus, 10 db is the surface, and 1 db is a standard height of ≈ 16 km, the

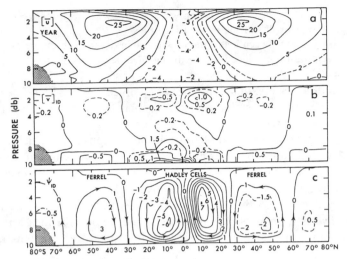

Figure 3. Zonal mean cross sections of (a) the east-west yearly-average wind in m/sec where negative values are from the east, (b) the north-south yearly-average wind in m/sec where negative values are towards the north, and (c) the yearly-average atmospheric mass flow in 10^{10} kg/sec. *Source*: Peixoto, J. P., and A. H. Oort. "Physics of Climate," *Rev. Mod. Phys.* 56:365–429 (1984).

approximate top of the troposphere. Note that the E-W velocities are from the east (easterly) in the tropics, from the west (westerly) at higher latitudes, and increase with height, reaching 25 m/sec in the upper troposphere of the midlatitudes (the jet stream). In contrast, the N-S velocities, though providing a more complex picture, are much weaker and seldom exceed 1 m/sec.

Because the average E-W winds are so strong, we can, as a first approximation, assume that an atmospheric tracer is well mixed in the E-W direction and focus on the slower transport that is driven by the time-averaged vertical and N-S winds. A standard two-dimensional meteorological picture of the time-averaged mass flow is shown in Figure 3c. A strong upward flux in the tropics and subsiding flux in the subtropics is produced by the thermally direct Hadley cells. The subtropics is that region lying between the tropics, a 20–30° latitude belt centered at the equator, and midlatitude. As a result of a statistical artifact, there is rising at midlatitudes and sinking in the subtropics. There is also very weak sinking at the Poles. This highly averaged picture, although statistically correct for the ensemble of atmospheric motions over the years, does not exist for the motion of a particular air parcel. Moreover, the time-averaged transport is even less complex. On the average, trace species are lifted in the tropics; they are carried poleward and descend in the subtropics, midlatitudes, and polar regions; and they are carried back to the tropics in the lower troposphere.

However, from these highly averaged wind fields, we can tell that

1. transport in the E-W direction will increase with height
2. air parcels will subside in the subtropics
3. air passing between hemispheres will rise in the tropics and be exposed to precipitation produced by the lifting of moist tropical air

Although soluble tracers, in the absence of chemical sinks, should be relatively long-lived in the dry and stable subtropics, their transport between the hemispheres is quite unlikely.

A more complex picture of the atmospheric motions is provided by Figures 4a and 4b, which show the seasonal average surface pressure and associated surface winds for January and July. Air flows from the subtropics into the tropics (i.e., converges in the tropics); areas of high pressure form in the subtropics and move north and south 10–15° with the sun. In the Northern Hemisphere winter, areas of high pressure also develop over the North American and Asian continents, while areas of low pressure form in the North

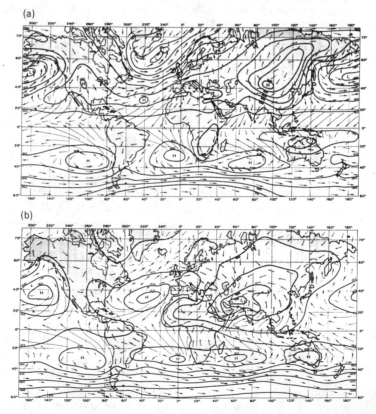

Figure 4.   Surface pressure (mb) and surface wind direction and speed. The wind flows with the arrows; the darker the arrows, the faster the wind. (a) January averages; (b) July averages. *Source*: Hasse, L., and F. Dobson. Introductory Physics of the Atmosphere and Ocean (Dordrecht, Netherlands: Reidel, 1986).

Atlantic and North Pacific. These low pressure areas are a time-averaged result of the dominant winter storm tracks off the eastern coasts of North America and Asia. The winter storms promote lifting out of the atmospheric boundary layer, the lowest layer (1–2 km thick) of the troposphere, and north-eastward transport of emissions from North America and Asia out over the ocean. There is a strong surface flow from India and Southeast Asia into the Southern Hemisphere and a strong westerly flow in the Southern Hemisphere midlatitudes.

In the northern summer, the flow in midlatitudes is weaker, and high surface pressure in both the subtropical Atlantic and Pacific becomes important. Rather than large-scale lifting of continental emissions by winter storms, the emissions are lifted into the free troposphere, that portion of the troposphere lying above the atmospheric boundary layer, by small-scale convection, both dry and wet. The surface flow in the Indian Ocean is now toward India and helps maintain that region's summer monsoons, while the converging flow in the rest of the tropics has moved north of the equator. A major feature of both seasons is this converging surface flow in the tropics that is coupled to a vigorous rising motion in the region and results in strong precipitation. This region is called the intertropical convergence zone (ITCZ) and forms a strong barrier to transport of soluble tracer between the hemispheres.

Time-Averaged Transport vs Time-Averaged Winds

Even the relatively complex picture presented by the time-averaged winds in Figure 4 fails to capture the time-averaged transport resulting from the time-dependent three-dimensional motions of the winds. As an example, winter averages of the simulated distributions of a water soluble tracer released in North America, in this case emissions of nitrogen oxides from fossil fuel combustion, are shown in Figures 5a and 5b.[3] In both cases there is a constant release of emissions in the two lowest model levels. In Figure 5a, the emissions are transported by constant winds that are the winter average of the time-dependent winds generated by the parent climate model, and the climate model's average precipitation is used in SINK. In Figure 5b, the transport is provided by the climate model's time-dependent winds, and the model's self-consistent and time-dependent precipitation is used in SINK.

Note that the transport by the time-averaged winds in Figure 5a corresponds to what might be expected, based on the previous section. In the boundary layer, the tracer is confined to the average North Atlantic storm track. It is confined to a midlatitude belt in the free troposphere, where, on average, the westerly winds are quite strong. In neither case is there transport to the Arctic. However, the actual average distribution—a result of time-dependent winds acting throughout the winter—is quite different (see Figure 5b). Although at the surface there is still a maximum along the North Atlantic storm track, the actual amount transported to Europe is much smaller. Furthermore, tracer is carried into the subtropics of both the Atlantic and Pacific, and there is

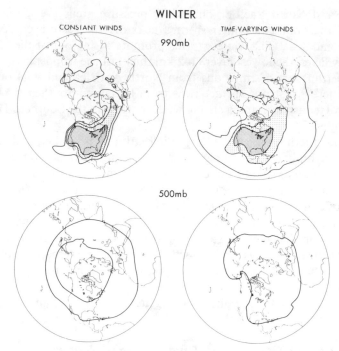

Figure 5. Average winter mixing ratios (ppbv) for a soluble tracer released at the surface in North America and transported by both (a) constant winter-average winds and (b) fully time-dependent winds. The clear contour contains mixing ratios in the range 0.01–0.1 ppbv, the dotted area is for the range 0.1–1.0 ppbv, and the dark area is 1.0–10.0 ppbv. *Source*: Levy and Moxim.[3]

significant transport into the Arctic. In the middle of the atmosphere, the tracer spreads throughout the Arctic, though the actual time-dependent winds do not sweep the tracer completely around the globe as was the case when they were time-averaged.

The inability of time-averaged winds to produce time-averaged transport is easily demonstrated by a simple thought experiment. Consider emissions in the eastern United States and a very simple wind pattern that blows from the west for half the year and from the east for the other half. Over the year, the winds over the eastern United States cancel and produce no transport by the average wind. The resulting distribution is a blob over the eastern United States. In fact, the actual transport would be towards the east for half the year and towards the west for the other half. The resulting average distribution would be a midlatitude belt of emissions around the globe, not a blob over the eastern United States.

Table 1. Percentage Contribution to NOy in the Arctic by Source Region.

Source	80 m	0.5 km	1.5 km	3.1 km
Eurasia N of 60° N	49	25	10	6
Europe S of 60° N	30	44	50	46
North America S of 60° N	10	15	26	37
Asia S of 60° N	11	16	14	11
Global	100	100	100	100

Source: Levy and Moxim.[3]

Regional and Global Transport

It should be clear from the preceding discussion that the average winds are only part of the story. For that reason, we have developed comprehensive global transport models that use time-dependent three-dimensional winds and require very powerful computers. Such a model, which has been discussed in considerable detail,[4] was used to simulate the examples of tracer transport presented in this chapter.

The wintertime transport of combustion products to the Arctic is simulated by a global transport model using time-dependent three-dimensional winds and a constant emission source in the lowest two model levels. The percentage contributions of different source regions to the mixing ratios at different heights in the Arctic atmosphere are summarized in Table 1. These results show the influence of two important transport mechanisms: the strong high pressure system over western Asia and the North Atlantic storm track. As a result of the northward transport of pollution by this Asian high pressure system, the surface concentration of combustion products in the Arctic is dominated by emissions from the European Arctic and northern Europe as a whole. The simulated distribution of the nitrogen component of arctic haze in Figure 6 shows a strong asymmetry between surface concentrations over the North American and European Arctic. North American emissions do become quite important in the free troposphere as they are lifted and carried northeastward by the prevailing storms over the North Atlantic (i.e., the North Atlantic storm track).

A summertime simulation, using the same global transport model and global emission source, finds that surface emissions from the southwestern United States are carried far out into the subtropical Pacific (see Figure 7). The observation of a similar plume of pesticides and other organics is reported in Chapter 11. This long-range transport in the free troposphere results from the subtropical high that moves north of Hawaii in the summer. The transport path is fed by emissions in the southwestern United States that are intermittently lifted from the surface by convection over the arid land. As a result, episodes of continental pollution are then observed in the eastern Pacific. A more complex transport process results in the transport of Asian emissions out over much of the North Pacific, particularly in the spring and the fall.[5]

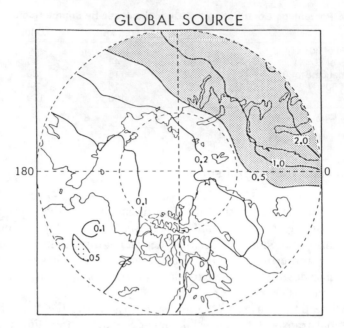

Figure 6. Simulated average winter mixing ratios (ppbv) for the nitrogen component of arctic haze in the atmospheric boundary layer. The light dots are for mixing ratios less than 0.05 ppbv and the dark shading is for mixing ratios that exceed 0.50 ppbv. Log contours 1–2–5 are used. *Source*: Levy and Moxim.[3]

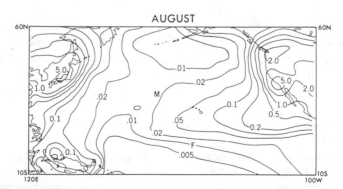

Figure 7. The simulated August average plume of the nitrogen component of North American pollution (ppbv) that has been transported out over the eastern North Pacific. M represents Midway Island; F, Fanning Island. The Hawaiian Islands are drawn on the map. *Source*: Levy and Moxim.[4]

REMOVAL

Sink

Pesticides and other toxics that react with and stick to surfaces are readily removed in the atmospheric boundary layer and must be lifted into the free troposphere for effective long-range transport. If these substances are also soluble, their rapid removal by intermittent precipitation inhibits transport out of the boundary layer. As a general rule, the long-range transport of "sticky" or soluble chemicals is greatly reduced by their short atmospheric lifetimes. The stable dry conditions found in the winter Arctic and discussed in the previous section are special cases that do promote long-range transport.

In Figure 8, we show the simulated yearly average surface distribution of an anthropogenic trace chemical — in this case, nitrogen oxides and nitrates — that is soluble and surface reactive. As expected for such a tracer, there is a strong correlation between the distribution maxima and the continental source regions. Transport from Asia out over the North Pacific storm track and from North America out over the North Atlantic storm track shows up clearly. With less than .005 ppbv over much of the southern ocean, it is clear that little is transported from the source regions of the Northern Hemisphere through the tropics and the ITCZ into the Southern Hemisphere.[4] We see downwind transport at southern midlatitudes due to the prevailing westerlies, and easterly transport in the tropics and subtropics.

Chemical Destruction

If the pesticides and other toxics are neither soluble nor surface reactive ("sticky"), their atmospheric lifetime is determined by their rate of chemical destruction. Methane, a chemical tracer with a moderately long (5–10 yr) lifetime, has been released for a long time in the atmosphere and has an

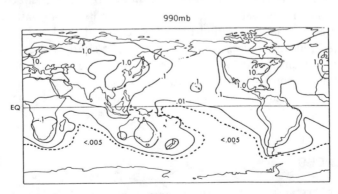

Figure 8.   Latitude-longitude plot of the yearly-average surface mixing ratio (ppbv) of a soluble tracer released by fossil fuel combustion. The contour intervals are 0.01, 0.1, 1.0, and 10.0. *Source*: Levy and Moxim.[3]

equilibrated distribution. With most of its source in either the tropics or the Northern Hemisphere, it has a significant interhemispheric gradient ($\approx 10\%$) and a small seasonal variation.[6] As a result of this interhemispheric difference, one can identify air masses that have been transported from one hemisphere to the other.

Carbon monoxide, an atmospheric trace gas with an even shorter lifetime (≈ 3 months), has much higher variability. Near the surface, its mole fraction or mixing ratio ranges from a high of 50 ppmv in severely polluted urban areas, to 0.15–0.25 ppmv in the background Northern Hemisphere, and to a low of 0.050 ppmv in the very clean Southern Hemisphere.[7] The range of variation is much less, 0.05–0.15 ppmv, in the middle of the troposphere. Carbon monoxide has frequently been used as a tracer of anthropogenic pollution[8] as well as a tracer of interhemispheric exchange. There are a number of very short-lived (i.e., highly reactive) hydrocarbons whose presence signifies a nearby source. These compounds are only found at very low levels in remote regions distant from their sources.

SUMMARY

Although time-averaged wind fields may help identify preferred paths of transport, they are not adequate to explain either simulated or observed average transport. Time-dependent meteorology, ranging from small-scale turbulence and convection to large storms, plays an important role in lifting surface emissions into the free troposphere, where long-range transport can readily occur. Once in the free troposphere, atmospheric transport is still strongly influenced by large-scale fluctuations in the time-dependent wind fields.

Even very long-lived trace chemicals will have a variable distribution and display identifiable transport events if their surface sources are nonuniform over the globe. Those tracers that are insoluble and nonreactive with surfaces will have their atmospheric lifetime determined by their chemical reactivity. The more chemically reactive the tracer, the more its mixing ratio distribution is influenced by the distribution of its sources, and the less likely it is to be transported long distances. Soluble tracers have short atmospheric lifetimes and are only transported long distances under stable dry conditions, such as those found in the winter Arctic and over the subtropical oceans. Due to removal by precipitation in the ITCZ, their transport between the hemispheres is quite unlikely.

REFERENCES

1. Cunnold, D. M., R. G. Prinn, R. A. Rasmussen, P. G. Simmonds, F. N. Alyea, et al. "Atmospheric Lifetime and Annual Release Estimates for $CFCl_3$ and CF_2Cl_2 from 5 Years of Data," *J. Geophys. Res.* 91:10797–817 (1986).

2. Prather, M. J. "Continental Sources of Halocarbons and Nitrous Oxide," *Nature* 317:221–25 (1985).

3. Levy, H., II and W. J. Moxim. "Examining the Global Impact of Local/Regional Air Pollution: The Role of Global Transport Models," in *17th International Technical Meeting on Air Pollution Modeling and Its Application,* H. van Dop, Ed., Vol. VII (New York, NY: Plenum Publishing Corporation, 1989).

4. Levy, H., II and W. J. Moxim. "Simulated Global Distribution and Deposition of Reactive Nitrogen Emitted by Fossil Fuel Combustion," *Tellus* 41B:256–71 (1989).

5. Levy, H., II and W. J. Moxim. "Influence of Long-Range Transport of Combustion Emissions on the Chemical Variability of the Background Atmosphere," *Nature* 338:326–28 (1989).

6. Steele, L. P., P. J. Fraser, R. A. Rasmussen, M. A. K. Khalil, T. J. Conway, et al. "The Global Distribution of Methane in the Troposphere," *J. Atm. Chem.* 5:125–71 (1987).

7. Logan, J. A., M. J. Prather, S. C. Wofsy, and M. B. McElroy. "Tropospheric Chemistry: A Global Perspective?" *J. Geophys. Res.* 86:7210–54 (1981).

8. Fishman, J. S., W. Seiler, and P. Haagensen. "Simultaneous Presence of O_3 and CO Bands in the Troposphere," *Tellus* 32:456–63 (1980).

## CHAPTER 7

The Calculated Effects of Nonexchangeable Material on the Gas/Particle Distributions of Organochlorine Compounds at Background Levels of Suspended Atmospheric Particulate Matter

James F. Pankow

INTRODUCTION

The global transport of chemical species through the atmosphere will depend on the manner in which those species are distributed between the gas and particulate phases. For example, the compound-dependent extent to which this partitioning occurs will have direct implications for the relative importance of processes such as the dry deposition of gaseous compounds, the dry deposition of compounds sorbed on particles, the precipitation scavenging of gaseous compounds, and the precipitation scavenging of compounds sorbed on particles.

Interest in the gas/particle distribution process was stimulated by Junge,[1] who carried out preliminary calculations based on a linear gas/solid sorption model. Yamasaki et al.[2] used the same type of model and also assumed that the particulate surface area concentration available for sorption is proportional to the level of TSP ($\mu g/m^3$). Using data collected in Tokyo, Yamasaki et al.[2] found that the partitioning of atmospheric polycyclic aromatic hydrocarbons (PAHs) could be described by a compound- and temperature-dependent constant of the form

$$K = \frac{A(TSP)}{F} = \frac{A}{F/(TSP)} \tag{1}$$

where A = atmospheric concentration measured in the gas phase (ng/m^3)

F = atmospheric concentration measured as associated with the particulate phase (ng/m^3)

F/TSP = concentration of the compound in/on the particles ($ng/\mu g$)

A/(F/TSP) = distribution coefficient between the gas and particle
           phases

Yamasaki et al.[2] developed Equation 1 by assuming that A and F are free from sampling artifacts. It is in this context that A and F will be used here.

Bidleman et al.[3] and Ligocki and Pankow[4] have found that Equation 1 is obeyed for PAHs in cities other than Tokyo, and that the compound-dependent values of K for several PAHs are about the same as for Tokyo. In addition, based on samples collected in several cities, Bidleman and co-workers[3,5,6] have found that a variety of organochlorine compounds also follow Equation 1. Based on these results, it may be concluded that the sorption properties of urban particulate matter tend to be similar independent of city and time.

As discussed by Pankow,[7] within a given compound class (e.g., PAHs, organochlorines, etc.), log K values at a given temperature correlate well with the vapor pressure values (p^o, torr) for the compounds at that temperature. If the compound is a solid at the temperature of interest, the vapor pressure of the subcooled liquid should be used. As discussed by Pankow[7,8], gas/solid sorption theory predicts that the dependence of K on p^o is given by

$$\log K = \log C + \log p^o \tag{2}$$

where C is a constant that depends on the chemical properties of the compound class, the properties of the aerosol particles, and the temperature T (°K).

Two important parameters that affect C are Q_1, the enthalpy of desorption from the surface of the particulate matter (kcal/mol), and Q_v, the enthalpy of vaporization of the liquid (or subcooled liquid) (kcal/mol). Using Pankow's estimates[7] for the other important parameters affecting C, at T = 293°K, it may be shown that

$$C = (5.6)^{-j} 5.5 \times 10^9 \tag{3}$$

where j represents the difference between Q_1 and Q_v in units of kcal/mol. As discussed by Pankow,[7] the organochlorine data of Bidleman et al.[3] for urban particulate matter are consistent with a j ≅ 1.5 kcal/mol. For this value, at 293°K we obtain log C = 8.6, and so for urban particulate matter

$$\log K = 8.6 + \log p^o \tag{4}$$

How well Equation 4 describes the sorption of organochlorine compounds to *nonurban* particulate matter is not known at the present time.

As has been discussed by Ligocki and Pankow[4] and Pankow,[8] a portion of any given compound can be bound within atmospheric particles in a manner that prevents exchange with the atmospheric gas phase, but can nevertheless be extracted during analysis. The presence of such bound material will affect the value of A(TSP)/F. In particular, the values of A and F will be less and greater, respectively, than what would be expected at full equilibrium. At a given temperature for a given compound, then, A(TSP)/F will be undervalued compared to what would be expected at equilibrium. The overall fraction ϕ associated with the particulate matter (which equals F/(A + F)) will be overvalued. Thus, even when C remains constant within a given compound class, this effect can cause actual field data for that class to exhibit significant negative deviation relative to what is predicted by Equation 2. When the level of sorbing particulate matter is low, as will be the case at remote sites, the presence of even a small amount of bound material can greatly affect the observed value of A(TSP)/F.

GOVERNING EQUATIONS

Following Pankow,[8] let x = the percentage of a given compound that is bound within the particulate matter and not available for exchange in the atmosphere. This fraction is assumed to be extractable during analysis. The percentage x is calculated over the total mass of material present in each m^3 of air. The value of x is likely to depend upon the compound as well as the source, nature, age, etc., of the particulate matter. The balance of the material will partition according to Equation 2. Pankow[8] then derives

$$\log \frac{A_T(TSP)}{F_T} = \log C + \log p^o + \log \frac{100 - x}{100 + Cp^o x/(TSP)} \tag{5}$$

$$\phi = \frac{\dfrac{(100 - x)}{(1 + Cp^o/TSP)} + x}{100} \tag{6}$$

where A_T is the concentration of the gaseous portion of the compound, and F_T is the total (i.e., exchangeable + bound) particulate-associated material. The subscript T is used here to indicate explicitly that the measured gaseous and particulate-associated fractions are affected by the presence of nonexchangeable material.

BEHAVIOR OF ORGANOCHLORINE COMPOUNDS AT LOW TSP LEVELS

In order to apply Equations 5 and 6 to predict the behavior of organochlorine compounds in remote air parcels as a function of p^o, we need to know

1. the value of C for the sorption of typical organochlorines on the type of particulate matter present in such air
2. how x varies with p^o

Since C values for nonurban particulate matter are not presently available, we will assume C = 8.6 in the calculations carried out here. Also, while Equations 5 and 6 may be used assuming any dependence of x on p^o, we will examine the behavior that results when x is assumed to take on various different *constant* values ranging between 0 and 20%.

When TSP = 30 $\mu g/m^3$, Figures 1 and 2 indicate that values of x that are even as low as 0.3% can have a major effect on the value of $A_T(TSP)/F_T$. Thus, while the assumption of complete equilibrium would predict that a compound with a p^o value of 10^{-4} torr would be < 0.08% in the particulate phase, at this level of TSP there would, in fact, be slightly more than 0.3% in that phase. The magnitudes of atmospheric loss mechanisms such as precipitation particle scavenging and dry particle deposition would therefore both be underestimated by a factor of about four. If other loss mechanisms such as those involving the gas phase portion of the compound are relatively unimportant, then major errors would result in any modeling effort designed to predict atmospheric lifetimes, total deposition rates, etc. As seen in Figures 3 and 4, when TSP is lowered to 3 $\mu g/m^3$, the effects of bound materials on the value of $A_T(TSP)/F_T$ are proportionally larger.

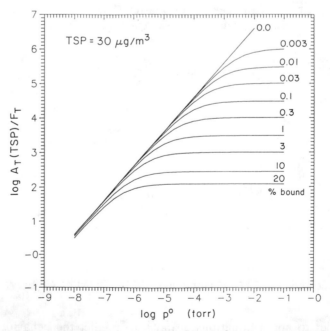

Figure 1. $A_T(TSP)/F_T$ vs log p^o at 293°K for TSP = 30 $\mu g/m^3$ with various percentages (x) of bound compound. The value of x is assumed to be independent of p^o. The energetics of sorption are assumed equal to give log C = 8.6.

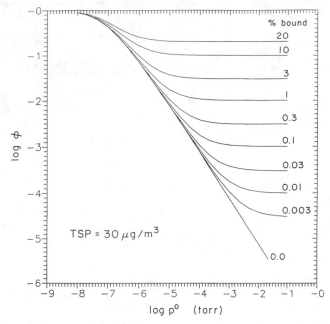

Figure 2.   Log ϕ vs log p^o at 293°K for TSP = 30 μg/m^3 with various percentages of bound compound. Other assumptions as in Figure 1.

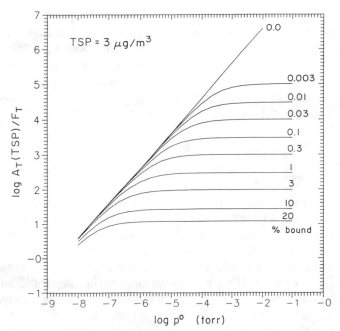

Figure 3. $A_T(TSP)/F_T$ vs log p^o at 293°K for TSP = 3 μg/m^3 with various percentages of bound compound. Other assumptions as in Figure 1.

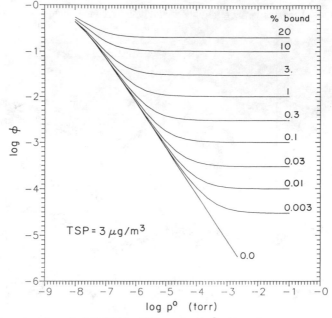

Figure 4. Log ϕ vs log p^o at 293°K for TSP = 3 $\mu g/m^3$ with various percentages of bound compound. Other assumptions as in Figure 1.

The case where x is independent of p^o is only one type of situation that can be examined using Equations 5 and 6. Some might expect that a more reasonable dependence of x would be a function that somehow increases as p^o decreases. Plots for that case and TSP = 30 and 3 $\mu g/m^3$ will nevertheless tend to be somewhat similar to Figures 1–4. For example, for small p^o, the proportion of material that will be bound to the particulate matter will always tend to be large, and therefore less affected by a nonzero value of x.[8]

CONCLUSIONS

The effects of a constant fraction of compound being bound within particulate matter on the parameters $A_T(TSP)/F_T$ and ϕ are readily calculated. For organochlorine compounds, the magnitude of the effect can be very significant even when the bound fraction is as low as a few percent. As the compound becomes more volatile and/or as TSP decreases, the magnitude of the effect increases. A knowledge of the compound-dependent magnitudes of this effect will be necessary for accurate predictive modeling of the long-range transport of organochlorine compounds in the atmosphere since the different processes and time scales govern the removal of gaseous and particle-associated contaminants.

REFERENCES

1. Junge, C. E. "Basic Considerations about Trace Constituents in the Atmosphere as Related to the Fate of Global Pollutants," in *Fate of Pollutants in the Air and Water Environments*, I. H. Suffet, Ed., Part I (New York, NY: J. Wiley and Sons, 1977), pp. 7–26.
2. Yamasaki, H., K. Kuwata, and H. Miyamoto. "Effects of Temperature on Aspects of Airborne Polycyclic Aromatic Hydrocarbons," *Env. Sci. Technol.* 16:189–94 (1982).
3. Bidleman, T. F., W. N. Billings, and W. T. Foreman. "Vapor-Particle Partitioning of Semivolatile Organic Compounds: Estimates from Field Collections," *Env. Sci. Technol.* 20:1038–43 (1986).
4. Ligocki, M. P., and J. F. Pankow. "Measurements of the Gas/Particle Distributions of Atmospheric Organic Compounds," *Env. Sci. Tech.* 23:75–83 (1989).
5. Foreman, W. T. and T. F. Bidleman. "An Experimental System for Investigating Vapor-Particle Partitioning of Trace Organic Pollutants," *Env. Sci. Technol.* 21:869–75 (1987).
6. Bidleman, T. F., and W. T. Foreman. "Vapor-Particle Partitioning of Semivolatile Organic Compounds," in *The Chemistry of Aquatic Pollutants*, R. A. Hites and S. J. Eisenreich, Eds., ACS Advances in Chemistry Series, No. 216 (New York, NY: American Chemical Society, 1987), pp. 27–56.
7. Pankow, J. F. "Review and Comparative Analysis of the Theories on Partitioning between the Gas and Aerosol Particulate Phases in the Atmosphere," *Atmos. Environ.* 21:2275–83 (1987).
8. Pankow, J. F. "The Calculated Effects of Non-Exchangeable Material on the Gas-Particle Distributions of Organic Compounds," *Atmos. Environ.* 22:1405–9 (1988).

## CHAPTER 8

Measurements of Selected Organochlorine Compounds in Air near Ontario Lakes: Gas-Particle Relationships

N. Douglas Johnson, Douglas A. Lane, William H. Schroeder, and William M. Strachan

INTRODUCTION

Atmospheric transport and deposition of toxic contaminants such as PCBs and various organochlorine pesticides is generally regarded as the major link to explain the occurrence and persistence of these organics in remote water bodies and in some of the Great Lakes. Despite their limited use in North America, hexachlorocyclohexanes (α- and γ-HCH) are often reported to be one of the most abundant chlorinated organic compounds found in the air of remote locales.[1-3] Some of the semivolatile chlorinated species are perceived to coexist in the atmosphere both in the vapor form and as adsorbed components of airborne particulate matter. Accordingly, the type and extent of atmospheric removal by processes such as wet and dry deposition will be partly governed by the vapor:particle ratio. For example, vapor scavenging at water surfaces and rainout washout can be expected to be the primary removal mechanisms for some of the more volatile lower molecular weight and soluble constituents such as lindane. Other substances, such as some PCB components, that may preferentially partition to the particle phase or for which Henry's law constants do not favor vapor scavenging, may be removed partly by dry particulate deposition or washout.[4] In addition, reevaporation or reentrainment of previously deposited material can be expected to oppose complete removal of these species from the atmosphere.

Because of the complex nature of airborne particulate matter and the various factors that can affect vapor/particle equilibrium, currently available sampling techniques may not adequately distinguish the physical forms of semivolatile components. For example, current sampling methods are usually based on the use of a filter followed by a vapor adsorber; large air volumes often are passed through the collection media at high face velocities in order to preconcentrate sufficient amounts of the contaminants for analytical determination.

Previous research[5-8] suggests that vapors could adsorb or desorb from collected particulate matter as an artifact of sampling, especially if the physical state of a compound changes during the collection period. Volatilization of several PAH compounds during sampling has been demonstrated in various studies[9,10] and by recent denuder difference measurements.[11] This implies that a more accurate means of determining gas and particle distributions of certain classes of semivolatile compounds is needed in an effort to address deposition mechanisms and environmental concerns regarding airborne toxic organic chemicals.

A prototype sampling system[12] has been developed for monitoring the occurrence and phase distribution in ambient air of selected compounds that are known to be contaminants of rainfall and surface waters in the Great Lakes. This gas and particle (GAP) sampler, using a denuder difference technique, has been evaluated in laboratory and field experiments for the target compounds HCB, α-HCH, and γ-HCH. Practically complete removal (i.e., greater than 98%) of vapors of these compounds by the denuder coating was demonstrated in laboratory tests at normal and cold temperature conditions.[12] Results of initial field measurements by this technique at a suburban location indicated that a substantially greater particulate fraction of HCB, for example, was present than found by conventional filter/adsorber measurements. The particle-associated fractions measured by denuder difference were regarded as upper limits only, since complete vapor removal efficiency by the denuder under field conditions was assumed. Subsequent field results (after improvements were made to the methodology) indicated that particle-associated fractions of the target compounds were lower than found earlier. The lowest particle-associated fractions were observed in the initial measurements at a remote sampling location in central Ontario (Turkey Lakes), where suspended particulate concentrations were also very low. Results of these and more recent seasonal field studies are presented here to further demonstrate the performance of the sampling system and to provide an indication of phase distributions of these compounds at various locales in Ontario.

EXPERIMENTAL

A schematic of the prototype GAP sampler is shown in Figure 1. This sampler comprises a multiannular diffusion denuder assembly, with a binary crushed Tenax/silicone gum coating (to trap the vapor-phase components of the target compounds), that is integrated with a filter/adsorber. The denuder is thermally conditioned between sampling intervals in order to desorb collected vapors from the coating. Packed Florisil (or Tenax) adsorbers (two in series) are positioned behind the glass-fiber filters to collect any material that may escape the filter, such as vapor desorbed from the particulate matter collected during filtration. This unit is operated in parallel with a conventional filter/adsorber sampler (reference unit) to obtain the total airborne constituent con-

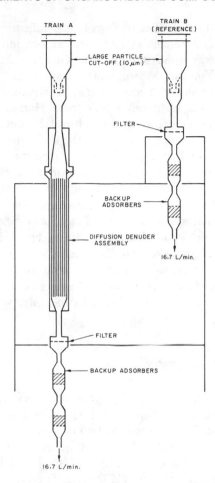

Figure 1.   Schematic diagram of the GAP sampler.

centration. Each unit was fitted with a 10-μm inlet to collect only inhalable particulate matter (determined gravimetrically on filters), and units were operated at flow rates of 16.7 L/min. The net difference in amounts of target compounds collected by filters and adsorbers of both units provides the constituent vapor fraction, and the combined filter/adsorber analyses of the denuder unit indicate the particulate-associated fraction. Equipment and experimental methods have been described in more detail elsewhere.[12]

Analytical methods to determine the target compounds involved extraction of filters and adsorbers with pesticide-grade hexane, solvent evaporation, sample cleanup through a Florisil column, further solvent evaporation, and quantification using dual capillary column gas chromatography with electron capture detection. In all instances, the lowest of the two column results was assigned, under the assumption that any positive differences in the values between columns represented analytical interferences.

In this study, field measurements (involving replicate sampling with two or three GAP samplers) were conducted at several locations in southern Ontario and at a remote (background) site near the Turkey Lakes watershed north of Sault Sainte Marie, Ontario. Measurement sets with duplicate units were done initially at a suburban site in Sheridan Park, Mississauga, approximately 3 km from Lake Ontario and in Stoney Creek (near Hamilton) at a site approximately 800 m downwind from an agricultural chemical formulation plant. Limited measurements were conducted at the shoreline of Lake Ontario near Niagara-on-the-Lake and aboard a Canada Centre for Inland Waters (CCIW) research vessel anchored approximately 5 km offshore. Similarly, onshore and offshore samples were taken at the eastern part of Lake Ontario near Sand Banks Provincial Park, and an integrated sample was collected on the ship while it traversed between the western and eastern sites. Most recently, sample sets (using triplicate GAP samplers) were also taken during each season at the remote Turkey Lakes area site near Lake Superior. Tenax adsorbers, used in earlier measurements, were subsequently replaced with Florisil adsorbers. Similarly, earlier measurements were conducted with a denuder/dichotomous sampler configuration and single filter/adsorber units were used in later measurements.

RESULTS AND DISCUSSION

Field studies took place in various periods from 1985 to 1987. Mean temperatures during sampling intervals ranged from –3.7 to 22.9°C, with hourly extremes from –11 to 34°C. Except on two occasions, under very hot summer conditions, temperature extremes were at or below temperature conditions selected in denuder performance laboratory tests in which practically complete vapor collection efficiency for HCB and lindane by the denuder was demonstrated.[12]

Results of individual field measurements to date, including ambient temperatures, airborne particulate concentrations, target compound concentrations, and the measured particulate-associated fractions by denuder difference at each site, are shown in Table 1. In most instances, values represent an average of two measurement results and appear in approximate chronological order. The general background nature of most sampling sites was evident from the low suspended particulate concentrations (usually less than 30 $\mu g/m^3$ for particles $< 10 \mu m$ in size). The lowest inhalable particulate loadings were encountered at the remote site, as might be anticipated. Only a small difference in the overall mean particle concentrations was found between the reference and GAP sampler units (about 4% difference based on approximately 40 comparative measurements). This confirmed earlier findings that no substantial particle losses occurred on average in the denuders.

Total airborne concentrations of HCB, as well as α- and γ-HCH, were detected at the sub-nanogram per cubic meter range, based on reference sam-

Table 1. Averaged Target Compound Concentrations and Particle-Associated Fractions Measured by Denuder Difference

Location	Season (Year)	Temp[a] (°C)	Part. Conc.[b] (μg/m³)	Total Conc. (pg/m³)[c] and Part. Fraction (%)[d] HCB	α-HCH	γ-HCH
Sheridan Park	Late Summer (1985)	22	22	—	550	—
(Mississauga)	Late Summer (1985)	21	32	720	480	40
	Late Summer (1985)	15	12	290	260	40
	Late Summer (1985)	16	55	300	450	30
	Early Autumn (1985)	14	14	100	270	20
	Early Autumn (1985)	8	26	280	270 (<1)	30
	Early Autumn (1985)	11	12	350	290 (14)	30
	Early Autumn (1985)	12	27	190	340 (~3)	30 (<1)
	Early Autumn (1985)	11	18	200	340 (<1)	20 (<1)
	Mid Autumn (1985)	3	16	230	410 (~7)	—
	Mid Autumn (1985)	5	17	110	510 (~6)	40 (<1)
	Early Autumn (1986)	18	8	210 (12)	360 (~8)	110 (~20)
	Early Autumn (1986)	8	36	150 (~1)	280 (~4)	30 (<1)
	Early Autumn (1986)	14	21	140 (~7)	270 (~1)	60
	Mid Winter (1987)	0	17	70	20	<20
	Mid Winter (1987)	−4	21	80 (<1)	120 (~7)	<20
	Mid Winter (1987)	−4	30	130 (~8)	110 (<1)	<20
	Mean		23	220	310	40
Stoney Creek	Late Autumn (1985)	3	50	210	130	360
(nr pesticide	Late Autumn (1985)	0	16	70 (31)	150 (<1)	490 (<1)
formulator)	Late Autumn (1986)	2	15	110	90	160
	Late Autumn (1986)	−1	12	250 (~30)	220 (<1)	110 (<1)
	Mean		23	160	150	280
Lake Ontario	Mid Summer (1986)	22	16	310	660	<30
W (nr Niagara)	Mid Summer (1986)	18	16	170 (5)	420 (12)	60 (20)
Central	Mid Summer (1986)	19	29	230 (<1)	490	60 (<1)
E (nr Sand Banks)	Mid Summer (1986)	23	31	260	270	<30
E (nr Sand Banks)	Mid Summer (1986)	20	30	280 (2)	500 (2)	70 (2)
	Mean		24	250	470	<50
Turkey Lakes	Early Autumn (1986)	15	9	160 (<1)	500 (<1)	40 (<1)
(nr Lake Superior)	Early Autumn (1986)	13	5	120 (<1)	270 (<1)	40 (<1)
	Early Autumn (1986)	8	5	120 (5)	300 (<1)	100 (<1)
	Mid Spring (1987)	12	12	110 (9)	300 (10)	100 (8)
	Mid Spring (1987)	9	10	80 (5)	270 (5)	80 (5)
	Mid Spring (1987)	10	13	70 (7)	310 (4)	120 (8)
	Mid Summer (1987)	19	6	80 (3)	480 (19)	50 (4)
	Mid Summer (1987)	18	6	90 (7)	450 (9)	40 (20)
	Mid Summer (1987)	18	14	30 (13)	450 (10)	40 (20)
	Mid Autumn (1987)	1	6	80 (<1)	200 (<1)	30 (<1)
	Mid Autumn (1987)	1	6	70 (<1)	210 (<1)	30 (<1)
	Mid Autumn (1987)	2	5	80 (<1)	180 (<1)	30 (<1)
	Mean		8	90 (4)	330 (6)	60 (6)

[a]Mean temperature during a 24- or 48-h sampling period.
[b]Concentration of airborne inhalable particulate matter (i.e., < 10μm particle size).
[c]Total constituent concentration derived from one or the average of two reference sampling units.
[d]Particulate-associated fractions of the total concentration are shown in parentheses and were derived by denuder-difference based on single measurements or an average of two or three replicate GAP sampling units.

Table 2.   Average Seasonal HCH Isomer Ratios

Season	Temperature Range (°C)	α-HCH:γ-HCH
Winter	−4 to 0	>6
Spring	9 to 12	3
Summer	15 to 23	11
Autumn	−1 to 18	9
Overall Average		~7

pler measurements performed at all of the sites. Although the mean HCB concentrations were generally similar for most of the southern Ontario sites, the average concentration measured at the Turkey Lakes area was somewhat lower. The higher concentrations might imply some influence of general urban air emission sources of HCB, such as coal combustion, incineration, chlorinated chemical manufacturing, and evaporation from contaminated water or soils. Lindane concentrations were usually an order of magnitude lower than α-HCH concentrations and were frequently at or near the detection limit of the analytical method used. Airborne lindane was greater when measured downwind of a pesticide formulation plant known to have prepared lindane formulations. The higher abundance of α-HCH relative to lindane may reflect environmental (perhaps photochemical) transformation of lindane to the more stable α-HCH isomer or different rates of atmospheric deposition between the isomers.[3,13-15] Average seasonal ratios of these components are shown in Table 2. The overall average ratio found in these studies compares favorably with averages determined from other measurements in northern climates.[15,16] Such ratios are reported to range from 1 to 4 in Europe, 7 to 10 in North America, and 10 to 100 under conditions of slow transport over long periods.[16] The lowest ratio in spring coincided with the expected time of pesticide application, but it must be noted that only limited spring measurements were available to deduce this ratio.

Concentrations of the target compounds as a function of temperature were also examined. For example, when all site data were combined, only a weak correlation of HCB concentrations with temperature occurred at different periods of time. The airborne concentrations and lack of a pronounced seasonal pattern of HCB were consistent with measurements at other remote or rural locales and implied a low-level, ubiquitous occurrence of this substance. This may also have reflected a rather constant emission rate to air during most of the year. However, ambient air HCB concentrations were found to be lowest during colder periods at the suburban location. A significant distinction between seasons was evident for the HCH compounds. For example, α-HCH increased substantially and consistently with increasing temperature. On the other hand, lindane concentrations at the Turkey Lakes site, where seasonal data are best compared, peaked during the spring period and decreased at both higher and lower temperatures. The increase of α-HCH relative to lindane in the summer may be related to such factors as

- increased volatilization of α-HCH from land or water surfaces at higher temperature
- greater resistance to atmospheric removal
- greater summer consumption of technical HCH (containing α-HCH) in distant areas with corresponding atmospheric transport
- increased lindane conversion to α-HCH in the summer

Representative particulate-associated fractions of these components, derived by denuder difference, are also shown in Table 1 for each site. From the results of these measurements, it appears that the target substances being examined occurred predominantly in the vapor phase, as has been found by other investigators. In some instances (especially earlier measurements), the particulate-associated fractions were found to be large, with some variability between days and between replicated measurements; on several occasions, significant amounts of the target compounds were detected in cartridges located after the denuder, presumably the result of partial volatilization from filtered particles during sampling. These compounds were rarely found by direct analysis of filters in the sampling system, except under very cold conditions. In retrospect, some of the disagreement between units might have been attributable to such factors as

- breakthrough of vapor-phase components from some denuders
- insufficient cleaning of impurities in some cartridges
- analytical imprecision at very low concentrations (especially in determining amounts in adsorbent media after the denuder, which often approached analytical detection limits)
- the convention of subtracting average blank cartridge results rather than individual blank values
- very fine particle penetration through filters of the collection system

For these reasons, the lowest value determined from replicated measurements by GAP samplers for a given day at each site is shown in Table 1, with the assumption that higher values represented measurement interferences. Both sampling and analytical improvements were implemented in later studies in an effort to improve accuracy and precision of these measurements. Nevertheless, these data suggested that the denuder, for the most part, efficiently removed the vapor fraction during sampling, as was demonstrated in earlier laboratory studies.

On several occasions, triplicate measurements of the target compound fractions associated with particulate matter were similar, as shown in Table 3. The selected results shown are based on single concurrent sets of measurements (48-hr sampling periods) at the remote Turkey Lakes area during each season. For example, generally favorable agreement between replicates was found especially during the autumn periods at this site. Surprisingly, these fractions were found to be highest for the spring and summer measurements and lowest for the autumn (cold temperature) measurements. However, the higher results in summer could have been related to partial breakthrough of vapors from

Table 3. Examples of Selected Replicated Measurements with the GAP Sampler at Turkey Lakes

Season	Total Airborne Conc. (pg/m³)			Particulate-Associated Fraction (%)		
	HCB	α-HCH	γ-HCH	HCB	α-HCH	γ-HCH
Autumn 1986	120	320	110	2	<1	<1
	120	270	90	7	<1	<1
				7	<1	10
Spring 1987	80	250	80	<1	2	10
	80	290	80	4	<1	<1
				8	11	9
Summer 1987	90	410	40	41	44	36
	90	490	40	5	10	12
				8	7	31
Autumn 1987	80	210	30	<1	<1	<1
	80	190	30	<1	<1	<1
				<1	<1	<1

Note: Each set of replicated measurements involved two reference (filter/adsorber) samplers, which were operated concurrently with three GAP (denuder/filter/adsorber) samplers. These data indicate replicated values for only one of three sets of samples taken during each season, which were typical of other sets.

denuders at higher summer temperatures (e.g., aging of denuders due to repeated use). On the other hand, the higher fractions in summer (especially α- and γ-HCH) could also be a function of component solubility in aqueous aerosols that penetrated the denuder under high summer water vapor conditions. If present in airborne water droplets, these compounds could be expected to readily strip from filters during sampling. In an examination of individual data, only a weak correlation between the amount of α-HCH in the particulate-associated fractions and ambient vapor pressure (i.e., higher atmospheric water vapor concentration in summer) was found, especially during foggy or hazy conditions. Furthermore, the differences between seasonal data could have reflected different types of airborne aerosols and different adsorptive properties. Nevertheless, it is noteworthy that these data agree in general with the gas/particle distributions for these compounds that have been reported by other investigators under moderate temperature conditions. Additional measurements over an extended period at a given site would be required to substantiate the apparent seasonal trends found in this study.

SUMMARY AND CONCLUSIONS

Field measurements utilizing a new sampling method for determining the vapor/particle distributions of semivolatile chlorinated organic compounds in ambient air—the GAP Sampler—were conducted at several sites in Ontario that were remote from known sources. The following are some of the major conclusions from this work:

- Particle losses in the denuder were small (about 4% loss on average for ambient particles < 10 μm in size) over an inhalable particulate concentration range of 5 to 55 $\mu g/m^3$.

- The three target compounds were essentially nondetectable in filtered particulate matter, as determined by direct filter analysis. This might be expected because of the low particulate loadings encountered and the potential for volatilization of these compounds from collected particles during sampling. The estimated measurement detection limits for HCB, α-HCH, and γ-HCH (using a 46-hr sample collection period) were 13, 3, and 12 pg/m^3, respectively.

- The ranges of total airborne concentrations of the compounds measured by the conventional (reference) sampling unit at these Ontario sites were 20–720 pg/m^3 HCB, 20–660 pg/m^3 α-HCH, and < 20–110 pg/m^3 γ-HCH. On a seasonal basis (when data from each measurement site were combined), HCB concentrations remained relatively constant, α-HCH varied directly with temperature (highest concentrations in summer), and lindane concentrations were highest in spring. Lindane was found to decrease in concentration at higher temperatures, possibly due to photomodification under warm, sunny conditions or other factors that might have affected lindane concentrations. The airborne concentrations, seasonal patterns, and α- to γ-HCH ratios were consistent with data reported by other investigators, implying that these substances are ubiquitous in occurrence.

- These organochlorine compounds were found to be primarily in the vapor phase at each of the sampling sites. At the remote Turkey Lakes location, the ranges of average particulate-associated fractions of the target compounds found by denuder difference (expressed as a percent of the total airborne concentration) were < 1–13% HCB, < 1–10% α-HCH, and < 1–20% γ-HCH. In several cases, there was favorable agreement between replicate measurements with the GAP sampler. Results were variable on some occasions (due, in part, to small amounts collected after denuders that were at or near detection limits). The largest fractions were unexpectedly found in the summer at this site. This might have been attributable to greater solubility of these species in aqueous airborne aerosols that penetrated the denuder or resulted from incomplete vapor collection by the denuder (breakthrough) under hot and humid conditions. Accordingly, the particulate-associated fractions measured during summer should be regarded as upper limits. The very low fractions found consistently in the cold autumn period at this site indicated that the denuders performed efficiently and that these constituents occurred predominantly in the vapor phase.

REFERENCES

1. Bidleman, T. F., and R. Leonard. *Atmos. Environ.* 16(5):1099–1107 (1982).
2. Oehme, M., and B. Ottar. *Geochem. Res. Lett.* 11(11):1133–36 (1984).
3. Atlas, E., and C. S. Giam. *Water Air Soil Poll.* 38(1,2):19–36 (1988).
4. Bidleman, T. F., W. N. Billings, and W. T. Foreman. *Environ. Sci. Technol.* 20(10):1038–43 (1986).
5. König, J., W. Funcke, E. Balfanz, G. Grosch, and F. Potts. *Atmos. Environ.* 14(5):609–13 (1980).
6. Broddin, G., W. Cautreels, and K. Van Cauwenberghe. *Atmos. Environ.* 14(8):895–901 (1980).
7. Grosjean, D. *Atmos. Environ.* 17(12):2565–73 (1983).
8. Spitzer, T., and W. Dannecker. *Anal. Chem.* 55(11):2226–28 (1983).
9. Van Vaeck, L., K. Van Cauwenberghe, and J. Janssens. *Atmos. Environ.* 18(2):417–30 (1984).
10. Peters, J., and B. Siefert. *Atmos. Environ.* 14(1):117–20 (1980).
11. Coutant, R. W., L. Brown, J. C. Chuang, R. M. Riggin, and R. G. Lewis. *Atmos. Environ.* 22(2):403–9 (1988).
12. Lane, D. A., N. D. Johnson, S. C. Barton, G. H. S. Thomas, and W. H. Schroeder. *Environ. Sci. Technol.* 22(8):941–47 (1988).
13. Malaiyandi, M., and S. M. Shah. *J. Environ. Sci. Health* A19(8):887–910 (1984).
14. Rapaport, R. A., and S. J. Eisenreich. *Environ. Sci. Technol.* 22(8):931–41 (1988).
15. Bidleman, T. F., U. Widequist, B. Jansson, and R. Soderlund. *Atmos. Environ.* 21(3):641–54 (1987).
16. Pacyna, J. M., and M. Oehme. *Atmos. Environ.* 22(2):243–57 (1988).

## CHAPTER 9

Atmospheric Reaction Pathways and Lifetimes for Organophosphorus Compounds

Arthur M. Winer and Roger Atkinson

INTRODUCTION

Although high atmospheric loadings of pesticides and other agricultural chemicals occur over localized areas from spraying and handling procedures, it is now well recognized that volatilization from soil and water surfaces is also an important mode of entry into the atmosphere.[1] Once present in the gas phase, organic chemicals are transformed into product species by photolysis and chemical reaction (mainly with OH and NO_3 radicals and O_3) and are removed by wet and dry deposition.[2] In order to determine the dominant loss process(es) and atmospheric lifetimes of the organophosphorus compounds emitted into the atmosphere, and hence the potential for human exposure to these chemicals and their transformation products, it is necessary to know the reaction rates for the important atmospheric removal processes. These data then allow evaluation of the potential for these compounds to undergo long-range transport and to impact remote global locations.

Despite the wide usage of organophosphorus compounds as insecticides and herbicides in agricultural operations[3] and the large database presently available for the kinetics and mechanisms of the gas-phase reactions of organic compounds with the OH radical,[4,5] O_3,[6] and the NO_3 radical,[7] until recently the only organophosphorus compound for which such data were available was trimethyl phosphate.[8] In order to begin to obtain a kinetic and mechanistic database concerning the atmospherically important gas-phase reactions of organophosphorus compounds, we have studied a series of simple, relatively volatile compounds containing the structural units

$$
\begin{array}{cc}
\underset{-X_3}{-X_4}\!\!\diagup\!\!\overset{X_1}{\underset{X_2-}{\diagdown}} P & \underset{-X_2}{-X_3}\!\!\diagup\!\!\overset{X_1}{\underset{NR_1R_2}{\diagdown}} P \\
A & B
\end{array}
$$

Table 1.   Rate Constants k for the Gas-Phase Reactions of NO_3 and OH Radicals and O_3 with a Series of Organophosphorus Compounds at 297 ± 3 K.

Organophosphorus Compound	k (cm^3/molecule/sec)		
	$10^{12} \times k_{OH}$[a]	$10^{14} \times k_{NO_3}$[b]	$10^{19} \times k_{O_3}$
$(CH_3O)_3PO$	7.37 ± 0.74[c]		<0.6[C]
$(CH_3O)_2P(O)SCH_3$	9.29 ± 0.68	<0.2	<2
$(CH_3S)_2P(O)OCH_3$	9.59 ± 0.75	≤0.3	<1
$(CH_3O)_3PS$	69.7 ± 3.9	<2	<3
$(CH_3O)_2P(S)SCH_3$	56.0 ± 1.8	<3	<2
$(C_2H_5O)_3PO$	55.3 ± 3.5		
$(CH_3O)_2P(O)N(CH_3)_2$	31.9 ± 2.4	<4.0	<2
$(CH_3O)_2P(S)N(CH_3)_2$	46.8 ± 1.4	3.2 ± 1.0	<2
$(CH_3O)_2P(S)NHCH_3$	232 ± 13	31 ± 4	<2
$(CH_3O)_2P(S)NH_2$	244 ± 9	40 ± 8	<2

[a]Indicated errors are two least-squares standard deviations. Placed on an absolute basis by use of rate constants k_2 for the reactions of the OH radical with cyclohexane, propene, 2-methyl-1,3-butadiene, and 2,3-dimethyl-2-butene as recommended by Atkinson.[5]
[b]Indicated errors are two least-squares standard deviations. Placed on an absolute basis as discussed in the text.
[c]From Tuazon et al.[8]

where X_1 through X_4 are O and/or S, and R_1 and R_2 are H or CH_3 (see Table 1 for structures). Examples of organophosphorus compounds containing these structural units that are used in agricultural operations[3] include demeton-O, demeton-S, ethoprophos, malathion, parathion, and phorate (structural unit A); and acephate and methamidophos (structural unit B) (Table 2). In addition, the trimethyl phosphorothioates studied occur as impurities in such pesticides as malathion and acephate.[9]

The kinetic and mechanistic data obtained from this study have allowed structure-reactivity relationships for the estimation of OH radical reaction rate constants, and hence of atmospheric lifetimes with respect to gas-phase reaction with the OH radical, to be extended to the classes of organophosphorus compounds containing the structural units A and B.[10]

EXPERIMENTAL METHODS

The experimental methods employed in this study have been described in detail previously[11-13] and are only briefly discussed here. Experiments were conducted in a 6400-L all-Teflon environmental chamber equipped with black-lamp irradiation. The OH radical reaction kinetics were determined using a relative rate technique in which the decay rates of the organophosphorus compounds and a reference organic (which does not photolyze and whose OH radical reaction rate constant is reliably known) were monitored in the presence of OH radicals. Hydroxyl radicals were generated at concentrations of $(2-10) \times 10^7$ molecules/cm^3 (a factor of 10-100 higher than present in the ambient atmosphere) from the photolysis of methyl nitrite (CH_3ONO) in air at wavelengths >300 nm.

Table 2. Examples of Agricultural Chemicals Containing Structural Units Investigated in This Study.

Name	Structure	$10^{12} \times k_{OH}$ (cm^3/molecule sec) [calculated]
Demeton-O	CH_3CH_2O, CH_3CH_2O bonded to P with =S and $OCH_2CH_2SCH_2CH_3$	128
Ethoprophos	CH_3CH_2OP with =O and two $SCH_2CH_2CH_3$	63
Acephate	CH_3O, CH_3S bonded to P with =O and $NHC(O)CH_3$	~ 51
Methamidophos	CH_3O, CH_3S bonded to P with =O and NH_2	26
Malathion	CH_3O, CH_3O bonded to P with =S and $SCHCOC_2H_5$ (with =O) and $C(O)OC_2H_5$	64
Parathion	CH_3CH_2O, CH_3CH_2O bonded to P with =S and O—C$_6$H$_4$—NO_2	92
Phorate	CH_3CH_2O, CH_3CH_2O bonded to P with =S and $PSCH_2SCH_2CH_3$	≥ 100

$$CH_3ONO + h\nu \rightarrow CH_3O + NO$$
$$CH_3O + O_2 \rightarrow HCHO + HO_2$$
$$HO_2 + NO \rightarrow OH + NO_2$$

NO was also included in the reactant mixtures to avoid the formation of O_3 and, hence, of NO_3 radicals. Additionally, irradiations were carried out in the absence of CH_3ONO to assess the contributions, if any, of photolysis as a removal process in these experiments for the organophosphorus compounds.

Propene, cyclohexane, isoprene (2-methyl-1,3-butadiene), or 2,3-dimethyl-2-butene were used as the reference organic for these irradiations. The approximate initial reactant concentrations were (in molecules/cm³) CH_3ONO (when present), 2.4×10^{14}; NO, 2.4×10^{14}; and reference organic and organophosphorus compound, $(5-25) \times 10^{12}$.

The experimental technique used to determine the NO_3 radical reaction rate constants was also a relative rate method,[14,15] in which the relative decay rates of the organophosphorus compounds and a reference organic (propene or *trans*-2-butene) were monitored in the presence of NO_3 radicals. NO_3 radicals were generated by the thermal decomposition of N_2O_5,

$$N_2O_5 \rightleftarrows NO_3 + NO_2$$

with up to six additions of N_2O_5 being made to the chamber during the reaction. Provided that both the organophosphorus compound (OPC) and the reference organic were consumed only by reaction with OH or NO_3 radicals, then

$$\ln\left\{\frac{[OPC]_{t_0}}{[OPC]_t}\right\} - D_t = \frac{k_1}{k_2}\left[\ln\left\{\frac{[\text{reference organic}]_{t_0}}{[\text{reference organic}]_t}\right\} - D_t\right] \tag{1}$$

where

$[OPC]_{t_0}$ = the concentration of the organophosphorus compound at time t_0

[reference organic]$_{t_0}$ = the concentration of the reference organic at time t_0

$[OPC]_t$ and [reference organic]$_t$ = the corresponding concentrations at time t

D_t = any dilution at time t caused by addition of reactants to the chamber (D_t was 0.0028 per addition of N_2O_5 to the chamber and zero for the OH radical reactions)

k_1 = the rate constant for the reaction
$$\left.\begin{array}{c}OH\\NO_3\end{array}\right\} + OPC \rightarrow \text{products}$$

k_2 = the rate constant for the reaction
$$\left.\begin{array}{c}OH\\NO_3\end{array}\right\} + \text{reference organic} \rightarrow \text{products}$$

Hence, plots of $\{\ln([OPC]_{t_0}/[OPC]_t) - D_t\}$ against $\{\ln([\text{reference organic}]_{t_0}/[\text{reference organic}]_t) - D_t\}$ should yield straight lines of slope k_1/k_2 and zero intercept.

Rate constants, or upper limits thereof, for the reactions of O_3 with the OPCs were obtained by monitoring the decay rates of the OPCs in the presence of known excess concentrations of O_3 over reaction times of approximately 3 hr, with ethane or cyclohexane being added to several of the reactant mixtures — at concentrations of $(2.4-24) \times 10^{14}$ molecules/cm^3 — to scavenge any OH radicals formed in these reactions.

All experiments were carried out at one atmosphere (740 torr) total pressure of dry air and at room temperature (297 ± 3 K). Ozone was monitored by a Monitor Labs chemiluminescence ozone analyzer, and the OPCs and the reference organics were monitored by gas chromatography with flame ionization detection (GC/FID), as described in detail by Goodman et al.[11,12] and Atkinson et al.[13]

The OPCs investigated in this study were either obtained commercially or synthesized, as described by Goodman et al.[11,12]

RESULTS

O_3 Reactions

The gas-phase concentrations of the organophosphorus compounds were monitored in the presence of an excess concentration, $(5-8) \times 10^{13}$ molecules/cm^3, of O_3 over periods of approximately 3 hr. In several of these experiments, ethane or cyclohexane was added to the reactant mixtures at concentrations sufficient to scavenge any OH radicals formed, since this method of investigating O_3 reactions is prone to the occurrence of secondary reactions involving radical chains,[6] with the radicals being formed from either the initial reaction of the organic compound with O_3 or from reactive impurities. As discussed in detail by Goodman et al.,[11,12] the experimental data showed no evidence for gas-phase reactions of O_3 with the organophosphorus compounds studied. The upper limits to the rate constants given in Table 1 were obtained from the O_3 concentrations and the observed slow decays of the organophosphorus compounds (or a maximum 10% loss over the duration of the experiment, set by the uncertainties in the concentration measurements of the organophosphorus compounds, whichever was greater). In all cases, the rate constants for reaction with O_3 were $<3 \times 10^{-19}$ cm^3/molecule sec at room temperature.

NO_3 Radical Reactions

The kinetics of the reactions of the NO_3 radical were studied with the trimethyl phosphorothioates, the dimethyl phosphoroamidate, and the dimethyl phosphorothioamidates shown in Table 1, with propene and/or *trans*-2 butene being used as the reference organic. No measurable consumptions of $(CH_3O)_3PO$, $(CH_3O)_2P(O)PSCH_3$, $(CH_3O)_2P(S)SCH_3$, or $(CH_3O)_2P(O)N(CH_3)_2$ were observed within the measurement uncertainties of 5–10%. Based upon a maximum loss of these compounds of 10% (due to the analytical uncertainties) and the observed amounts of reaction of the reference organics, upper limits to the rate constant ratios k_1/k_2 were obtained.

For $(CH_3S)_2P(O)OCH_3$, a 16–18% disappearance from the gas phase was observed over reaction times of approximately 3 hr, irrespective of a variation in the integrated NO_3 radical concentration of a factor of 30, depending on whether propene or *trans*-2-butene was used as the reference organic. These data suggest that this loss of $(CH_3S)_2P(O)OCH_3$ from the gas phase was due to adsorption at the chamber walls and not to gas-phase reaction with the NO_3 radical. Accordingly, an upper limit to the rate constant ratio was derived from the observed loss rates of this trimethyl phosphorothioate and the reference organics.

For the three dimethyl phosphorothioamidates studied, reaction was observed and the experimental data are plotted in Figure 1 in accordance with Equation 1. Rate constant ratios k_1/k_2 were obtained from the slopes of these plots by least-squares analyses. Rate constants k_1 were obtained from the rate constant ratios k_1/k_2 by use of rate constants for the reference organics of $k_2(\text{propene}) = (9.4 \pm 1.2) \times 10^{-15}$ cm³/molecule sec and $k_2(\textit{trans}\text{-2-butene}) = (3.87 \pm 0.45) \times 10^{-13}$ cm³/molecule sec.[7,16,17] The resulting rate constants are given in Table 1.

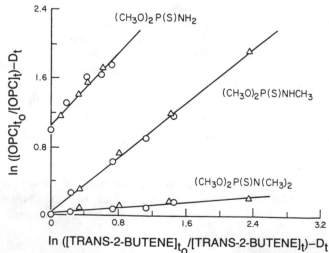

Figure 1. Plots of Equation 1 for the reactions of the NO_3 radical with $(CH_3O)_2P(S)N(CH_3)_2$, $(CH_3O)_2P(S)NHCH_3$ and $(CH_3O)_2P(S)NH_2$, with *trans*-2-butene as the reference organic. The data for $(CH_3O)_2P(S)NH_2$ have been displaced vertically by 1.0 unit for clarity. Initial concentrations of NO_2: (○) 1.2×10^{14} molecules/cm³; (△) 2.4×10^{14} molecules/cm³.

OH Radical Reactions

A series of $CH_3ONO-NO$–organophosphorus compound–reference organic
–air and NO-organophosphorus compound-reference organic–air irradiations
were carried out, and the experimental data for four of these compounds are
plotted in accordance with Equation 1 in Figures 2 and 3. The NO-air irradiations of $(CH_3O)_2P(O)SCH_3$, $(CH_3S)_2P(O)OCH_3$, $(CH_3O)_2P(O)N(CH_3)_2$, and
$(CH_3O)_2P(S)N(CH_3)_2$ showed no evidence for photolysis under the experimental conditions employed. Furthermore, although a disappearance of
both the organophosphorus compound and the reference organic was
observed in the NO-air irradiations of $(CH_3O)_3PS$, $(CH_3O)_2P(S)SCH_3$,
$(CH_3O)_2P(S)NHCH_3$, and $(CH_3O)_2P(S)NH_2$, the data were consistent with this
disappearance being due to the presence of OH radicals, at concentrations an
order of magnitude lower than in the corresponding CH_3ONO-NO-air irradiations. Thus, the slopes of the plots of Equation 1 yield the rate constant ratios
k_1/k_2, and the rate constants k_1 obtained by use of rate constants k_2, taken
from the review of Atkinson,[5] are given in Table 1.

DISCUSSION

The rate constants, or upper limits to the rate constants, given in Table 1 for
the gas-phase reactions of OH and NO_3 radicals and O_3 with the organophosphorus compounds studied, can be combined with measured or estimated
ambient atmospheric concentrations of OH and NO_3 radicals and O_3 to calculate the atmospheric lifetimes of these organophosphorus compounds with

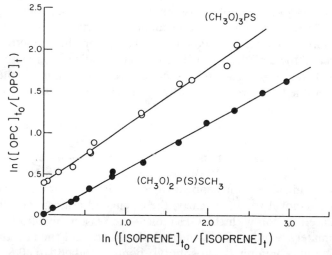

Figure 2. Plots of Equation 1 for the data obtained from NO-air and CH_3ONO-NO-air irradiations of $(CH_3O)_3PS$ and $(CH_3O)_2P(S)SCH_3$, with isoprene as the reference organic.
The data for $(CH_3O)_3PS$ have been displaced vertically by 0.1 unit for clarity.

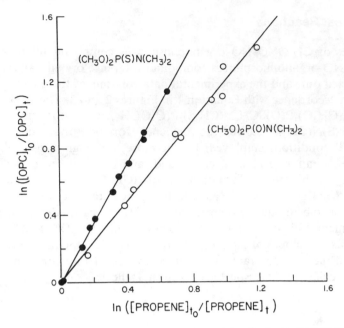

Figure 3. Plots of Equation 1 for the reactions of the OH radical with $(CH_3O)_2P(O)N(CH_3)_2$ and $(CH_3O)_2P(S)N(CH_3)_2$, with propene as the reference organic.

respect to these potential gas-phase reactions. Using ambient atmospheric concentrations of 1.5×10^6 molecules/cm³ of OH radicals during a 12-hr daytime period,[18] 2.4×10^8 molecules/cm³ of NO_3 radicals during a 12-hr nighttime period[19,20] and 7×10^{11} molecules/cm³ of O_3 during a complete 24-hr day,[21] the calculated lifetimes given in Table 3 are obtained. These data show that the O_3 reactions are of negligible importance as an atmospheric loss process for the organophosphorus compounds studied. Similarly, for the trimethyl phosphorothioates and the dimethyl phosphoroamidate studied, no reactions with the NO_3 radical were observed, and the calculated lower limits to the atmospheric lifetimes due to this reaction pathway indicate that these reactions can be neglected as an atmospheric loss process.

However, all of the organophosphorus compounds studied react with the OH radical and, except possibly for $(CH_3O)_2P(S)NHCH_3$ and $(CH_3O)_2P(S)NH_2$, this reaction is the dominant gas-phase atmospheric removal pathway. For $(CH_3O)_2P(S)NHCH_3$ and $(CH_3O)_2P(S)NH_2$, the OH radical and NO_3 radical reactions appear to be of relatively similar importance as atmospheric removal routes. Clearly, the calculated atmospheric lifetimes due to gas-phase reaction with the OH radical are reasonably short, ranging from approximately two days for trimethyl phosphate and the two trimethyl phosphorothioates containing phosphoryl bonds studied to 0.8 hr for $(CH_3O)_2P(S)NHCH_3$ and $(CH_3O)_2P(S)NH_2$.

These data for the OH radical rate constants have been incorporated[10] into

Table 3.   Calculated Atmospheric Lifetimes of a Series of Organophosphorus Compounds Due to Gas-Phase Reaction with OH and NO$_3$ Radicals and with O$_3$.

Organophosphorus Compound	Lifetimes Due to Reaction with		
	OH[a]	NO$_3$[b]	O$_3$[c]
(CH$_3$O)$_3$PO[d]	2.1 days		>275 days
(CH$_3$O)$_2$P(O)SCH$_3$	1.7 days	>48 days	> 83 days
(CH$_3$S)$_2$P(O)OCH$_3$	1.6 days	≥32 days	>165 days
(CH$_3$O)$_2$PS	2.7 hr	>4.8 days	> 55 days
(CH$_3$O)$_2$P(S)SCH$_3$	3.3 hr	>3.2 days	> 83 days
(C$_2$H$_5$O)$_3$PO	3.3 hr		
(CH$_3$O)$_2$P(O)N(CH$_3$)$_2$	5.8 hr	>2.4 days	> 83 days
(CH$_3$P)$_2$(S)N(CH$_3$)$_2$	4.0 hr	3.0 days	> 83 days
(CH$_3$P)$_2$(S)NHCH$_3$	0.8 hr	3.7 hr	> 83 days
(CH$_3$P)$_2$(S)NH$_2$	0.8 hr	2.9 hr	> 42 days

[a]For a 12-hour daylight average OH radical concentration of 1.5 × 10^6 molecules/cm^3 (Prinn et al[18]).
[b]For a 12-hour nighttime average NO$_3$ radical concentration of 2.4 × 10^8 molecules/cm^3 (Platt et al.[19]; Atkinson et al.[20]).
[c]For a 24-hour average O$_3$ concentration of 7 × 10^{11} molecules/cm^{-3} (Logan[21]).
[d]From Tuazon et al.[8]

an estimation method[4,22] for the calculation of OH radical reaction rate constants for organic compounds. In this structure-reactivity relationship, the overall OH radical reaction rate constant is obtained by summing up the individual contributions from the four processes:

$$k_{total} \quad = \quad k(\text{H-atom abstraction from C-H and O-H bonds})$$
$$+ \quad k(\text{OH radical addition to} > C = C < \text{ double bonds})$$
$$+ \quad k(\text{OH radical addition to aromatic rings})$$
$$+ \quad k(\text{OH radical reaction with N-, S-, and P-containing groups})$$

As discussed in detail by Atkinson,[4,10,22] the calculation of the overall H-atom abstraction rate constant is based on the estimation of -CH$_3$, -CH$_2$-, >CH-, and O-H group rate constants. The -CH$_3$, -CH$_2$-, and >CH-group rate constants depend on the identity of the substituents around these groups, with

$$k(CH_3\text{-}X) \quad = \quad k_{prim} \, F(X)$$

$$k(X\text{-}CH_2\text{-}Y) \quad = \quad k_{sec} \, F(X)F(Y)$$

$$k(X\text{–}CH \, {Y \atop Z}) \quad = \quad k_{tert} \, F(X)F(Y)F(Z)$$

where k_{prim},
k_{sec} and
k_{tert} = the rate constants per -CH$_3$, -CH$_2$-, and >CH- group for a standard substituent (chosen to be a -CH$_3$ group)

X, Y and Z = substituent groups
F(X), F(Y)
   and F(Z) = the corresponding substituent factors

At 298°K the values of k_{prim}, k_{sec}, and k_{tert} are (in units of 10^{-12} cm^3/molecule sec) 0.144, 0.838 and 1.83, respectively, and F(-CH$_2$-) = F(>CH-) = F(>C<) = 1.29.[4,10,22]

The experimental data obtained for the organophosphorus compounds allow the following relevant group rate constants and substituent factors to be derived:[10]

$$F(-OP-) = F(-SP-) = 20$$
$$k_{P=O} = 0$$
$$k_{P=S} = 5.5 \times 10^{-11} \text{ cm}^3/\text{molecule sec}$$

all at room temperature. Furthermore, the kinetic data for the dimethyl phosphoroamidate and dimethyl phosphorothioamidates studied are reasonably consistent with the factors obtained from the aliphatic amines and related compounds of

$$k_{-NH2} = 2.0 \times 10^{-11} \text{ cm}^3/\text{molecule sec}$$
$$k_{>NH} = k_{>N-} = 6.0 \times 10^{-11} \text{ cm}^3/\text{molecule sec}$$
$$F(-NH_2) = F(>NH) = F(>N-) = 10$$

Use of these group rate constants and substituent factors, together with the other parameters derived by Atkinson[22] for H-atom abstraction from C-H and O-H bonds and OH radical addition to aromatic rings, then allows the room temperature rate constants for the gas-phase reactions of the OH radical with the more complex organophosphorus compounds encountered in agricultural operations to be calculated; selected examples are given in Table 2. The organophosphorus compounds used in commercial operations all appear to be highly reactive, with atmospheric lifetimes during daytime of a few hours or less. Thus, this class of organic chemicals will not undergo long-range transport, and their effects will be limited to local or, at most, regional scales.

A recent product study[23] has shown that (CH$_3$O)$_3$PS and (CH$_3$O)$_2$P(S)SCH$_3$ react with the OH radical to form (CH$_3$O)$_3$PO and (CH$_3$O)$_2$P(O)SCH$_3$, with yields of 0.28 and 0.13, respectively, at room temperature and atmospheric pressure. These experimental observations are consistent with the expected reaction of the OH radical with the thiophosphoryl bond and with laboratory and ambient observations of a rapid transformation of organophosphorus compounds containing thiophosphoryl bonds to, at least in part, the corresponding compounds containing a phosphoryl bond — for example, the conversion of parathion to paraoxon (see Woodrow et al.[24] and references therein).

Although the present investigation has provided much new and interesting data concerning the atmospheric reactivity and potential reaction products of simple organophosphorus compounds, further research is needed to identify the reaction products formed under atmospheric conditions and to experimentally extend this work to the more complex compounds used in commercial applications.

ACKNOWLEDGMENTS

We gratefully acknowledge Janet Arey, Sara M. Aschmann, Mark A. Goodman, Patricia McElroy, and Ernesto C. Tuazon for carrying out various portions of this experimental program. We thank Professor Roy Fukuto and Dr. S. Keadtisuke of the UCR Department of Entomology for valuable assistance in the syntheses of several of the organophosphorus compounds employed in this research. The financial support of the University of California, Riverside, Toxic Substances Research and Training Program is gratefully acknowledged.

REFERENCES

1. Jury, W. A., A. M. Winer, W. F. Spencer, and D. D. Focht. "Transport and Transformations of Organic Chemicals in the Soil-Air-Water Ecosystems," *Rev. Env. Contam. and Toxicol.* 99:120–64 (1987).
2. Bidleman, T. F. "Atmospheric Processes," *Environ. Sci. Technol.* 22:361–67 (1988).
3. Worthing, C. R. *The Pesticide Manual,* 7th ed. (Croydon, UK: British Crop Protection Council, 1983).
4. Atkinson, R. "Kinetics and Mechanisms of the Gas-Phase Reactions of the Hydroxyl Radical with Organic Compounds under Atmospheric Conditions," *Chem. Rev.* 86:69–201 (1986).
5. Atkinson, R. "Kinetics and Mechanisms of the Gas-Phase Reactions of the Hydroxyl Radical with Organic Compounds," *J. Phys. Chem. Ref. Data*, Monograph 1 (1989), pp. 1–246.
6. Atkinson, R., and W. P. L. Carter. "Kinetics and Mechanisms of the Gas-Phase Reactions of Ozone with Organic Compounds under Atmospheric Conditions," *Chem. Rev.* 84:437–70 (1984).
7. Atkinson, R., S. M. Aschmann, and J. N. Pitts, Jr. "Rate Constants for the Gas-Phase Reactions of the NO_3 Radical with a Series of Organic Compounds at 296 ± 2 K," *J. Phys. Chem.* 92:3454–57 (1988).
8. Tuazon, E. C., R. Atkinson, S. M. Aschmann, J. Arey, A. M. Winer, and J. N. Pitts, Jr. "Atmospheric Loss Processes for 1,2-Dibromo-3-chloropropane and Trimethyl Phosphate," *Environ. Sci. Technol.* 20:1043–46 (1986).
9. Fukuto, T. R. "Toxicological Properties of Trialkyl Phosphorothioate and Dialkyl Alkyl-and Arylphosphorothioate Esters," *J. Environ. Sci. Health* B18:89–117 (1983).
10. Atkinson, R. "Estimation of Gas-Phase Hydroxyl Radical Rate Constants for Organic Chemicals," *Environ. Toxicol. Chem.* 7:435–42 (1988).
11. Goodman, M. A., S. M. Aschmann, R. Atkinson, and A. M. Winer. "Kinetics of the Atmospherically Important Gas-Phase Reactions of a Series of Trimethyl Phosphorothioates," *Arch. Environ. Contamin. Toxicol.* 17:281–88 (1988).
12. Goodman, M. A., S. M. Aschmann, R. Atkinson, and A M Winer. "Atmospheric

Reactions of a Series of Dimethyl Phosphoroamidates and Dimethyl Phosphorothioamidates," *Environ. Sci. Technol.* 22:578–83 (1988).

13. Atkinson, R., S. M. Aschmann, M. A. Goodman, and A. M. Winer. "Kinetics of the Gas-Phase Reactions of the OH Radical with $(C_2H_5O)_3PO$ and $(CH_3O)_2P(S)Cl$ at 296 ± 2 K," *Int. J. Chem. Kinet.* 20:273–81 (1988).

14. Atkinson, R., C. N. Plum, W. P. L. Carter, A. M. Winer, and J. N. Pitts, Jr. "Rate Constants for the Gas-Phase Reactions of Nitrate Radicals with a Series of Organics in Air at 298 ± 1 K," *J. Phys. Chem.* 88:1210–15 (1984).

15. Atkinson, R., J. N. Pitts, Jr., and S. M. Aschmann. "Tropospheric Reactions of Dimethyl Sulfide with NO_3 and OH Radicals," *J. Phys. Chem.* 88:1584–87 (1984).

16. Ravishankara, A. R., and R. L. Mauldin, III. "Absolute Rate Coefficient for the Reaction of NO_3 with *trans*-2-Butene," *J. Phys. Chem.* 89:3144–47 (1985).

17. Dlugokencky, E. J., and C. J. Howard. "Studies of NO_3 Radical Reactions with Some Atmospheric Organic Compounds at Low Pressures," *J. Phys. Chem.* 93: 1091–96 (1989).

18. Prinn, R., D. Cunnold, R. Rasmussen, P. Simmonds, F. Alyea, A. Crawford, P. Fraser, and R. Rosen. "Atmospheric Trends in Methylchloroform and the Global Average for the Hydroxyl Radical," *Science* 238:945–50 (1987).

19. Platt, U. F., A. M. Winer, H. W. Biermann, R. Atkinson, and J. N. Pitts, Jr. "Measurement of Nitrate Radical Concentrations in Continental Air," *Environ. Sci. Technol.* 18:365–69 (1984).

20. Atkinson, R., A. M. Winer, and J. N. Pitts, Jr. "Estimation of Night-Time N_2O_5 Concentrations from Ambient NO_2 and NO_3 Radical Concentrations and the Role of N_2O_5 in Night-Time Chemistry," *Atmos. Environ.* 20:331–39 (1986).

21. Logan, J. A. "Tropospheric Ozone: Seasonal Behavior, Trends, and Anthropogenic Influence," *J. Geophys. Res.* 90:10463–82 (1985).

22. Atkinson, R. "A Structure-Activity Relationship for the Estimation of Rate Constants for the Gas-Phase Reactions of OH Radicals with Organic Compounds," *Int. J. Chem. Kinet.* 19:799–828 (1987).

23. Atkinson, R., S. M. Aschmann, J. Arey, P. A. McElroy, and A. M. Winer. "Product Formation from the Gas-Phase Reactions of the OH Radical with $(CH_3O)_3PS$ and $(CH_3O)_2P(S)SCH_3$," *Environ. Sci. Technol.* 23:243–44 (1989).

24. Woodrow, J. E., D. G. Crosby, and J. N Seiber. "Vapor Phase Photochemistry of Pesticides," *Residue Reviews* 85:111–25 (1983), and references therein.

## CHAPTER 10

Global Monitoring of Organochlorine Insecticides—An 11-Year Case Study (1975–1985) of HCHs and DDTs in the Open Ocean Atmosphere and Hydrosphere

Ryo Tatsukawa, Yukihiko Yamaguchi, Masahide Kawano, Narayanan Kannan, and Shinsuke Tanabe

INTRODUCTION

In the past few decades, agricultural, public health, and industrial activities around the world have contributed to the widespread contamination of global ecosystems with persistent organochlorine compounds such as HCHs, DDTs, chlordanes, and PCBs.[1-5] Their continued use in many developing countries has created concerned debates on aspects such as environmental conservation versus economical development. Investigation is still going on regarding the geochemical and ecotoxicological meaning of these persistent and bioaccumulative pollutants. These contaminants are used in recent years as model compounds for our general understanding of biogeochemical dynamics of persistent chemicals and their possible long-term effects to humans and wildlife.[6-8]. Through a global monitoring program, we have been observing the environmental fate and distribution of HCHs, DDTs, chlordane compounds, and PCBs. In this chapter, an eleven-year case study on the levels of organochlorine insecticides, HCHs, and DDTs in air and water media of the global environment is described.

MATERIALS AND METHODS

Sampling

Air and water samples were collected from various parts of the world, covering different seas and oceans, from 1975 to 1985 on 16 survey cruises, as indicated in Table 1 and Figures 1 and 2.

Table 1.   Details of Samples Collected During the Survey Cruises Between 1975 to 1985.

Cruise No.	Date	Samples	References
1	May 1975	air	a
		water	
2	July–Aug. 1976	air	a
		water	
3	Oct. 1976–Jan. 1977	air	a
		water	
4	Sept.–Nov. 1977	air	a
5	Jan.–Feb. 1978	air	a
		water	
6	June–July 1979	air	a
		water	
7	July 1979	air	a
		water	
8	Nov. 1980–Mar. 1981	air	b
		water	
9	Jan. 1981–Jan. 1982	air	b
10	Feb.–Mar. 1982	air	c
11	July 1982	air	unpublished
		water	
12	Dec. 1982	air	unpublished
		water	
13	Nov. 1983–Jan. 1984	air	d
		water	
14	July 1984	air	unpublished
15	June–July 1985	air	unpublished
16	Sep.–Oct. 1985	air	unpublished
		water	

[a]Tanabe S., and R. Tatsukawa. "Chlorinated Hydrocarbons in the North Pacific and Indian Oceans," *J. Oceanogr. Soc. Japan* 36:217–26 (1980).
[b]Tanabe, S., R. Tatsukawa, M. Kawano, and H. Hidaka. "Global Distribution and Atmospheric Transport of Chlorinated Hydrocarbons: HCH (BHC) Isomer and DDT Compounds in the Western Pacific, Eastern Indian, and Antarctic Oceans," *J. Oceanogr. Soc. Japan* 38:137–48 (1982).
[c]Tanabe, S., H. Hidaka, and R. Tatsukawa. "PCBs and Chlorinated Hydrocarbon Pesticides in Antarctic Atmosphere and Hydrosphere, *Chemosphere* 12:277–88 (1983).
[d]Kawano, M., S. Tanabe, T. Inoue, and R. Tatsukawa. "Chlordane Compounds found in the Marine Atmosphere from the Southern Hemisphere," *Trans. Tokyo Univ. Fisher.* 6:59–66 (1985).

Air samples were collected by high-volume and low-volume air samplers packed with polyurethane foam plugs, following the method of Simon and Bidleman.[9] About 1000 m^3 of air was drawn at a flow rate of 550–650 L/min for DDTs, and ~200 m^3 at 25–27 L/min for HCHs. During our earlier surveys (cruises 1–7, see Figure 1), air samples were collected using glycerine-coated Florisil columns for both HCHs and DDTs. The volume of air samples collected was about 20 m^3 at a flow rate of 7 L/min.

A metal bucket was used to collect the surface water. In some cases, surface water was also collected from the tap on board the research vessel. About 100–500 L of water was collected and immediately passed through Amberlite (Rohm and Haas Co., Philadelphia, PA) XAD-2 resin columns.

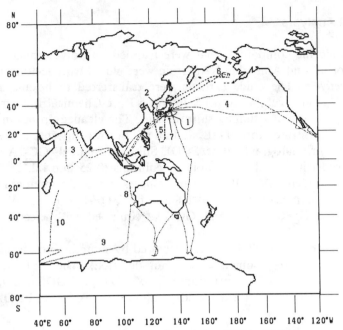

Figure 1. Route map of first 10 survey cruises during 1975 to 1981.

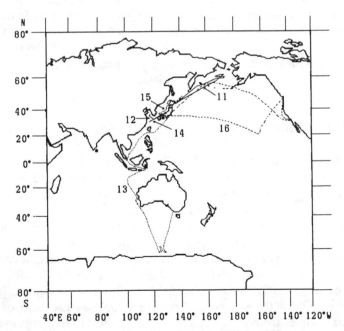

Figure 2. Route map of survey cruises 11–16 during 1982 to 1985.

Chemical Analysis

Organochlorine compounds that were trapped on polyurethane foam (as well as Florisil) and Amberlite XAD resin were eluted with acetone and ethanol, respectively. The crude extracts were transferred to hexane and then cleaned using silica gel (Wako gel S-1, Wako Pure Chemical Industries Ltd., Osaka, Japan) and 5% fuming sulfuric acid. The cleaned-up hexane extract was microconcentrated to 100 mL, and an aliquot was injected into electron-capture gas chromatographs (GC/ECD), Shimadzu models GC-7A and GC-9A, equipped with fused silica capillary columns (0.25 mm i.d. × 25 m) of chemically bonded OV-1701. Earlier samples were chromatographed on packed glass columns (2 mm i.d. × 1.8 m) of 2% QF-1 + 1.5% OV-17, using Shimadzu models GC-4BM and GC-5A, equipped with electron-capture detectors.

The concentration of ΣHCH was calculated by summing up the concentrations of α-, β-, and γ-isomers (δ was not detected). The concentration of ΣDDT is the sum of the concentrations of p,p'-DDT, p,p'-DDE and o,p'-DDT isomers (p,p'-DDD was not detected). Laboratory recoveries of DDTs and HCHs were more than 90%.

Data Analysis and Management

The analytical data obtained during a period of eleven years were processed using a computer program specially developed for this purpose in our laboratory. This "data analysis and management system" was written in BASIC, and computer graphics, were obtained with a NEC model PC-9801V personal computer. Data plotted in the figures were irrespective of temporal or spatial variations in concentrations and compositions.

RESULTS AND DISCUSSION

The atmospheric distribution of HCH isomers (ΣHCH) and DDT compounds (ΣDDT) on a global basis is illustrated in Figures 3 and 4. Their concentration profiles on a latitudinal basis are given in Figures 5 and 6. These figures clearly indicate that the highest concentration of these chlorinated pollutants are in the mid–Northern Hemisphere where several industrially developed nations are situated. High consumption of these compounds in the past in those countries contributed to this situation. The environmental source of HCHs is mainly technical BHC and lindane. The former formulation contains all the major HCH isomers, namely, α-, β-, and γ-isomers; lindane is composed mainly of the γ-isomer. We have calculated the ratio of α- to γ-HCH in air samples from various locations and plotted the results in Figure 7. Their latitudinal distribution is given in Figure 8. It is apparent from these figures that the atmospheric samples from the Northern Hemisphere are high

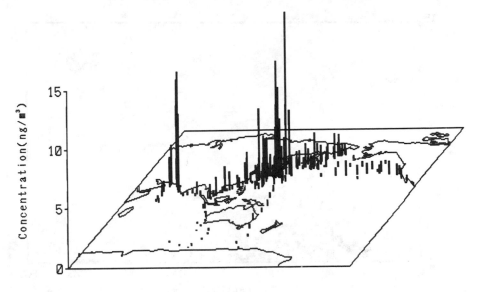

Figure 3. Atmospheric distribution of ΣHCH.

in α-HCH content. On the other hand, the low ratio of α- to γ-HCH in southern latitudes likely reflects the larger usage of lindane. However, the characteristic enrichment of α-HCH in Northern Hemispheric air needs some careful explanation. Technical BHC could be a possible source of α-HCH. However, the α-HCH content in that formulation is not more than 5 times that of γ-HCH (the α- and γ-HCH comprise about 70% and 13% of technical BHC, respectively).[10] Interestingly, the atmospheric ratio of α- to γ-HCH was > 5 in many locations and clearly increased from south to north (Figures 7 and 8). A recent report of Kurtz and Atlas (see Chapter 11) confirms that

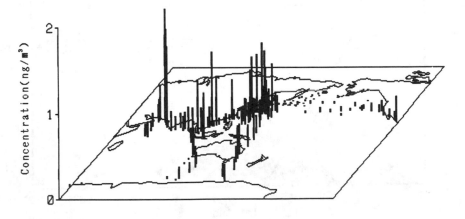

Figure 4. Atmospheric distribution of ΣDDT.

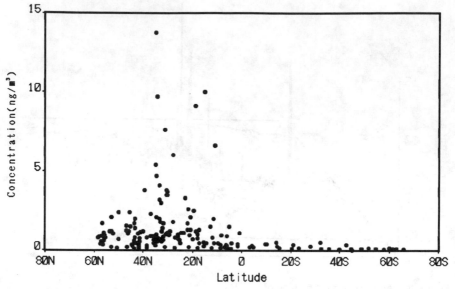

Figure 5. Atmospheric distribution of ΣHCH on a latitudinal basis.

atmospheric α-HCH is extremely high not only in Northern Hemisphere, but also southern, including tropical areas.

The temporal trend of ΣHCH and ΣDDT within a span of eleven years is plotted in Figures 9 and 10. A declining pattern in the concentrations of these persistent pollutants is seen, probably reflecting the effect of a ban on these

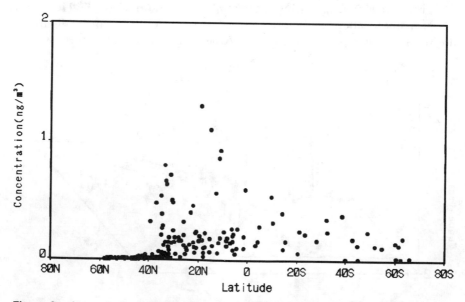

Figure 6. Atmospheric distribution of ΣDDT on a latitudinal basis.

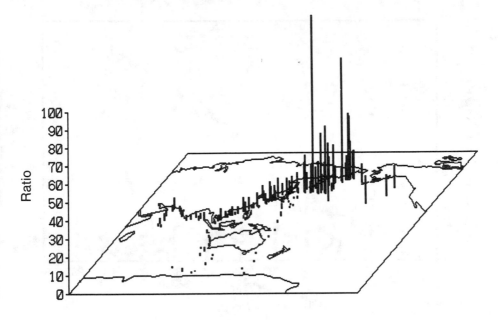

Figure 7. Distribution profiles of α-HCH: γ-HCH in air.

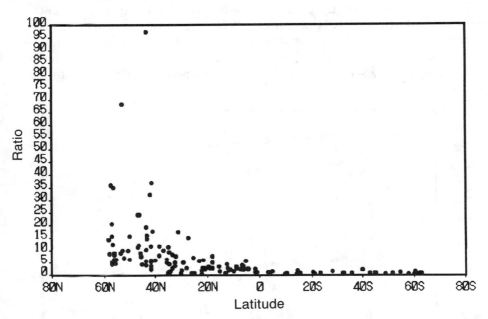

Figure 8. Latitudinal distribution pattern of α-HCH:γ-HCH in air.

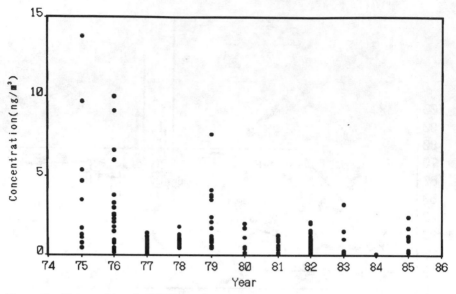

Figure 9.   Temporal trend of ΣHCH in air.

chemicals in many countries, including some developing nations. The generally higher levels of ΣHCH in air than ΣDDT may arise out of the fact that HCH isomers have higher vapor pressures than DDT compounds. The temporal trends of the α- to γ-HCH ratio for the Northern and Southern Hemispheres are also given in Figures 11 and 12, respectively. Compared to the

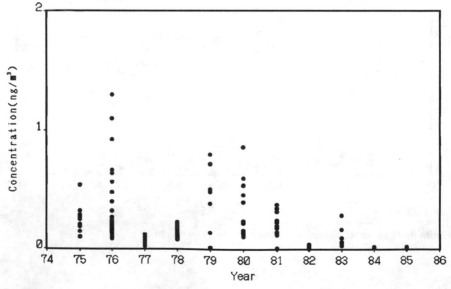

Figure 10.   Temporal trend of ΣDDT in air.

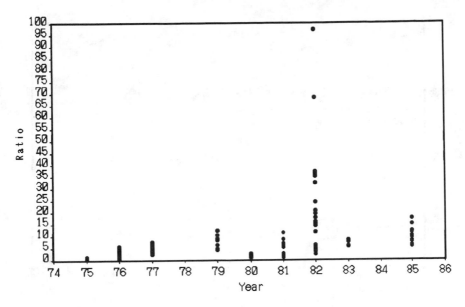

Figure 11. Temporal trend of α-HCH:γ-HCH in Northern Hemispheric air.

discernible trend of ΣHCH concentrations, the isomer pattern does not vary as much. However, it may be safely mentioned that the α:γHCH ratio in the Southern Hemisphere is gradually increasing year by year, probably implying that a change in the usage pattern from lindane to technical BHC is occurring in southern countries.

The global distributions of ΣHCH and ΣDDT in surface seawater are presented in Figures 13 and 14, as well as on a latitudinal basis in Figures 15 and 16. Since the major input source of persistent pollutants to open ocean water is the atmosphere, these figures agree with the observation that Northern Hemisphere samples are more contaminated than those from the Southern Hemisphere. The temporal trend of ΣHCH and ΣDDT, as seen in Figures 17 and 18, is also similar to the trend noticed in air samples (see Figures 9 and 10). The ratio of α-HCH to γ-HCH in water samples is presented both on a latitudinal basis (Figure 19) and on a temporal basis (Figures 20 and 21). The observation that technical BHC was possibly consumed at higher rates in countries in the Northern Hemisphere than in the Southern Hemisphere is further substantiated from Figure 19. Like the change in the atmospheric composition of HCH isomers, the subtle change in the α:γ ratio in surface seawater is visible in the Southern Hemisphere (Figure 21), but not in the Northern Hemisphere (Figure 20).

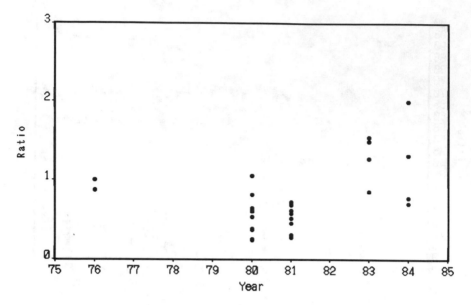

Figure 12. Temporal trend of α-HCH:γ-HCH in Southern Hemispheric air.

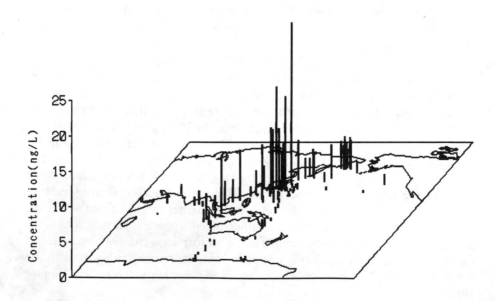

Figure 13. Distribution of ΣHCH in surface seawater.

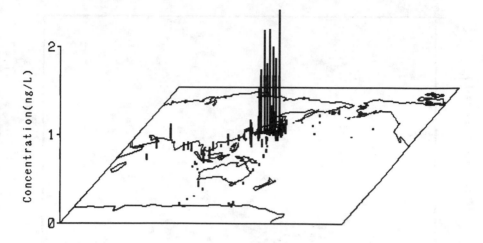

Figure 14. Distribution of ΣDDT in surface seawater.

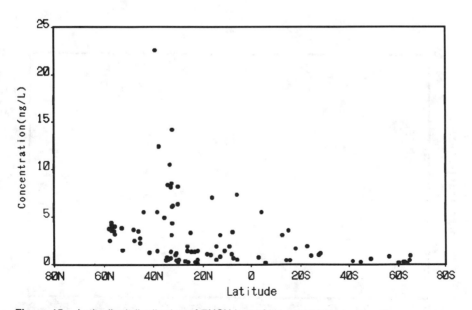

Figure 15. Latitudinal distribution of ΣHCH in surface seawater.

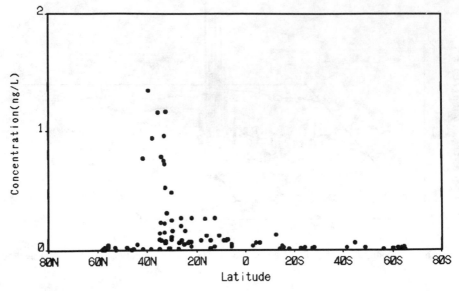

Figure 16.   Latitudinal distribution of ΣDDT in surface seawater.

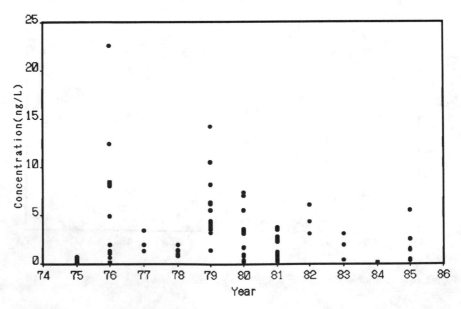

Figure 17.   Temporal trend of ΣHCH in surface seawater.

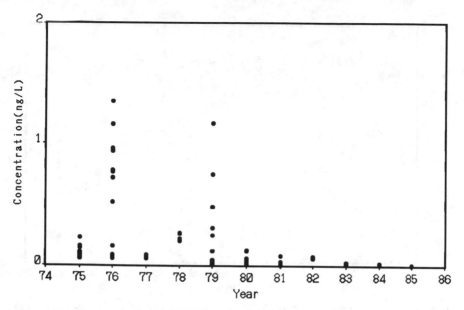

Figure 18.   Temporal trend of ΣDDT in surface seawater.

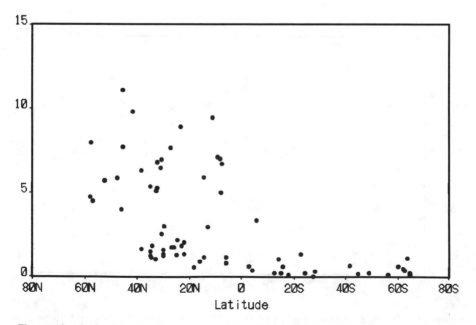

Figure 19.   Latitudinal distribution pattern of α-HCH:γ-HCH in surface seawater.

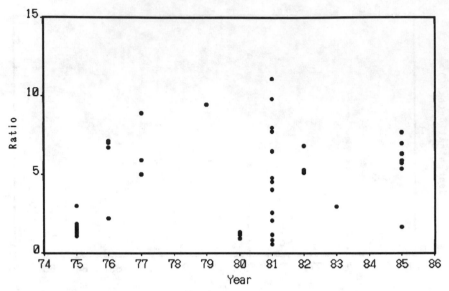

Figure 20.   Temporal trend of α-HCH:γ-HCH in surface seawater of the Northern Hemisphere.

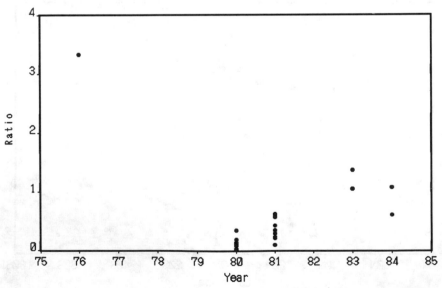

Figure 21.   Temporal trend of α-HCH:γ-HCH in surface seawater of the Southern Hemisphere.

ACKNOWLEDGMENTS

The authors are grateful to the Ocean Research Institute, the University of Tokyo, Tokyo University of Fisheries, the National Institute of Polar Research, Nagasaki University, Shimonoseki University of Fisheries, the Fisheries Agency of Japan, and the Toyota Foundation for providing the opportunity to participate in their cruises. They would also like to thank Mr. N. Hamada, Faculty of Agriculture, Tokyo University, for his help in developing the computer graphics programs.

REFERENCES

1. Atlas, E., T. Bidleman, and C. S. Giam. "Atmospheric Transport of PCBs to the Oceans," in *PCBs and the Environment*, Vol. 1, John S. Waid, Ed. (Boca Raton, FL: CRC Press, 1986), pp. 79–100.
2. Tanabe, S., and R. Tatsukawa. "Distribution, Behavior, and Load of PCBs in the Oceans," in *PCBs and the Environment*, Vol. 1, John S. Waid, Ed. (Boca Raton, FL: CRC Press, 1986), pp. 143–61.
3. Norstrom, R. J., M. Simon, D. C. G. Muir, and R. E. Schweinsburg. "Organochlorine Contaminants in Arctic Marine Food Chains: Identification, Geographical Distribution, and Temporal Trends in Polar Bears," *Environ. Sci. Technol.* 22:1063–71 (1988).
4. Muir, D. C. G., R. J. Norstrom, and M. Simon. "Organochlorine Contaminants in Arctic Marine Food Chains: Accumulation of Specific Polychlorinated Biphenyls and Chlordane-Related Compounds," *Environ. Sci. Technol.* 22:1071–79 (1988).
5. Kawano, M., T. Inoue, T. Wada, H. Hidaka, and R. Tatsukawa. "Bioconcentration and Residue Patterns of Chlordane Compounds in Marine Animals: Invertebrates, Fish, Mammals, and Seabirds," *Environ. Sci. Technol.* 22:792–97 (1988).
6. Tateya, S., S. Tanabe, and R. Tatsukawa. "PCBs on the Globe: Possible Trend of Future Levels in the Open Ocean Environment," in *Toxic Contamination in Large Lakes*, Vol. 3, Norbert W. Schmidtke, Ed. (Chelsea, MI: Lewis Publishers, 1988), pp. 237–81.
7. Clark, T., K. Clark, S. Paterson, D. Mackay, and R. J. Norstrom. "Wildlife Monitoring, Modeling, and Fugacity," *Environ. Sci. Technol.* 22:120–27 (1988).
8. Tanabe, S. "PCB Problems in the Future: Foresight from Current Knowledge," *Environ. Poll.* 50:5–28 (1988).
9. Simon, C. G., and T. Bidleman. "Sampling Airborne Polychlorinated Biphenyls with Polyurethane Foam—Chromatographic Approach to Determining Retention Efficiencies," *Anal. Chem* 51:1110–13.
10. Tatsukawa, R., T. Wakimoto, and T. Ogawa. "BHC Residues in the Environment," *Environmental Toxicology of Pesticides*, F. Matsumura and G. M. Boush, Eds. (San Diego, CA: Academic Press, 1972), pp. 229–238.

## CHAPTER 11

Distribution of Hexachlorocyclohexanes in the Pacific Ocean Basin, Air and Water, 1987

David A. Kurtz and Elliot L. Atlas

INTRODUCTION

Farm-applied pesticides are known to be distributed throughout the surface of the earth.[1-5] Virtually every geographical region, even the most remote, has been shown to be impacted by pesticides or other synthetic organic compounds. These compounds are transported to all regions of the earth by exchange processes between reservoirs of pesticides in the land, water, and air, and by movement of the pesticides within these reservoirs themselves.[6-8] Thus, anthropogenic compounds are distributed in the environment by a combination of atmospheric transport,[2,9-13] river drainage,[14-16] and aquifer flow.[17-18] However, various modeling efforts have suggested that atmospheric transport is the major pathway of input of synthetic organic compounds to most of the global ocean.[19] Once pesticides reach the marine environment, transport of compounds between air and water and between water and sediment can occur by evaporation,[20-24] precipitation,[25-27] distillation and condensation,[28-29] and occlusion and adsorption processes.[30-31]

Though the motion of the atmosphere and the oceans contains a complex vertical as well as horizontal component, our study focuses only on the region adjacent to the air-ocean interface. Measurements of trace chemicals in the atmospheric boundary layer and in surface ocean waters can be used to obtain useful inferences on the flux of chemicals across this interface. Unfortunately, the data available to evaluate pesticide inputs to the oceans are very limited. Studies have been carried out in several areas of the Atlantic,[10,32-34] Pacific,[2,5,7,13,35-37] and Indian Ocean,[7,29,36-39] but the geographic and temporal coverage of the measurements is sparse.

In the summer of 1987, we were able to participate in a circum-Pacific cruise on the NOAA research vessel *Oceanographer*. This cruise allowed sampling of chlorinated pesticides and other organic compounds in both air and water of two hemispheres and near three continents. This report presents results of the analysis of these samples for hexachlorocyclohexanes (HCHs). The HCHs were the most abundant organochlorine pesticides measured during the cruise and have been shown to be a major contaminant in other studies of the marine environment. It should be pointed out that other literature may refer to the HCHs by the often used misnomer BHCs (referring to benzene hexachlorides).

The commercial HCH pesticide can contain a number of isomers of slightly varying composition in the different formulations. The main pesticidal component is the γ-isomer, also known as lindane. In the technical mixture of HCH, an approximate composition is 70% α-HCH, 7% β-HCH, 13% γ-HCH, 5% δ-HCH, and 5% other isomers.[40-41] In addition, pure lindane, or a pesticide mixture fortified with lindane, is now applied in many parts of the world. In the environment, the composition of HCH is observed to differ from the pesticide mixture originally used. Source variation, metabolism, and physical and chemical degradation are processes that can alter the composition of these HCH residues. In some cases, change in composition can be used to trace the sources of HCH to more remote areas of the earth.[42]

This chapter will examine the concentration and composition of HCH in the air and water of diverse geographic areas of the Pacific Ocean. These data allow regions of the Pacific to be characterized in terms of sources and inputs of pesticides (HCH) and in terms of potential exchange processes between the atmosphere and the ocean.

MATERIALS AND METHODS

Cruise Track

Figure 1 shows the cruise track of the NOAA R/V *Oceanographer* from May 19 to August 29, 1987. Sampling locations for both air and water are shown. Air samples required up to two days for collection; locations for these samples are depicted as long bands. Water samples required only two hours to collect; these samples integrate surface water from typically < 20 nautical miles. Locations for water samples are indicated by squares on Figure 1.

Sample Collection and Analysis

Air

Samples were collected by drawing air through a 15-cm column of a Mg-silicate adsorbent material (Florisil, 16–30 mesh) contained in a 6 mm i.d. glass tube. Air flow rates were measured with a mass flow meter. Typical flow rates through the column were 10–15 Lpm, and total air volumes were 25–50 m³. The tubes were mounted on a metal holder on the bow mast of the vessel. Polyethylene tubing connected the sampler to an oilless vacuum pump located on the deck immediately below the bow. In preparation for sampling, all

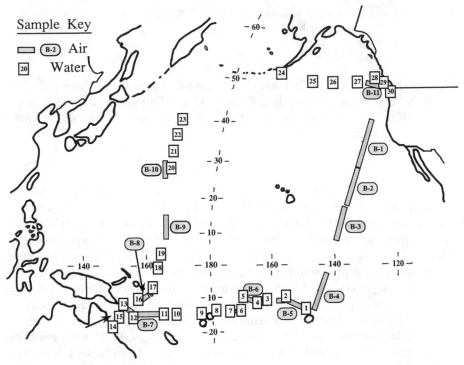

Figure 1. Cruise track and sampling locations on R/V *Oceanographer,* Summer 1987.

adsorbent columns were prepacked, combusted for 16 hours at 420°C, and sealed with aluminum foil and Teflon tape. After collection the samples were resealed with aluminum foil and Teflon tape and stored refrigerated or frozen until later analysis in the lab. Unfortunately, only a limited number of air samples could be collected during the cruise because of power restrictions on the bow of the ship.

In the laboratory the Florisil sample tubes were spiked with 8.1 ng of dibromooctafluorobiphenyl and 2.7 ng decachlorobiphenyl to serve as internal standards. The samples were then eluted from the Florisil column with 300 mL of 20% (v/v) methylene chloride in pentane. The extracts were reduced in volume in a Kuderna-Danish evaporative concentrator fitted with a three-ball Snyder condenser column. Final extracts were reduced to ~1 mL after solvent exchange to hexane. Also, several samples were analyzed for breakthrough of the analytes. No breakthrough of analytes was found in the back section of the trap.

Identification and quantification of the samples was accomplished by capillary gas chromatography using electron capture detection and computerized data collection/integration. A Varian Model 3500 GC (Varian Instrument Division, Walnut Creek, CA) with a Model 651 Data System was used. Chromatographic conditions were

- Oven program: 1 min hold at 100°C, 3°C/min to 250°C, 10°C/min to 300°C, and final hold 1 min
- Injector: 275°C
- Detector: 325°C

The column used was a DB-1 30 m, 0.25 mm i.d., fused silica column, 0.25 μm film thickness (J & W Scientific, Folsom, CA). Helium was used as a carrier gas with a linear velocity of approximately 30 cm/sec. A typical chromatogram is shown in Figure 2.

Water

Surface seawater was collected by pumping water from an intake located 3 m below the surface at the bow of the ship. Experiments conducted before the cruise showed that the ship's pumping system did not introduce artifacts into the samples that would interfere with the analysis of organochlorine pesticides. Seventy-liter water samples obtained from the pumping system were subjected to continuous liquid-liquid extraction utilizing an all-glass apparatus (the Goulden Large Sample Extractor).[43] Filtered water was continuously extracted by 200 mL methylene chloride, which was reserved for later analysis in our home laboratory. Prior experiments demonstrated good recovery of organochlorine compounds with this apparatus. Recovery of HCH with the apparatus was > 96% in laboratory tests when compared to in-bottle liquid-liquid extractions. Other uses and recovery experiments are detailed in recent publications.[44-45]

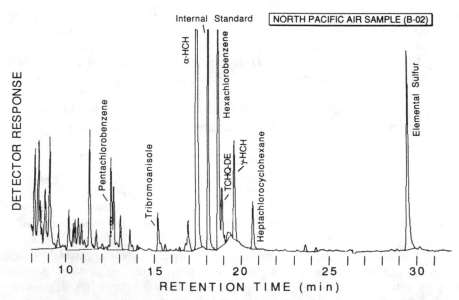

Figure 2. Capillary electron capture gas chromatogram of air sample B-02. Conditions of analysis are given in text. (TCHQ-DE is tetrachlorohydroquinone dimethyl ether.)

The methylene chloride extracts were analyzed by careful rotary evaporation of the solvent, exchange of solvent to hexane, Florisil fractionation, and capillary gas chromatography quantification with electron capture detection.[46]

The Florisil fractionation utilized 3 g of deactivated Florisil in a 7 mm i.d. column. Deactivation was accomplished with 3 mL of a petroleum ether solution containing 2.67% methanol and 0.67% methylene chloride. The fraction eluted by 15 mL petroleum ether contained most PCBs, some p,p'-DDE, and the unsaturated chlordane compounds. The fraction eluted by 20 mL of 20% methylene chloride in petroleum ether included the HCH isomers, the remaining p,p'-DDE and other DDT analogs, and the remaining chlordane compounds. A final fraction containing dieldrin, endosulfans, endrin, and heptachlor epoxide was eluted with 50% methylene chloride/0.35% acetonitrile in petroleum ether. Florisil cleanup procedures were similar to those used by FDA.[47] Evaporative concentration was carefully done to avoid loss of the HCH compounds through volatilization.

Method recoveries of the HCH compounds at 3–6 ng charge levels averaged between 91–107%. Quality control was achieved with method control recoveries run every five samples.

Chromatographic analysis was performed on a Varian 3600 gas chromatograph (Varian Instrument Division, Walnut Creek, CA) equipped with an automatic injector. A 60 m SPB-5 fused silica column, 0.32 mm i.d., 0.25 μm film thickness (Supelco, Inc., Bellefonte, PA), was used for compound separation. The column was operated under the following conditions: 40°C (2 min), 5°C/min to 250°C, final hold for 11 min. Helium was used as a carrier gas at a flow rate of 11 mL/min and a flow velocity of 23 cm/sec. Detector response was recorded electronically with a Shimadzu Chromatopac CR4A Integrator (Giancarlo Scientific, Pittsburgh, PA). Control chart GC standards were run every 6–12 injections. Chromatograms of representative fractions of HCHs in water are shown in Figure 3. This figure demonstrates the contrast between the lowest level samples in the South Pacific and the high concentration samples of the North Pacific.

RESULTS

Air Samples

α-HCH was found to have the highest concentration in air of all the pesticides investigated (Table 1). It ranged in concentration from 8 to 300 pg/m³ in the Pacific Ocean atmosphere. γ-HCH was found in concentrations ranging from <2 to 44 pg/m³. The highest concentrations of HCHs were observed in the Northeast Pacific, and samples from the Southern Hemisphere contained the smallest amounts of HCH. Intermediate concentrations were measured in the central Northwest Pacific (Figures 4 and 5). On average the concentration of HCHs was 16 times higher in the Northern Hemisphere than in the South-

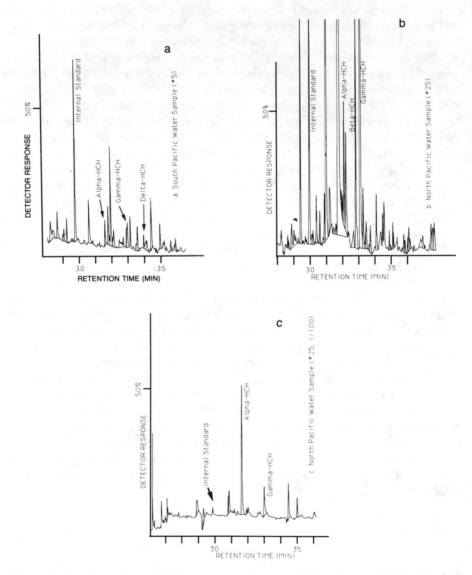

Figure 3. Capillary electron capture gas chromatograms of (a) water sample 5 from tropical South Pacific, (b) sample 25 from North Pacific showing the typically very high HCH concentrations, and (c) sample 25 diluted 1/100.

ern Hemisphere. This interhemispheric difference is similar to results obtained at other sites in the Pacific.[48]

The ratio of $\alpha{:}\gamma$ HCH in the air ranged from 4 to 33. The values around 4 were found in the Southern Hemisphere, those around 7–13 were in the NE

Table 1.   Concentrations of Hexachlorocyclohexanes in the Pacific Ocean Atmosphere, Summer 1987

| Sample | Mid-Date | Concentration (pg/m^3) | | Ratio |
		α-HCH	γ-HCH	$\alpha{:}\gamma$
B-01	May 22	300	44	6.8
B-02	May 24	260	29	9.0
B-03	May 27	121	12	10.1
B-04	May 31	12	3	4.0
B-05	June 8	14	<2	<7
B-06	June 14	8	<2	<4
B-07	June 30	17	4	4.3
B-09	July 31	130	4	33
B-10	August 5	132	5	26
B-11	August 28	280	21	13.3

Note: See Figure 1 for sample locations.

Pacific, and those from 26–33 were located in the central NW Pacific (Table 1).

Water Samples

α-HCH had the highest concentration and widest range of all compounds found in the water samples (Table 2 and Figure 4). Surprisingly, the highest concentration of α-HCH was found in waters of the Coral Sea off Australia, 1100–5100 pg/L. In most other areas of the tropical South Pacific there was very little α-HCH; concentrations ranged from 5–81 pg/L in the waters between Tahiti and Samoa. Closer to the Fiji Islands, the concentrations were higher, 330–1180 pg/L. In the NW Pacific the concentration of α-HCH ranged from 970 to 2050 pg/L. The NE Pacific region between Dutch Harbor (Alaska) and Seattle contained concentrations of α-HCH from 2000 to 2800

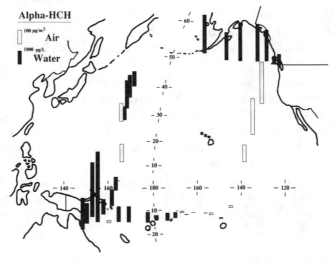

Figure 4.   Concentrations of α-HCH found over and in the Pacific Ocean, 1987.

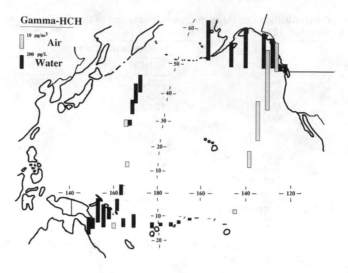

Figure 5. Concentrations of γ-HCH found over and in the Pacific Ocean, 1987.

pg/L, with lower concentrations observed in the areas close to Vancouver Island and the Strait of Juan de Fuca.

γ-HCH was found to vary between <4 to 600 pg/L (Table 2 and Figure 5). Minimum concentrations were found in the South Pacific between Tahiti and Fiji. Concentrations of γ-HCH were higher, 150–330 pg/L, in the Coral Sea near Australia and New Guinea. However, the highest concentrations of γ-HCH, up to 600 pg/L, were found in the NE Pacific near Alaska.

β-and δ-HCH were also identified in the water samples, though not in the air samples. β-HCH closely tracked the concentration of γ-HCH in the samples. With the exception of a few samples in the South Pacific and one in the Strait of Juan de Fuca, the concentration of β-HCH was about half the concentration of γ-HCH.

The composition of HCH in surface waters, as indicated by the α:γ ratio, appears to have significant geographic differences (see Table 2 and Figure 6). The α:γ HCH ratio in Alaskan samples all fall in the range of 4.6 to 6.7 (avg. = 5.5). Samples from the NW Pacific were higher toward the equator, to a maximum of 11.9. The samples from the Coral Sea had the highest ratio (max. = 15.3) and the highest average ratio (avg. = 11.4). The more remote regions of the South Pacific had variable ratios caused by the inaccurate determination at very low concentrations of HCHs in the waters. Excluding samples 1–5, the average α:γ HCH ratio was found to be 7.4 in the tropical South Pacific waters.

Table 2. Concentrations of Hexachlorocyclohexane Isomers in Pacific Ocean Surface Waters, Summer 1987

Sample	Date	Concentration (pg/l)				Ratio	
		α-HCH	γ-HCH	β-HCH	δ-HCH	$\alpha{:}\gamma$	$\beta{:}\gamma$
1	June 7	81	16	35	0	5	2.2
2	June 8	23	1	31	0	16	22
3	June 11	7	13	16	0	0.56	1.3
4	June 12	33	36	13	0	0.90	0.35
5	June 15	5	4	0	4	1.1	—
6	June 17	390	53	26	0	7.4	0.5
7	June 19	550	68	32	0	8.1	0.48
8	June 21	330	47	0	0	6.9	—
9	June 22	750	83	39	0	9.0	0.47
10	June 25	1180	194	73	11	6.1	0.38
11	June 27	1120	127	55	0	8.8	0.43
12	June 30	5100	330	111	13	15.3	0.33
13	July 2	4000	350	118	12	11.3	0.33
14	July 5	2300	250	96	9	9.2	0.39
15	July 14	1480	153	63	7	9.7	0.41
16	July 16	1100	170	93	7	6.5	0.55
17	July 19	1190	170	66	0	7.0	0.39
18	July 21	990	144	56	4	6.8	0.39
20	August 5	970	81	29	0	11.9	0.35
21	August 8	2050	210	119	5	9.6	0.56
22	August 10	1620	290	129	9	5.5	0.44
23	August 12	1010	260	114	7	3.9	0.44
24	August 23	2800	590	220	14	4.6	0.37
25	August 24	2000	360	171	14	5.6	0.48
26	August 25	2800	600	240	18	4.6	0.39
27	August 26	2700	520	200	19	5.1	0.39
28	August 27	2300	430	186	14	5.3	0.43
29	August 28	580	87	34	0	6.7	0.40
30	August 28	890	136	114	4	6.6	0.84
	Avg. blank	<4	<4	<4	<4	—	—

Note: See Figure 1 for sample locations.

DISCUSSION

Ocean Concentration Regions

Based on our measurement of the concentration and composition of HCH in Pacific Ocean surface waters, we can identify four distinct geographic regions:

1. NE Pacific
2. NW Pacific
3. Coral Sea
4. Tropical South Pacific

In Table 3 we compare concentrations in air and water from these areas with other data from the Pacific Ocean, excluding those samples taken near Japan and the Asian coastline.[2,7,37,48-50,57] These latter samples are strongly influenced by emissions from Asia and Japan. In spite of the difficulty of making com-

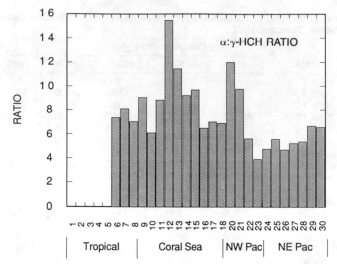

Figure 6.   Ratio of concentrations of α:γ-HCH in Pacific Ocean surface waters. *Note*: Concentrations of one or both isomers in samples 1–5 were so low that the ratios calculated from them have been omitted from the figure.

parisons between samples taken at different locations at different times and analyzed by different laboratories, we are surprised at the overall comparability of the data collected in different regions over more than a decade.

NE Pacific

Observations of both α-and γ-HCH in the remote areas of the NE Pacific atmosphere, and indeed from throughout the Northern Hemisphere, typically have shown concentrations of several hundred pg/m^3. Atmospheric concentrations near source areas have shown much higher concentrations ($> 10,000$ pg/m^3).[13,51] Our data suggest that the values found in 1987 in the atmosphere of the NE Pacific region fell within the range of the variation of total HCHs seen in the past decade. HCH usage appears to be stable, although relative amounts of the α-and γ-isomers have shifted from 1987 to 1988 (the slight drop in α-HCH concentration and the large increase in γ-HCH may indicate a shift in land usage from the technical mixture to pure lindane). Significant changes in atmospheric HCH over days or months can be found and may be due to the variation in sources of the pesticide and its transport through the marine atmosphere.[48] Larger data sets and longer time series measurements would be required to determine if there might be a temporal trend in the concentration or composition of HCHs in the Northern Hemispheric atmosphere during the last decade.

The surface waters of the NE Pacific, Gulf of Alaska, and the Bering Sea showed both a high and constant level of HCHs over the last decade. Furthermore, the ratios of α:γ-HCH also appear to have been relatively stable (~5-6) over this same time period (Figure 6). We would expect the concentrations in

Table 3. Average Concentrations of HCH in Pacific Ocean Air and Water, 1976 to Present

Region	Date	Location	α-HCH	γ-HCH	Σ-HCH	Reference
		Air				
NE Pacific	9/77	40–50°N	—	—	400	37
	7/79	Bering Sea	—	—	980	37
	8/87	50° N	280	21	300[a]	this work
	7/88	Bering Sea	200	130	330[a]	49
NW Pacific	5/75	20–40° N	—	—	1910	37
	9/77	30–40° N	—	—	420	37
	5/79	12° N	250	15	265	2
	3/81	0–30° N	370	200	700	7
	6/86	35–45° N	310	28	340	50
	8/87	15–40° N	131	5	136[a]	this work
Coral Sea	2/81	Coral Sea	120	210	380	7
	6/85	New Zealand	25	1.3	26	48
	7/87	Coral Sea	17	4	21	this work
Tropical S Pacific	1,7/81	Samoa	32	2	34[a]	48
	3/81	Peru	10	<1	≤ 11	48
	6/87	0–15° S	11	1	12[a]	this work
		Water				
NE Pacific	7/79	Bering Sea	—	—	3900	37
	7/81	Bering Sea	2800	650	3400	57
	8/87	40–50° N	2500	500	3200	this work
	7/88	Bering Sea	2400	520	2900[a]	49
NW Pacific	3/81	20–30° N	500	280	950	7
	6/86	35–40° N	850	200	1050	50
	8/87	28–40° N	1410	210	1700	this work
Coral Sea	3/81	0–30° S	210	590	820	7
	7/83	New Zealand	450[b]	—	>450[b]	48
	7/87	0–18° S	2180	220	2500	this work
Tropical S Pacific	1,7/81	Samoa	320[b]	30[b]	350[b]	48
	6/87	0–15° S	241	36	300	this work

Note: Excludes measurements taken near Japan and the Asian coastline. Concentrations are reported in pg/m^3 for air and pg/L for water.
[a] α-HCH and γ-HCH only.
[b] Rain water

surface waters to change more slowly than air concentrations because of the slower movement and higher mass.

NW Pacific

Regions of the NW Pacific are known to be subjected to periodic pulses of trace chemicals from the Asian continent.[52-54] Areas closer to an Asian source of pesticides might be subjected to considerable variability in the input of HCHs to this area. In fact, reported data indicate more variation in the concentration and composition of HCH in the atmosphere for this area than the other described areas, as shown in Table 3. Average atmospheric concentrations of HCH in this region ranged from 136 to 1910 pg/m^3, nearly a factor of 15 difference. Composition of the HCH mixture also was different, largely as a result of changes in the γ-HCH content. The measurements made during our cruise were among the lowest total HCH and highest α:γ HCH ratio reported for the marine atmosphere. These lower concentrations and high α:γ ratios may result from degradation and removal of HCH in the atmosphere.[42] Water samples from this region contained ~ 1000–2000 pg/L total HCHs, about one-half that observed in the northern Pacific areas.

Coral Sea

Surprisingly, the highest marine values of α-HCH we measured were in the Coral Sea near Australia. One sample at the SE tip of Papua New Guinea reached over 5000 pg/L. Of the five samples in the Coral Sea, three samples had α-HCH concentrations in the range of 1100–2300 pg/L, and two samples were a factor of 2–5 times higher (4000–5100 pg/L). These samples also had the highest α:γ HCH ratio observed during the cruise. These results are an interesting contrast to measurements taken off the east coast of Australia during 1981 by Tanabe et al.[7] They reported concentrations of ΣHCH 5–10 times lower than reported here, but they also found that γ-HCH was the most abundant isomer. Our measurements showed that α-HCH is the most abundant isomer and the total HCH concentrations in air are comparable to samples obtained from New Zealand.[48] Thus, if these measurements are accurate, HCHs in surface waters near Australia appear to be responding very quickly to changing sources of HCH. The anomalous concentration of HCHs may have been produced by surface runoff or improper disposal of pesticide stocks or waste products. Also, atmospheric or ocean transport mechanisms may contribute to the burden of HCHs in the Coral Sea. The mechanisms responsible for the high concentrations of HCH in this area should be investigated.

Tropical South Pacific

The region of the South Pacific around 10–15° S contained only trace residues of HCH in both the air and the water. Interhemispheric atmospheric transport of HCH from the Northern to the Southern Hemisphere is limited by the relatively short lifetime of HCH in the atmosphere, and by the removal of HCH in tropical rain systems located near the high usage land masses to the west of these sea areas. Thus, concentrations of HCH and other organochlorine compounds in this region[48] are among the lowest anywhere in the world. It is of interest to note that these waters contained nearly the same concentrations of HCH as measured in rain in the South Pacific.[48]

In Table 4 we summarize the characteristics and potential sources of HCH in the different regions of the Pacific sampled during 1987. Potential sources are

Table 4. Characteristics of HCHs in Pacific Ocean Air and Water in This Study

Region	Water ΣHCH (ng/L)	Water α:γ Ratio	Air Σ-HCH (pg/m³)	Air α:γ Ratio	Potential Sources
NE Pacific	3–4	4–7	200–300	10–15	N. America/USSR
NW Pacific	1–2	4–11	100–300	Variable	Asia
Coral Sea	2–5	10–15	20–30	10–15	Aust./New Guinea
Tropical S. Pacific	<1	Variable	10–20	10–15	Diffuse

assessed by considering expected air trajectories and surface water currents to the different regions.[50]

AIR-SEA EXCHANGE PROCESSES

Transport of atmospheric HCH across the air-sea interface can occur by several processes. While the dry deposition of particles containing adsorbed HCH to the water phase is one possible process, the most prominent have been shown recently to be both gas exchange and wet deposition through rainfall.[19,55] No precipitation was collected during this cruise, but reports of HCH in marine rain have been made elsewhere.[48,56] Gas exchange as the other major process can be evaluated with data from our cruise. Though the process of gas exchange of these higher molecular weight compounds, the organochlorines, is complex,[2,19] the basic parameter that determines the direction and magnitude of gas flux is the difference between equilibrium and ambient concentrations of the compound in the surface seawater. The equation to describe the flux from air to water[19] can be simply written as

$$F = K_w (C_a RT/H - C_w)$$

where
F = flux of HCH into the ocean
K_w = overall exchange velocity of HCH
C_a = concentration of HCH in air
C_w = concentration of HCH in water
R = gas constant
T = temperature (°K)
H = Henry's law constant

Thus, comparison of the concentration of HCH in the air and the water, using an appropriate Henry's law constant, can determine the potential for gas exchange processes to drive HCH into or out of the ocean. The rate of exchange is determined by the magnitude of disequilibrium and the exchange velocity K_w.

In Figure 7 we have calculated, from the concentrations of HCH found in water, the observed concentrations that would be expected to be in air under equilibrium conditions. The calculations used a dimensionless Henry's law

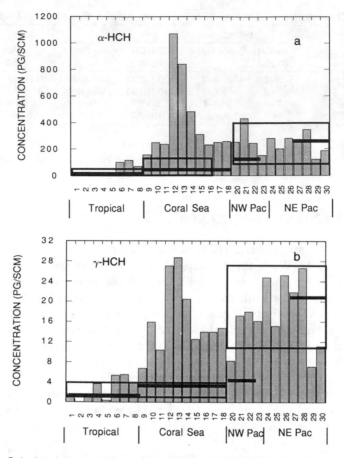

Figure 7. Calculated concentrations of (a) α-HCH and (b) γ-HCH in the atmosphere based on equilibrium with existing levels of α-HCH in the surface ocean (vertical bars). Heavy horizontal lines are ambient concentrations measured during the cruise. Heavy boxes encompass the range of atmospheric concentrations previously found in the North and South Pacific. Vertical bars extending above the lines or boxes indicate supersaturation of surface seawater and potential exchange of HCH from the water to the atmosphere. Bars within the boxes suggest equilibrium conditions between the atmosphere and the ocean.

constant at 24°C of 2.1×10^{-4} for α-HCH and 8.2×10^{-5} for γ-HCH.[56] Though most water temperatures were near 24°C, we did apply a small temperature correction when the temperature was near 30°C or < 20°C. Figure 7 compares the values in air measured during the cruise (heavy horizontal lines) with the equilibrium air concentrations calculated (vertical bars). If the calculated concentrations are close to the measured values, the air and water are close to equilibrium. Calculated concentrations above the measured air concentrations indicate a water-driven system (water to air transport). Conversely, calculated concentrations below the measured atmospheric levels indicate an air-driven system. This analysis shows that during this cruise most surface waters of the North Pacific could be a source of HCH found in the atmosphere.

The one region where waters appear to be highly supersaturated is the Coral Sea area adjacent to Australia and New Guinea. In this area it is possible that there are sources other than the atmosphere that are controlling the concentration of HCHs in the water. As noted earlier, this area deserves more attention to determine the source, variation, and behavior of HCHs.

ACKNOWLEDGMENTS

The authors thank Kevin Lloyd for sample collection, Sue Schauffler for assistance in air sample analysis, and Robert Bond for analysis of the water samples. We gratefully acknowledge the assistance of the officers, crew, and technical staff of the R/V *Oceanographer* during the cruise. Appreciation is given for helpful suggestions from Norman Macdonald for meteorological matters. This work was supported, in part, by NSF grants OCE-84–15726 and ATM-87–170002, and by a grant of equipment from Supelco, Inc. This paper is published as Journal Series paper No. 8260 of the Pennsylvania Agricultural Experiment Station.

REFERENCES

1. George, J. L., and D. E. H. Frear. "Pesticides in the Antarctic," *J. Appl. Ecol.* 3(suppl.):155–66 (1966).
2. Atlas, E., and C. S. Giam. "Global Transport of Organic Pollutants: Ambient Concentrations in the Remote Marine Atmosphere," *Science* 211:163–65 (1981).
3. Ballschmiter, K., and M. Zell. "Baseline Studies of Global Pollution: I. Occurrence of Organohalogens in Pristine European and Antarctic Aquatic Environments," *Intern. J. Environ. Anal. Chem.* 8:15–35 (1980).
4. Giam, C. S., R. L. Richardson, D. Taylor, and M. K. Wong. "DDT, DDE, and PCBs in the Tissues of Reef Dwelling Groupers (Serranidae) in the Gulf of Mexico and the Grand Bahamas," *Bull. Environ. Contam. Toxicol.* 11:189–92 (1974).
5. Tanabe, S., T. Mori, R. Tatsukawa, and N. Miyazaki. "Global Pollution of Marine Animals by PCBs, DDTs, and HCHs (BHCs)," *Chemosphere* 12:1269–75 (1983).
6. Atlas, E. L., T. Bidleman, and C. S. Giam. "Atmospheric Transport of PCB's to the Oceans," in *PCB's and the Environment,* J. Waid, Ed. (Boca Raton, FL: CRC Press, 1986), pp. 79–100.
7. Tanabe, S., R. Tatsukawa, M. Kawano, and R. Hidaka. "Global Distribution and Atmospheric Transport of Chlorinated Hydrocarbons: HCH (BHC) Isomers and DDT Compounds in the Western Pacific, Eastern Indian and Antarctic Oceans," *J. Oceanogr. Soc. Japan* 38(3):137–48 (1982).
8. Villeneuve, J. P., and C. Cattini. "Input of Chlorinated Hydrocarbons through Dry and Wet Deposition to the Western Mediterranean," *Chemosphere* 15(2):115–20 (1986).
9. Risebrough, R. W., R. J. Huggett, J. J. Griffin, and E. D. Goldberg. "Pesticides: Transatlantic Movements in the Northeast Trades" *Science* 20:1233 (1968).

10. Bidleman, T. F., E. C. Christensen, W. N. Billings, and R. Leonard. "Atmospheric Transport of Organochlorines in the North Atlantic Gyre," *J. Marine Res.* 39:443–64 (1981).
11. Villeneuve, J. P., and E. Holm. "Atmospheric Background of Chlorinated Hydrocarbons Studied in Swedish Lichens," *Chemosphere* 13:1133–38 (1984).
12. Rapaport, R. A., N. R. Urban, P. D. Capel, B. B. Looney, and S. J. Eisenreich. "New DDT Inputs to North America Atmospheric Deposition," *Chemosphere* 14(9):1167–74 (1985).
13. Tanabe, S., M. Kawano, and R. Tatsukawa. "Chlorinated Hydrocarbons in the Antarctic, Western Pacific Ocean, and Eastern Indian Ocean." *Trans. Tokyo Univ. Fish.* 5(5):97–110 (1982).
14. Baker, D. B. "Sediment, Nutrient, and Pesticide Transport in Selected Great Lakes Tributaries," Great Lakes National Program Office, EPA-905/4-88-001 (1988).
15. Duinker, J. C., and M. T. J. Hillebrand. "Behavior of PCB, Pentachlorobenzene, Hexachlorbenzene, α-HCH, β-HCH, γ-HCH, Dieldrin, Endrin, and p,p'-DDT in the Rhine-Meuse Estuary and the Adjacent Coastal Area," *Neth. J. Sea Res.* 3(2):256–81 (1979).
16. Kurtz, D. A. "Residues of Polychlorinated Biphenyls, DDT, and DDT Metabolites in Pennsylvania Streams, Community Water Supplies, and Reservoirs: 1974–1976," *Pest. Monit. J.* 11:190–98 (1978).
17. Kurtz, D. A., and R. R. Parizek. "Complexity of Contaminant Dispersal in a Karst Geological System," in *Evaluation of Pesticides in Groundwater*, W. Y. Garner, R. C. Honeycutt, and H. N. Nigg, Eds., Symposium Series #315 (Washington, DC: American Chemical Society, 1986), Chapter 13.
18. Pacenka, S., K. S. Porter, R. L. Jones, Y. B. Zecharias, and H. B. F. Hughes. "Changing Aldicarb Residue Levels in Soil and Ground Water, Eastern Long Island, NY," *J. Contam. Hydrology* 2:73–91 (1987).
19. Atlas, E. L., and C. S. Giam. "Sea-Air Exchange of High Molecular Weight Synthetic Organic Compounds," in *The Role of Air-Sea Exchange in Geochemical Cycling,* P. Buat-Menard, Ed. (Dordrecht, Netherlands: Reidel, 1986), 295–330.
20. Arthur, R. D., J. D. Cain, and B. F. Barrentine. "Atmospheric Levels of Pesticides in the Mississippi Delta," *Bull. Environ, Contam. Toxicol.* 15(2):129–34 (1976).
21. Spencer, W. F. "Movement of DDT and Its Derivatives into the Atmosphere," in *Residue Reviews. Residues of Pesticides and Other Contaminants in the Total Environment,* F. A. Gunther, Ed. (New York, NY: Springer-Verlag, 1975), pp. 91–117.
22. Murphy, T. J., L. J. Formanski, B. Brownawell, and J. A. Meyer. "Polychlorinated Biphenyl Emissions to the Atmosphere in the Great Lakes Region USA, Canada Municipal Landfills, and Incinerators," *Environ. Sci. Technol.* 19(10):942–46 (1985).
23. Larsson, P. "Change in Polychlorinated Biphenyls Clophen A-50 Composition When Transported from Sediment to Air in Aquatic Model Systems," *Environ. Pollut. Ser. B Chem. Phys.* 9(2):81–94 (1985).
24. Jury, W. A., W. F. Spencer, and W. J. Farmer. "Behavior Assessment Model for Trace Organics in Soil. 1. Model Description," *J. Environ. Qual.* 12(4):558–64 (1983).
25. Pankow, J. F., L. M. Isabelle, and W. E. Asher. "Trace Organic Compounds in Rain. 1. Sampler Design and Analysis by Adsorption, Terminal Desorption," *Environ. Sci. Technol.* 18(5):310–18 (1984).

26. Murphy, T. J., and A. W. Schinsky. "New Atmospheric Inputs of Polychlorinated Biphenyls to the Ice Cover on Lake Huron, USA, Canada," *J. Great Lakes Res.* 9(1):92–96 (1983).

27. Atkins, D. H. F., and A. E. J. Eagleton. "Studies of Atmospheric Washout and Deposition of γ-BHC, Dieldrin and *p,p'*-DDT Using Radiolabelled Pesticides," in *International Atomic Energy Agency Proceedings Series. Nuclear Techniques in Environmental Pollution,* C. N. Welsh, Ed. (New York, NY: Unipub, Inc., 1971), pp. 521–33.

28. McNeeley, R., and W. D. Gummer. "A Reconnaissance Survey of the Environmental Chemistry in East-Central Ellesmere Island, Northwest-Territories, Canada," *Arctic* 37(3):210–23 (1984).

29. Peel, D. A. "Organochlorine Residues in Antarctic Snow," *Nature* 254:324–25 (1975).

30. Offstad, E. B., G. Lunde, and H. Drangsholt. "Chlorinated Organic Compounds in the Fatty Surface Film on Water," *Int. J. Environ. Anal. Chem.* 6(2):119–32 (1979).

31. Eisenreich, S. J., G. J. Hollod, and T. C. Johnson. "Accumulation of Polychlorinated Biphenyls in Surficial Lake Superior Sediments from Atmospheric Deposition," *Environ. Sci. Technol.* 13(5):569–73 (1979).

32. Harvey, G. R., and W. G. Steinhauer. "Biogeochemistry of PCB and DDT in the North Atlantic Ocean," in *Environmental Biogeochemistry,* J. O. Nriagu, Ed., Vol. I (Ann Arbor, MI: Ann Arbor Science, 1976), pp. 203–21.

33. Harvey, G. R., and W. G. Steinhauer. "Transport Pathways of PCB in the Atlantic Ocean," *J. Marine Res.* 34:561–75 (1976).

34. Knap, A. H., and K. Binkley. "The Flux of Synthetic Organic Chemicals to the Deep Sargasso Sea," *Nature* 319:572–74 (1986).

35. Atlas, E. L., K. Sullivan, and C. S. Giam. "Widespread Occurrence of Polyhalogenated Anisoles and Related Compounds in the Marine Atmosphere," *Atmos. Environ.* 20:1217–20 (1986).

36. Tanabe, S, and R. Tatsukawa. "Vertical Transport and Residence Time of Chlorinated Hydrocarbons in the Open Ocean Water Column," *J. Oceanogr. Soc. Japan* 39(2):53–62 (1983).

37. Tanabe, S, and R. Tatsukawa. "Chlorinated Hydrocarbons in the North Pacific and Indian Oceans," *J. Oceanogr. Soc. Japan* 36(4):217–26 (1980).

38. Tanabe, S., H. Hidaka, and R. Tatsukawa. "PCB's and Chlorinated Hydrocarbon Pesticides in Antarctic Atmosphere and Hydrosphere," *Chemosphere* 12:277–88 (1983).

39. Kawano, M., S. Tanabe, T. Inoue, and R. Tatsukawa. "Chlordane Compounds Found in the Marine Atmosphere from the Southern Hemisphere," *Trans. Tokyo Univ. of Fisheries* 6:59–66 (1985).

40. Tatsukawa, R., T. Wakimoto, and T. Ogawa. "BHC Residues in the Environment," in *Environmental Toxicology of Pesticides,* F. Matsumura, C. M. Boush, and T. Misato, Eds. (New York, NY: Academic Press, 1972), pp. 229–38.

41. Baumann, K., J. Angerer, R. Heinrich, and G. Lehnert. "Occupational Exposure of Hexachlorocyclohexane. 1. Body Burden of HCH Isomers," *Int. Arch. Occup. Environ. Health* 47:119–27 (1980).

42. Pacyna, J. M., and M. Oehme. "Long-Range Transport of Some Organic Compounds to the Norwegian Arctic," *Atmos. Environ.* 22:243–57 (1988).

43. Goulden, P. D., and D. H. J. Anthony. "Design of a Large Sample Extractor for

the Determination of Organics in Water," Analytical Methods Division, National Water Research Institute, Canada Centre for Inland Waters, Burlington, Ont., unpublished document (1987).

44. Afghan, B. K., J. Carron, P. D. Goulden, J. Lawrence, D. Leger, F. Onuska, J. Sherry, and R. Wilkinson. "Recent Advances in Ultra Trace Analysis of Dioxins and Related Halogenated Hydrocarbons," *Can. J. Chem.* 65:1086–97 (1987).

45. Stevens, R. J. J., and M. A. Neilson. "Inter- and Intralake Distribution of Trace Organic Contaminants in Surface Waters of the Great Lakes," *J. Great Lakes Res.* 15(3):377 (1989).

46. Kurtz, D. A., and E. L. Atlas. "Distribution of Predominant Organochlorine Pesticides throughout the Pacific Ocean in Air and Water samples: 1987," *Environ. Sci. Technol.* (in preparation).

47. Mills, P. A., B. A. Bong, LaV. R. Kamps, and J. A. Burke. "Elution Solvent System for Florisil Column Cleanup in Organochlorine Pesticide Residue Analyses," *J. Assoc. Off. Anal. Chem.* 55:39–43 (1972).

48. Atlas, E. L., and C. S. Giam. "Sea-Air Exchange of High Molecular Weight Synthetic Organic Compounds — Results from the SEAREX Program," in *Chemical Oceanography,* J. P. Riley and R. A. Duce, Eds. Academic Press, San Diego, CA.

49. Hinckley, D., and T. Bidleman. Personal communication (February 1989).

50. Chapter 12.

51. Bidleman, T. F., and R. Leonard. "Aerial Transport of Pesticides over the Northern Indian Ocean and Adjacent Seas," *Atmos. Environ.* 16:1099–1107 (1982).

52. Uematsu, M., R. A. Duce, and J. M. Prospero. "Deposition of Atmospheric Mineral Particles in the North Pacific Ocean," *J. Atmos. Chem.* 3:123–38 (1985).

53. Prospero, J. M. "Mineral Aerosol Transport to the North Atlantic and the North Pacific: The Impact of African and Asian Sources," in *The Large Scale Atmospheric Transport of Natural and Contaminated Substances,* A. Knap, Ed., NATO ASI Series (in press).

54. Arimoto, R., R. A. Duce, B. J. Ray, and C. K. Unni. "Atmospheric Trace Elements at Enewetak Atoll," *J. Geophys Res.* 90:2391–2408 (1985).

55. Bidleman, T. F. "Atmospheric Processes," *Environ. Sci. Technol.* 22:361–67 (1988), and "Errata," *Environ. Sci. Technol.* 22:726–27 (1988).

56. Fendinger, N. J., and D. E. Glotfelty. "A Laboratory Method for the Experimental Determination of Air-Water Henry's Law Constants for Several Pesticides," 194th American Chemical Society Meeting, New Orleans, LA, August 1987. AGRO 155.

## CHAPTER 12

Concentration and Variation of Trace Organic Compounds in the North Pacific Atmosphere

Elliot L. Atlas and S. Schauffler

INTRODUCTION

Studies over the last several decades have revealed the presence of organic pollutants in a wide variety of environmental systems, even from the most remote areas of the earth. Synthetic chlorinated hydrocarbon pesticides and industrial chemicals have been identified in arctic ecosystems,[1,2] in the antarctic environment,[3-6] in deep-sea organisms,[7,8] and in a variety of other areas. Evidence has been accumulating to support the idea that atmospheric transport and deposition are the major processes for distributing many organic pollutants over the earth.[9-25] However, for most organic compounds and over large areas of the earth's surface, data coverage is very sparse. Few measurements are available over long-enough time scales to evaluate the variability of the concentration (and subsequent deposition) of synthetic organic compounds in remote ocean areas.

As part of the SEAREX (Sea-Air Exchange) program, we have been investigating the transport and deposition of several classes of organic compounds to the Pacific Ocean. The major emphasis has been on high molecular weight organochlorine compounds such as pesticides and PCBs. These types of compounds have been found widely dispersed in the global troposphere,[1-25] and generally higher concentrations are observed in the Northern Hemisphere than the Southern Hemisphere.[12,22] It is expected that a major source of anthropogenic compounds is associated with industrial and agricultural activity in the midlatitudes of the Northern Hemisphere. Thus, midlatitude westerlies may transport significant amounts of synthetic material to the North Pacific Ocean. Such transport has been well documented for aluminosilicate dust,[26,27] various trace elements,[28,29] and epicuticular plant waxes.[30]

During April-July 1986, shipboard experiments were conducted in the mid-Pacific to evaluate tropospheric transport associated with the westerlies. Air samples collected during the cruise were analyzed for high molecular weight organochlorines and other classes of organic compounds, including polycyclic aromatic hydrocarbons (PAH), $> C_3$ alkyl nitrates, $> C_{15}$ n-alkanes, and several low molecular weight organohalogen compounds. Other chemical measurements obtained during the experiment included atmospheric radon, which is an indicator of air masses with a recent continental origin. In addition, subsequent meteorological analyses were performed to calculate 10-day back trajectories of air masses sampled during the cruise.[31] These trajectory analyses have proved to be powerful in interpreting changes in the composition and concentration of chemicals in the atmosphere.[32] This chapter examines the variability of different compounds and discusses concentration changes in the marine atmosphere that are related to long-range transport of organic pollutants from continents. It presents a broad overview of different classes of organic compounds; a more detailed evaluation of the data is in preparation.

EXPERIMENTAL

Cruise Track

The research cruise track is shown in Figure 1. The cruise contained two segments, originating and ending in Hawaii. The first segment (April 28-June 3, 1986) departed Hawaii and maintained a generally northward heading to Kodiak, Alaska; the track then followed a southwesterly heading into the central Pacific between 35–45° N. The second segment (June 9-July 14, 1986) returned to the central Pacific and remained in roughly a 5° square area centered around 175° W and 38° N.

Sampling

Air samples were collected from a specially constructed air sampling tower on the bow of the R/V *Moana Wave*. The sample collectors were mounted on

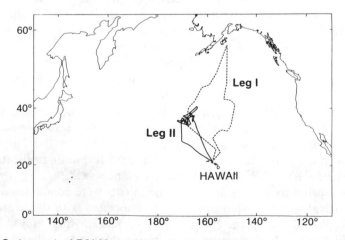

Figure 1.   Cruise track of R/V *Moana Wave*, April-July 1986.

a platform approximately 7 m above the bow, and pumps and flow meters were located in a storage van on the deck downwind of the tower. To avoid any contamination from the ship itself, air samples were collected only when the wind was blowing over the bow of ship. Extreme care was taken to ensure that samples were representative of only the ambient marine atmosphere, without artifacts from the ship platform or from ship traffic. Radar and visual observations were used to detect nearby ship traffic. In addition, condensation nuclei (CN) were continuously monitored, and the sampling system was programmed to shut down if CN were greater than ~750/cc. Radon was collected several times daily, and samples were analyzed by counting Rn daughter products.[33]

For high molecular weight organic compounds, high-volume samples were collected by drawing air through a glass-fiber filter to collect particles, then through a solid adsorbent (Florisil or polyurethane foam) to collect gas-phase organic compounds. Two gas-phase adsorbent traps were used in series to monitor breakthrough of compounds in the sample train. A third trap was placed immediately downstream of the sampler to protect the sample adsorbents from contamination by back diffusion of compounds from the tubing and/or pump. For the compounds reported here, Florisil was used to collect alkanes, PAH, aldehydes, and most chlorinated hydrocarbons. Because of some breakthrough of HCHs on a number of Florisil traps, polyurethane foam was used for quantification of HCH compounds. In cases where both foam and Florisil quantitatively collected chlorinated hydrocarbons, the agreement between analyses was good. Prior to the cruise, all polyurethane foam plugs (15 cm × 7.5 cm) were exhaustively extracted with a series of solvents including water, acetonitrile, methylene chloride, acetone, and pentane. After drying in a vacuum desiccator, the foam plugs were sealed in precombusted glass jars. Florisil (16–30 mesh) was precombusted in foil-covered glass jars at 420°C for 16 hr and sealed. Glass-fiber filters were combusted at 450°C in individually sealed aluminum foil packets.

High-volume samples were collected at flow rates of about 25 m^3/hr. Typical sample volumes were 1000–1500 m^3. After sampling, the filters and adsorbents were returned to precombusted containers, sealed, and frozen. The samples remained frozen, except for one day in transit from the ship to the home laboratory.

Low-volume samples were collected for analyses of bromoform, perchlorethylene, and alkyl nitrates by drawing air through preextracted microcharcoal cartridges. The cartridges contain ~5 mg of charcoal. They were precleaned by extraction with methanol, acetone, and benzene. After sampling from 10–300 L of air at 200–250 mL/min, the cartridges were spiked with 3 μL internal standard and extracted three times with a total of 30 μL benzene. Samples were analyzed immediately on board ship. Analysis was by capillary gas chromatography with electron capture detection.

Chemical Analysis

In the home laboratory, high-volume samples were extracted overnight with methylene chloride in a Soxhlet extractor. Sorbent traps on the vent of the extraction apparatus were used to limit exposure of the sample to contamination by laboratory air. Prior to extraction, the samples were spiked with an internal standard containing deuterated alkanes (C_{12} and C_{20}), deuterated PAH (naphthalene, phenanthrene, chrysene, and perylene), dibromooctafluorobiphenyl, and decachlorobiphenyl. Sample extracts were concentrated in a Kuderna-Danish apparatus and the solvent exchanged to hexane. Analysis of aliphatic and polycyclic aromatic hydrocarbons was performed on the unseparated extract using capillary gas chromatography and mass selective detection. Analysis of chlorinated hydrocarbon pesticides was performed first on raw extracts; then the analysis was repeated after separation of the compound classes on deactivated Florisil.[34] Compound quantification was by the internal standard technique.

RESULTS

Air Mass Trajectories and Radon Concentrations

During the cruise, conditions favorable for long-range transport of continental material to the sampling sites occurred only sporadically. Trajectory analyses[35] showed that favorable conditions for transport were more frequent in the first segment of the cruise. Typical trajectories for different segments of the cruise are shown in Figure 2. The 10–12 day history of air masses sampled during the cruise showed trajectories crossing Asia, North America, or both continents; other trajectories were totally marine. In some cases, the trajectories indicated a complex air-mass history. Studies have shown that a combination of trajectory analysis and tracer measurements can be used to indicate source areas for specific chemicals in the marine atmosphere.[32] During this research cruise, an unambiguous signal of Asian source material was sampled on board the research vessel and on mid-Pacific island stations. This signal was due to radionuclides from the Chernobyl accident in April 1986, and these chemical observations were supported by trajectory analyses indicating transport from the Asian continent.[36]

Measurement of atmospheric radon is another unambiguous indicator of air recently over continental areas, though it cannot directly specify the continental source region. Continual outgassing of radon from soils provides a radioactive source signal in an air mass that originates over a continent. The radon signal then decays in the atmosphere with a half-life of ~4.5 days. Thus, air that has recently been over continental soils and that is rapidly transported to the marine environment will retain a relatively high radon signal. Air that has been stagnant over the oceans for several Rn half-lives will show a much

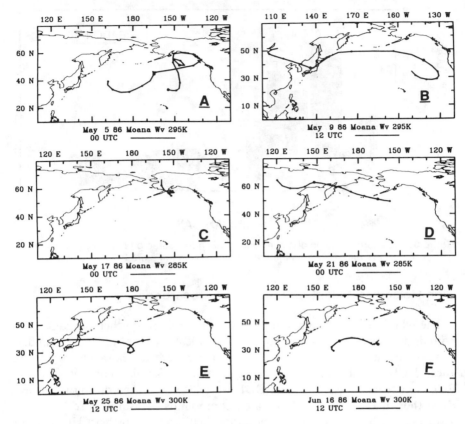

Figure 2. Calculated back trajectories indicating possible origins of different air masses sampled during the cruise. Time and day refer to Greenwich Mean Time. (Courtesy of J. Merrill, University of Rhode Island.)

diminished radon signal. Typical Rn concentrations over continental regions are ~200 pCi/m³; background concentrations of Rn in remote ocean atmospheres is < 2 pCi/m³. Figure 3 shows Rn concentrations during the cruise (averaged over time intervals of the high-volume air samples). These data are consistent with predictions of the meteorological and trajectory analyses. The first leg of the cruise has the highest Rn and most often indicated transport from continental areas. The highest concentration of Rn is observed at the highest latitude sampling locations, and it appears to be correlated with air trajectories direct from Alaska. Lower Rn concentrations typical of "background" levels are observed during the second leg of the cruise, and trajectory analysis indicated primarily oceanic trajectories during this time period. Thus, for purposes of comparison to other chemicals, the pattern of temporal variation in Rn concentration can serve as a template to indicate periods of relatively rapid long-range transport of continental air masses reaching the ocean.

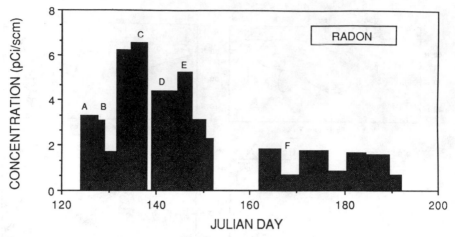

Figure 3. Temporal variation of radon concentration measured on board R/V *Moana Wave*. Letters refer to different air trajectories shown in Figure 2.

High Molecular Weight Chlorinated Hydrocarbons

The main groups of compounds discussed here are synthetic organic compounds used as pesticides and industrial chemicals. Two of the most abundant compound classes measured during the cruise were the hexachlorocyclohexanes (HCHs) and chlorinated benzenes (CBs). Smaller concentrations of PCBs, chlordanes, dieldrin, and DDE were also measured during the cruise. Even though most of the chlorinated compounds measured here have been associated with the midlatitude "pollution belt" in the Northern Hemisphere, there was relatively little variation observed in their concentrations. Furthermore, the pattern of variation did not parallel that of Rn, even though chlorinated hydrocarbons also are emitted as gases at ground level. The concentrations of the chlorinated compounds measured during the cruise fall in the range expected for remote areas of the Northern Hemisphere (Table 1).[37-41]

Of the suite of isomers in the commercial HCH pesticide mixtures, the most abundant and commonly measured isomers are α- and γ-HCH. These two HCH isomers account for over 50% of the total high molecular weight organochlorines we measured in the North Pacific atmosphere. It is interesting to examine not only the total concentration of HCHs, but also the relative composition of HCH in the atmosphere. The HCH mixture used for pesticide application contains α-HCH and γ-HCH in ratios from < 1 to 6, depending on the formulation. Thus, a change in α:γ ratio can be associated with transport from separate geographical regions where different mixtures are used (or have been used).[5,42] Low α:γ ratios in atmospheric samples indicate transport of HCH from regions where the common pesticide mixture is pure lindane (γ-HCH) or a mixture fortified in the γ-isomer. Away from source regions the α:γ HCH ratio is often higher than in most commercial mixtures. Our data show α:γ ratios of 10–15 in most regions of the central Pacific, and similar ratios

Table 1.   Concentration of Selected High Molecular Weight Chlorinated Hydrocarbons in the Marine and Remote Atmosphere of the Northern Hemisphere.

	Concentration (pg/scm)						
	HCB	ΣHCH	ΣDDT	Chlordane	Dieldrin	ΣPCB	Reference
Atlantic/Arctic							
Barbados	—	—	4	9	5	57	15
North Atlantic	150	390	6	30	20	690	10
North Atlantic	133	386	—	8[a]	8	—	37
Baltic	291	553	—	—	—	—	38
North Sea	125	130	—	—	—	—	38
Bear Island	40	390	—	—	—	17	39
Bear Island	111	237	—	1[a]	—	—	39
Canadian Arctic	—	—	—	3.9	—	—	40
Ellesmere Island	147	385	5.2	6.3	0.6	17	2
Sweden	64	489	7.2	8.4	—	165	41
North Pacific							
Enewetak Atoll	100	260	<6	13	10	110	10,13
Central Pacific	—	—	—	—	—	43	21
Bering Sea	—	—	—	—	—	41	21
West Pacific	—	550	280	—	—	300	22
	108	365	0.4	6.8	1.9	32	this work

[a]Alpha chlordane only.
[b]p,p'-DDE only.

have been reported in other areas.[2,25,43] Processes that can lead to a relative increase in α-HCH are photoisomerization of γ- to α-HCH, enhanced washout and removal of γ-HCH due to differences in Henry's law constants, and volatilization of already altered and/or degraded HCHs from treated soils and surfaces.

The concentration of HCHs was relatively uniform during each leg of the cruise (Figure 4). There were sporadic pulses of HCH during the second leg of the cruise, but the composition of HCH did not change during these episodes (see days 162 and 172, Figure 4). In general, the *composition* of HCH varied less than the *concentration* of HCH. The observations of HCH suggest a relatively diffuse and slowly varying source of this pesticide to the marine atmosphere. No transport of HCH from large source areas or differences in HCH emissions were evident during the period of the cruise. The data do suggest that the relative proportion of lindane was smaller in the summer period. The average α:γ ratio increased from 9.6 during the first leg of the cruise to 14.7 during the second half of the cruise. Similar changes in HCH composition from early to late summer periods have been observed in the Arctic.[2] The reason for this slow (seasonal?) change in HCH composition is not known at present.

The chlorobenzenes also were among the most abundant chlorinated hydrocarbons measured during the cruise. Hexachloro- and pentachlorobenzene (HCB and PeCB) are stable molecules in the atmosphere, and they have been found to be distributed fairly uniformly throughout the Northern Hemi-

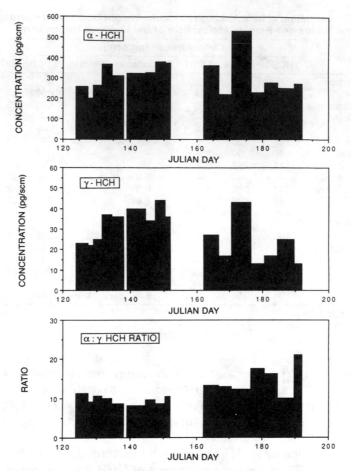

Figure 4.   Temporal variation in concentration of α-HCH and γ-HCH and their ratio in atmospheric samples from the North Pacific.

sphere.[2,12,43] In the North Pacific atmosphere, these two compounds account for ~25% of the total high molecular weight hydrocarbons. The concentrations of HCB and PeCB showed even less variation than the HCH compounds (Figure 5). This is not surprising given the long lifetimes of HCB and PeCB in the atmosphere. However, the tri- and tetrachlorinated benzenes showed greater variability. The most abundant chlorobenzene compound measured here was 1,2,4-trichlorobenzene. The mean concentration of total Cl_3-benzenes during the cruise was 5900 pg/m³, over 50 times greater than HCB. Maximum concentrations exceeded 20,000 pg/m³. Interestingly, the Cl_3-benzene variation is well correlated with the Rn distribution, suggesting that there is a strong source of Cl_3 benzenes in continental areas (Alaska?) also associated with Rn emissions. Trichlorobenzenes have been reported to average between 50–100 ng/m³ in ambient air in the Netherlands, but maxima can

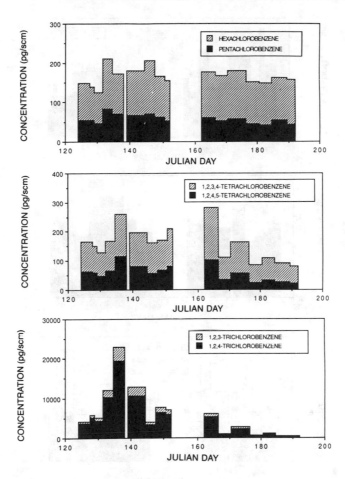

Figure 5. Temporal variation in concentration of chlorobenzenes in atmospheric samples from the North Pacific. Trichlorobenzenes show a variation similar to that observed for radon; hexachlorobenzene and pentachlorobenzene are nearly constant. Compound concentrations shown on single graph are stacked.

be up to 50 times higher.[44] Thus, urban sources are likely for this compound, but other sources may be possible. Also, the large range of concentration observed in trichlorobenzene make this compound an excellent candidate for tracing polluted air masses to remote ocean regions.

The other chlorinated compounds measured during the cruise were fairly constant throughout the experiment. Variation in PCBs, dieldrin, and chlordane compounds is illustrated in Figure 6. There is some indication that there is a shift in the relative composition of chlordane constituents from late spring to early summer. The ratio of *trans-* (γ-) to *cis-* (α-) chlordane showed a gradual decline from 0.75 in early May to 0.35 in mid-July. A general depletion of *trans*-chlordane (relative to its content in the technical chlordane mixture) also has been noted in atmospheric samples by others.[2,40] As with changes in

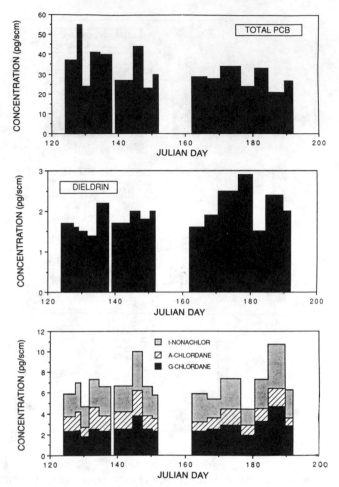

Figure 6. Temporal variation in concentration of PCB, dieldrin, and chlordanes in atmospheric samples from the North Pacific. Compound concentrations shown on single graph are stacked.

HCH composition, the variation in chlordane components may be due to several factors that cannot be specified now.

Polynuclear Aromatic Hydrocarbons (PAH)

The source of PAH in the atmosphere is primarily combustion processes. Combustion of petroleum fuels, coal, wood, grasses, etc., can produce a range of 2- to 6-ring polynuclear aromatic compounds. These compounds are relatively reactive and are expected to have a short lifetime in the atmosphere.[45,46] Thus, the PAH represent a class of compounds with an anthropogenic origin, but they have source characteristics and atmospheric chemistry different from the chlorinated organic compounds discussed above.

In samples from this cruise, we were able to analyze ≥ 3-ring PAH. More volatile PAH were not trapped in the sampling system. We found that the 3- and 4-ring PAH were primarily in the gas phase, and 5-ring and larger PAH were associated with particles. On average the total gas-phase PAH concentration (≥ 3-ring) was three times the total particulate PAH concentration. The samples from the North Pacific contained much lower concentrations of PAH than other marine or remote areas (Table 2).[30,47-49] Concentrations of PAH in these samples ranged from ≥ 1 to 35 pg/m³. In samples closer to a European source, Marty et al.[50] reported "total gaseous PAH" of 7–18 ng/m³, nearly three orders of magnitude greater than observed in the central North Pacific. Masclet et al.[49] found that naphthalene and methyl naphthalenes were major PAH in the Mediterranean Sea. Including all PAH, the range of gaseous PAH they reported was from 24 to 76 ng/m³. Excluding naphthalene and methylated compounds reduces the concentrations found by Masclet et al.[49] to 4.9–9.3 ng/m³, similar to the range found by Marty et al.[50] Total particulate PAH over the Mediterranean also was several orders of magnitude higher than in the central North Pacific atmosphere. Masclet et al.[49] report an average concentration of 1.6 ng/m³, and Sicre et al.[48] report an average of 320 pg/m³ total particulate PAH. The average concentration of particulate PAH from this cruise was 12 pg/m³. These reports suggest that PAH can be concentrated near large industrialized areas (Western Europe) but can rapidly decrease in the atmosphere away from source areas. In the Pacific, Ohta and Handa[51] report a rapid decrease in concentration of particulate PAH in the atmosphere away from the Asian/Japanese coastlines. Other reports in the Pacific atmosphere also show very low concentrations of PAH.[52]

As an example of variation in PAH, we show the concentrations of gas-phase and particulate pyrene (Figure 7). There is a reasonable correlation of gas-phase pyrene (and other PAH) with Rn concentrations and air mass trajectories, though the correlation is not nearly as good as seen in other classes of compounds. Higher concentrations are observed during the first leg of the cruise; generally low concentrations are found during the second leg. A pulse of high pyrene occurs on days 176–180, but this is not observed for all PAH and may be an artifact. Particulate PAH shows a different variation, with a maximum occurring for all particulate PAH between days 144 and 147. A small increase in Rn is noted for this period, but it is not the maximum concentration. We believe this maximum is due to sources of PAH in Japan. Trajectories from this period showed low-level transport directly over Japan. It is interesting, too, that the next highest concentration of particulate pyrene (and the highest gas-phase pyrene concentration) also were associated with trajectories over Japan. Low concentrations of PAH are associated with oceanic trajectories (also low Rn) and with trajectories coming from the Alaska/Siberia region (higher Rn). This observation suggests that the arctic regions are not a significant source of PAH for the marine atmosphere during this time of the year. Overall, it appears that there is a complex relationship between concentration of PAH and their transport to the marine atmosphere. Differ-

Table 2. Selected Gas-Phase and Particulate Polycyclic Aromatic Hydrocarbons in the Marine and Remote Atmosphere.

	Particulate Conc. (pg/scm)					Gaseous Conc. (pg/scm)	
	Barrow, AK (Ref. 47)	Medit. Sea (Ref. 48)	(Ref. 49)	Enewetak (Ref. 30)	North Pacific (This Work)	Medit. Sea (Ref. 49)	North Pacific (This Work)
Fluorene	—	—	—	<5	0.1	2360	16
Phenanthrene	17	52	—	<5	2.2	3414	14
Fluoranthene	28	82	429	<5	1.3	549	2.0
Pyrene	29	70	229	<5	1.1	411	3.0
Chrysene	5	42	76	<5	1.8	19	1.0
Benzofluoranthenes	—	35	128	<5	2.1	—	—
Benzo(a)pyrene	10 [a]	8	21	<5	0.4	6	—
Benzo(e)pyrene	—	21	186	<5	1.0	—	—
Benzo(ghi)perylene	10	11	224	<5	1.0	—	—
Indeno(c,d)pyrene	—	6	87	<5	0.7	—	—
Total PAH	99	327	1598	—	12	7192	36
Oxygenated PAH							
Fluorenone	—	—	—	—	9.0	—	33
Dibenzofuran	—	—	—	—	—	—	248

[a]Below detection.

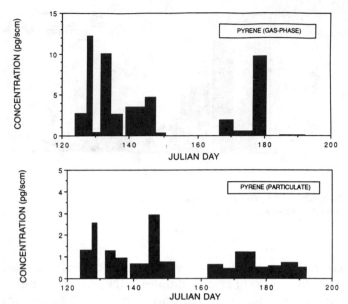

Figure 7. Temporal variation in concentration of gas-phase and particulate pyrene in atmospheric samples from the North Pacific.

ences in particle and gas transport and scavenging processes may contribute to the complexity. Still, it is encouraging to be able to interpret variation in PAH at the very low concentrations observed during the cruise. Interesting variations related to long-range transport of PAH are found to occur in the remote ocean atmosphere, even at concentrations orders of magnitude lower than that found in other ocean regions.

Organic Nitrates

Aliphatic nitrates are formed in the atmosphere as a result of hydrocarbon oxidation in the presence of NO_x.[53] It has been shown that this oxidation pathway for hydrocarbons becomes increasingly important for hydrocarbons $\geq C_3$.[53,54] Also, model studies have suggested that these $> C_3$ alkyl nitrates are formed primarily in atmospheres relatively rich in hydrocarbons and NO_x, i.e., urban atmospheres.[55] These nitrates are sufficiently long-lived in the atmosphere to be transported to remote areas. Thus, variation in the concentration of alkyl nitrates should reflect urban emissions in much the same way as PAH. One important difference is that PAH are primary emissions, but organic nitrates are secondary photochemical products in the atmosphere.

On this cruise we were able to identify C_4 and C_5 nitrates (Figure 8).[56] Figure 9 shows that there is a strong correlation between organic nitrates and radon. At this time it is not clear why there should be such a good correlation. We might expect relatively higher levels of alkyl nitrates when transport is associated with air masses from the industrialized areas of Japan and China, and

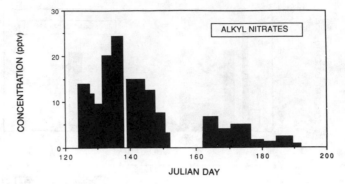

Figure 8. Temporal variation in concentration $C_4 + C_5$ alkyl nitrates in atmospheric samples from the North Pacific.

lower concentrations in air masses originating over Siberia and Alaska. However, recent measurements of organic nitrates in the arctic region show concentrations over 10 times that observed on the cruise.[57] Cooler regions of the Arctic may be a reservoir of organic nitrates. Thus, air masses originating in higher latitudes could be a strong source of both Rn and organic nitrates. This suggestion needs to be tested by additional measurements. At this time it seems that organic nitrates may be another class of compound that can serve as a sensitive tracer of polluted air masses over the ocean.

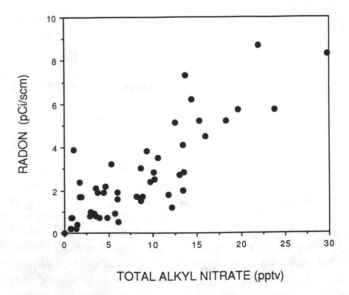

Figure 9. Correlation between atmospheric radon and total alkyl nitrates during SEAREX North Pacific experiment.

Aliphatic Hydrocarbons

Aliphatic hydrocarbons are ubiquitous chemicals in the atmosphere.[58-61] Gas-phase hydrocarbons from C_1 to C_{25} have been measured over the oceans.[59-61] The distribution of C_2-C_6 hydrocarbons over the oceans has been shown to be influenced by anthropogenic emissions.[59,60] Fewer measurements of $> C_6$ hydrocarbons have been reported. Data summarized by Duce and Gagosian[58] show that the higher molecular weight ($> C_{15}$) gas-phase hydrocarbons are more abundant in the North Atlantic than in the Pacific atmosphere. This concentration difference may be related to industrial hydrocarbon emissions from Europe and North America. Also, detailed studies of the concentration and composition of *particulate* alkanes ($> C_{25}$) have shown that these compounds can be used as molecular markers to trace regional sources of continental emissions to the oceans.[32]

During the cruise we measured concentrations of C_{15}-C_{25} alkanes. The average concentration of C_{15}-C_{25} alkanes was 3.2 ng/m³ with a range from 2.2 to 7.1 ng/m³. These concentrations are somewhat higher than similar measurements in other Pacific areas[12] but are lower than 80 ng/m³ reportedly representative of the North Atlantic.[58] The temporal variation of alkanes is not similar to any other class of compound measured during the cruise, and it is difficult to interpret the variation at this point (Figure 10).

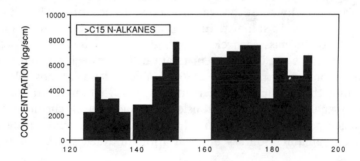

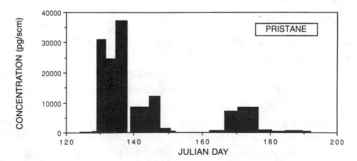

Figure 10. Temporal variation in concentration of > C15 n-alkanes and pristane in atmospheric samples from the North Pacific.

Table 3. Concentrations of Tetrachloroethylene and Bromoform in the Marine Troposphere of the Northern Hemisphere.

Location	Concentration (pptv)		Reference
	Tetrachloroethylene	Bromoform	
North Atlantic	15	0.6	64
North Atlantic	—	0.85	65
North Atlantic	29	—	66
North Pacific	—	3.1	67
North Pacific	12	0.94	68
North Pacific	15	0.89	this work

A different variation is observed, though, for one individual hydrocarbon compound—pristane (Figure 10). In several samples this one hydrocarbon compound exceeded the total of the other alkanes. A maximum concentration of 38 ng/m³ was measured. Pristane is a branched hydrocarbon strictly of biogenic origin; it is a significant hydrocarbon component in the lipid of several types of marine copepod. It is interesting, then, that the temporal variation of this compound seems to follow the Rn "template," which indicates transport from continental areas. Obviously continental sources do not control the distribution of this marine lipid material, and other explanations are required. In this case we suggest that high concentrations of pristane (and other biogenic compounds) are associated with emissions from areas of high biological productivity. Such areas are known to occur in high latitudes in the late spring and early summer. Either the sampling platform was directly in these areas, or air masses traveled over high productivity areas before reaching the ship. In any case, the apparent relationship between Rn and marine biogenic emissions underscores the necessity of considering all possible sources of chemicals in the marine atmosphere and suggests that interpretations based on trajectory analysis be carefully considered in light of all chemical data available. When possible, a variety of chemical and meteorological data is most useful in evaluating sources and variation of trace species in the marine atmosphere.

Light Halocarbons

The final group of compounds we will discuss consists of two very different chemicals with different sources in the marine atmosphere. The compounds are tetrachloroethylene and bromoform. Tetrachloroethylene is a widely distributed solvent with sources primarily in industrialized and urbanized areas.[44] Bromoform has a marine source and is associated with production from several types of algae.[62] Reports suggest that bromoform production by algae in the arctic region is a significant source of organic bromine in the atmosphere.[63] Comparison of the measurements here with other reports[64-68] of these compounds in the remote atmosphere is given in Table 3.

The temporal variation in bromoform (Figure 11) is similar to that of organic nitrates and pristane. Pulses of bromoform up to 2 pptv (~20 ng/m³)

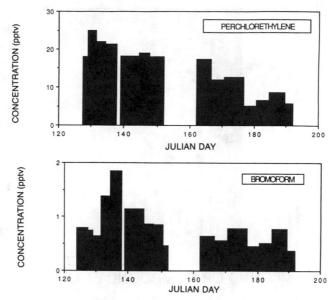

Figure 11.   Temporal variation in concentration of tetrachloroethylene (perchlorethylene) and bromoform in atmospheric samples from the North Pacific.

appear over a background concentration of ~0.6 pptv. We interpret these pulses as indicators of either arctic air masses or biological productivity near the sampling platform.

Tetrachloroethylene shows yet another pattern of temporal change (Figure 11). A gradual decrease in concentration is observed during the entire cruise. Concentrations decreased from ~25 pptv (~170 ng/m^3) in the early part of May to ~8 pptv by mid-July. We had anticipated that tetrachloroethylene might be the most sensitive indicator of transport of polluted air masses to the mid-Pacific, but this expectation proved incorrect. In our data it is difficult to discern distinct periods of relatively high concentrations of tetrachloroethylene associated with transport episodes. Other studies have suggested a seasonal change in tropospheric levels of tetrachloroethylene.[69] Higher concentrations in the winter compared to the summer may be due to a seasonally varying source or increased oxidation of tetrachloroethylene in the summer atmosphere. Our measurements suggest that the variation in tetrachloroethylene is primarily related to long-term seasonal effects rather than short-term transport phenomena.

SUMMARY

The concentration of different organic compounds measured during this study range over five orders of magnitude. Figure 12 shows the average concentration of different groups of compounds (including some not discussed in

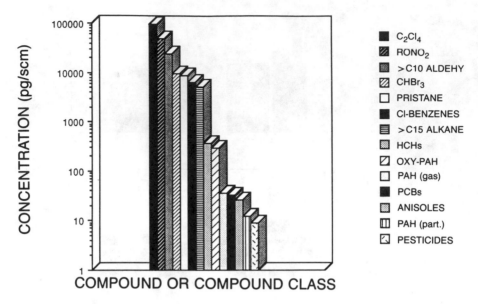

Figure 12.   Average concentration of trace organic gases in the North Pacific atmosphere.

this chapter). The most abundant compounds were tetrachloroethylene and alkyl nitrates. These compounds are both exclusively "pollutant" in origin, but they have temporal variations that are distinctly different. The differences may be attributed to different regional source strengths for each compound and different reactivity in the atmosphere. Our data suggest that organic nitrates are more reactive than tetrachloroethylene. This suggestion is consistent with recent data on the kinetics of oxidation and photolysis of these compounds.[70,71]

The concentration of other classes of pollutant compounds in the North Pacific showed different relationships compared to marine or remote regions elsewhere. PAH compounds were one to two orders of magnitude lower than in other marine locations. However, major chlorinated compounds, e.g., HCB and HCH, have similar concentrations in most marine sites over the Northern Hemisphere. Differences in the behavior of PAH and chlorinated hydrocarbons also should be related both to regional source strength and chemical stability in the atmosphere. The data shown here are consistent with chlorinated hydrocarbons having a greater stability in the atmosphere than PAH.

We observed several types of temporal variation in the data. The concentration of most high molecular weight chlorinated organic compounds was relatively constant over the three-month period. This type of behavior suggests a diffuse source of pesticides to the marine atmosphere and fairly long residence times for these compounds in the atmosphere. There was some suggestion of a seasonal trend in the composition of HCH and chlordanes. Aliphatic hydrocarbons displayed a unique temporal variability apparently unrelated to transport pathways. Changes in the sources and emission rates of these compounds are likely causes of the observed changes, but this suggestion cannot be evaluated without additional information. PAH showed a complex relationship between gas and particle transport, but there is evidence that periodic pulses of "polluted" air masses from the Asian coastline can have an impact on the background concentrations of PAH in the North Pacific atmosphere. Similarly, alkyl nitrate species and trichlorobenzenes appear to be well correlated with continental air mass transport over the Pacific. These compounds may serve as effective tracers of anthropogenic influence in remote areas. However, longer term measurements at suitable marine and remote sites are necessary to test these relationships. Finally, we observed variation in chemicals related to marine biological productivity. This variation was noted for pristane and bromoform.

ACKNOWLEDGMENTS

This work was supported by grants OCE-84-15726 and ATM-8717002 of the National Science Foundation. We thank the officers and the crew of the R/V *Moana Wave* for their help during the cruise. We also gratefully acknowledge coworkers in the SEAREX program for their valuable assistance and discussions. In particular, we thank J. Merrill of the University of Rhode Island for trajectory calculations, and members of the Center for Atmospheric Chemistry Studies for making data available.

REFERENCES

1. Hargrave, B. T., W. P. Vas, P. E. Erickson, and B. R. Fowler. "Supply of Atmospheric Organochlorines to Food Webs in the Arctic Ocean," Sixth Int. Symp. of the Comm. for Atm. Chem. and Global Poll. on Global Atmospheric Chemistry, Peterborough, Ont., August 23–29, 1987.
2. Patton, G. W., D. A. Hinckley, M. D. Walla, T. F. Bidleman, and B. T. Hargrave. "Airborne Organochlorines in the Canadian High Arctic," *Tellus* 41B(3):243–55 (1989).
3. Giam, C. S., R. L. Richardson, M. K. Wong, and W. M. Sackett. "Polychlorinated Biphenyls in Antarctic Biota," *Antarctic J. U.S.* 8:303–5 (1974).
4. Risebrough, R. W., and G. M. Carmignani. "Chlorinated Hydrocarbons in Antarctic Birds," in *Proceedings of the Colloquium, Conservation Problems in Antarctica*, B. C. Parker, Ed. (Lawrence, KS: Allen Press, 1972), pp. 63–78.
5. Tanabe, S., and R. Tatsukawa. "Chlorinated Hydrocarbons in the Southern Ocean," *Memoirs of National Institute of Polar Research Special Issue No. 27, Proceedings of BIOMASS Colloquium in 1982*, (1983), pp. 64–76.
6. Bacci, E., D. Calamari, C. Gaggi, R. Fanelli, S. Focarki, and M. Morosini. "Chlorinated Hydrocarbons in Lichen and Moss Samples from the Antarctic Peninsula," *Chemosphere* 15(6):747–54 (1986).
7. Barber, R. T., and S. M. Warlen. "Organochlorine Insecticide Residues in Deep

Sea Fish from 2500 m in the Atlantic Ocean," *Environ. Sci. Technol.* 13:1146–48 (1979).

8. Ballschmiter, K., H. Buchert, S. Bihler, and M. Zell. "Baseline Studies of Global Pollution: IV. The Pattern of Pollution by Organochlorine Compounds in the North Atlantic as Accumulated by Fish," *Fres. Z. Anal. Chem.* 306:323–39 (1981).

9. Atlas, E. "Long-Range Transport of Organic Compounds," in *Long-Range Atmospheric Transport of Natural and Contaminant Substances,* A. H. Knap, Ed., NATO ASI Series (Dordrecht, The Netherlands: Kluwer Academic Publishers, 1990), pp. 105–35.

10. Atlas, E. L., and C. S. Giam. "Global Transport of Organic Pollutants: Ambient Concentrations in the Remote Marine Atmosphere," *Science* 211, 163–65 (1981).

11. Atlas, E. L., and C. S. Giam. "Sea-Air Exchange of High Molecular Weight Synthetic Organic Compounds," in *The Role of Air-Sea Exchange in Geochemical Cycling,* P. Buat-Menard, Ed. (Dordrecht, Netherlands: Reidel, 1986), pp. 295–330.

12. Atlas, E. L., and C. S. Giam. "Sea-Air Exchange of High Molecular Weight Synthetic Organic Compounds — Results from the SEAREX Program," in *Chemical Oceanography,* J. P. Riley, R. Chester, and R. Duce, Eds., Vol. 10 (San Diego, CA: Academic Press, 1989), pp. 340–78.

13. Atlas, E. L., T. Bidleman, and C. S. Giam. "Atmospheric Transport of PCB to the Oceans," in *PCB's and the Environment,* J. S. Waid, Ed., Vol. 1 (Boca Raton, FL: CRC Press, 1986), pp. 79–100.

14. Knap, A. H., K. Binkley, and R. Artz. "The Occurrence and Distribution of Trace Organic Compounds in Bermuda Precipitation," *Atmos. Environ.* 22:1411–23 (1987).

15. Bidleman, T. F., and R. Leonard. "Aerial Transport of Pesticides over the Northern Indian Ocean and Adjacent Seas," *Atmos. Environ.* 16:1099–1107 (1982).

16. Harvey, G. R., and W. G. Steinhauer. "Atmospheric Transport of Polychlorobiphenyls to the North Atlantic," *Atmos. Environ.* 8:777–82 (1974).

17. Eisenreich, S., Ed. *Atmospheric Pollutants in Natural Waters* (Ann Arbor, MI: Ann Arbor Science, 1981).

18. Rice, C. P., P. J. Samson, and G. R. Noguchi. "Atmospheric Transport of Toxaphene to Lake Michigan," *Environ. Sci. Technol.* 20(11):1109–16 (1986).

19. Tateya, S., S. Tanabe, and R. Tatsukawa. "PCBs in the Globe: Possible Trend of Future Levels in Open Ocean Environment," submitted to *J. Great Lakes Res.*

20. Tanabe, S., and R. Tatsukawa. "Chlorinated Hydrocarbons in the North Pacific and Indian Oceans," *J. Oceanogr. Soc. Japan* 36(1):217–26 (1980).

21. Tanabe, S., and R. Tatsukawa. "Distribution, Behavior, and Load of PCB's in the Oceans," in *PCB's and the Environment,* J. S. Waid, Ed., Vol. 1 (Boca Raton, FL: CRC Press, 1986).

22. Tanabe, S., R. Tatsukawa, M. Kawano, and H. Hidaka. "Global Distribution and Atmospheric Transport of Chlorinated Hydrocarbons: HCH (BHC) Isomers and DDT Compounds in the Western Pacific, Eastern Indian and Antarctic Oceans," *J. Oceanogr. Soc. Japan* 38:137–48 (1982).

23. Tanabe, S., H. Hidaka, and R. Tatsukawa. "PCBs and Chlorinated Hydrocarbon Pesticides in Antarctic Atmosphere and Hydrosphere," *Chemosphere* 12(2):277–88 (1983).

24. Tanabe, S., H. Tanaka, and R. Tatsukawa. "Polychlorobiphenyls, ΣDDT, and

Hexachlorocyclohexane Isomers in the Western North Pacific Ecosystem," *Arch. Environ. Contam. Toxicol.* 13:731–38 (1984).

25. Bidleman, T. F., and E. J. Christensen. "Atmospheric Removal Processes for High-Molecular Weight Organochlorines," *J. Geophys. Res.* 84:7857 (1979).

26. Uematsu, M., R. A. Duce, and J. M. Prospero. "Deposition of Atmospheric Mineral Particles in the North Pacific Ocean," *J. Atmos. Chem.* 3:123–38 (1985).

27. Prospero, J. M. "Mineral Aerosol Transport to the North Atlantic and the North Pacific: The Impact of African and Asian Sources," in *Long-Range Atmospheric Transport of Natural and Contaminant Substances,* A. H. Knap, Ed., NATO ASI Series (Dordrecht, The Netherlands: Kluwer Academic Publishers, 1990), pp. 59–86.

28. Arimoto, R., R. A. Duce, B. J. Ray, and C. K. Unni. "Atmospheric Trace Elements at Enewetak Atoll," *J. Geophys. Res.* 90:2391–2408 (1985).

29. Patterson, C. C., and D. M. Settle. "Review of Data on Eolian Fluxes of Industrial and Natural Lead to the Lands and Seas in Remote Regions on a Global Scale," *Mar. Chem.* 22:137–62.

30. Gagosian, R. B., E. T. Peltzer, and O. C. Zafiriou. "Atmospheric Transport of Continentally-Derived Lipids to the Tropical North Pacific," *Nature* 291:312–14 (1981).

31. Merrill, J. T. "Atmospheric Pathways to the Oceans," in *The Role of Air-Sea Exchange in Geochemical Cycling,* P. Buat-Menard, Ed. (Dordrecht, Netherlands: Reidel, 1986), pp. 35–63.

32. Gagosian, R. B., E. T. Peltzer, and J. T. Merrill. "Long-Range Transport of Terrestrially Derived Lipids in Aerosols from the South Pacific," *Nature* 325:800–803 (1987).

33. Pszenny, A. Manuscript in preparation.

34. Chang. L. W., E. L. Atlas, and C. S. Giam. "Chromatographic Separation and Analysis of Chlorinated Hydrocarbons and Phthalate Esters from Ambient Air Samples," *J. Intern. Environ. Anal. Chem.* 19:145 (1985).

35. Merrill, J. M. Personal communication.

36. Uematsu, M., and R. A. Duce. "Tracking the Chernobyl Plume Across the Pacific," *Maritimes* 30:1–4 (1986).

37. Atlas, E., and S. Schauffler. Unpublished data.

38. Reinhart, K. H., and D. Wodarg. "Transport of Selected Organochlorine Compounds over the Sea," presented at European Aerosol Conference, Lund, Sweden, August 30-September 2, 1988.

39. Oehme, M., and H. Stray. "Quantitative Determination of Ultra-Traces of Chlorinated Compounds in High-Volume Air Samples from the Arctic Using Polyurethane Foam as Collection Medium," *Fres. Z. Anal. Chem.* 311:665–73 (1982).

40. Hoff, R. M., and K. W. Chan. "Atmospheric Concentrations of Chlordane at Mould Bay, N.W.T., Canada," *Chemosphere* 15(4):449–52 (1986).

41. Bidleman, T. F., U. Wideqvist, B. Jansson, and R. Soderlund. "Organochlorine Pesticides and Polychlorinated Biphenyls in the Atmosphere of Southern Sweden," *Atmos. Environ.* 21:641–54 (1987).

42. Pacyna, J. M., and M. Oehme. "Long-Range Transport of Some Organic Compounds to the Norwegian Arctic," *Atmos. Environ.* 22:243–57 (1988).

43. Oehme, M., and B. Ottar. "The Long Range Transport of Polychlorinated Hydrocarbons to the Arctic," *Geophys. Res. Lett.* 11(10):1133–36 (1984).

44. Guicherit, R., and F. L. Schulting. "The Occurrence of Organic Chemicals in the

Atmosphere of the Netherlands," *The Science of the Total Environment* 43:193–95 (1985).

45. Atkinson, R. "Kinetics and Mechanisms of the OH Radical Reactions of the Hydroxyl Radical with Organic Compounds under Atmospheric Conditions," *Chem. Rev.* 85:69–201 (1985).

46. Atkinson, R., S. M. Aschmann, A. M. Winer, and J. N. Pitts, Jr. "Atmospheric Gas Phase Loss Processes for Chlorobenzene, Benzotrifluoride, and 4-Chlorobenzotrifluoride, and Generalization of Predictive Techniques for Atmospheric Lifetimes of Aromatic Compounds," *Arch. Environ. Contam. Toxicol.* 14:417–25 (1985).

47. Daisey, J. M., R. J. McCaffrey, and R. A. Gallagher. "Polycyclic Aromatic Hydrocarbons and Total Extractable Particulate Organic Matter in the Arctic Aerosol," *Atmos. Environ.* 15(9):1353–63 (1981).

48. Sicre, M. A., J. C. Marty, A. Saliot, X. Aparicio, J. Grimalt, and J. Albaiges. "Aliphatic and Aromatic Hydrocarbons in the Different Sized Aerosols over the Mediterranean Sea, Occurrence and Origin," *Atmos. Environ.* 12:2247–59 (1987).

49. Masclet, P., P. Pistikopoulos, S. Beyne, and G. Mouvier. "Long-Range Transport and Gas/Particle Distribution of Polycyclic Aromatic Hydrocarbons at a Remote Site in the Mediterranean Sea," *Atmos. Environ.* 22:639–50 (1988).

50. Marty, J. C., M. J. Tissier, A. Saliot. "Gaseous and Particulate Polycyclic Aromatic Hydrocarbons (PAH) from the Marine Atmosphere," *Atmos. Environ.* 18(10):2183–90 (1984).

51. Ohta, K., and N. Handa. "Organic Components in Size-Separated Aerosols from the Western North Pacific," *J. Oceanogr. Soc. Japan* 41:25–32 (1985).

52. Gagosian, R. B., O. C. Zafiriou, E. T. Peltzer, and J. B. Alford. "Lipids in Aerosols from the Tropical North Pacific: Temporal Variability," *J. Geophys. Res.* 87(C):11133–44 (1982).

53. Atkinson, R. A., and A. C. Lloyd. "Evaluation of Kinetic and Mechanistic Data for Modelling of Photochemical Smog," *J. Phys. Chem. Ref. Data* 13:315–444 (1984).

54. Atkinson, R. A., S. M. Aschmann, W. P. L. Carter, A. M. Winer, and J. N. Pitts, Jr. "Alkyl Nitrate Formation from the NO_x-Air Photooxidation of C_2-C_8 Alkanes," *J. Phys. Chem.* 86:4563–69 (1982).

55. Calvert, J. G., and S. Madronich. "Theoretical Study of the Initial Products of the Atmospheric Oxidation of Hydrocarbons," *J. Geophys. Res.* 92:2211–20 (1987).

56. Atlas, E. "Evidence for $\geq C_3$ Alkyl Nitrates in Rural and Remote Atmospheres," *Nature* 331:426–28 (1988).

57. Atlas, E., G. Patton, J. Bottenheim, and L. Barrie. Unpublished data.

58. Duce, R. A., and R. B. Gagosian. "The Input of Atmospheric n-C_{10} to n-C_{30} Alkanes to the Ocean," *J. Geophys. Res.* 87(C9):7192–7200 (1982).

59. Greenberg, J. P., and P. R. Zimmerman. "Nonmethane Hydrocarbons in Remote Tropical, Continental, and Marine Atmospheres," *J. Geophys. Res.* 89(D3):4767–78 (1984).

60. Bonsang, B., and G. Lambert. "Nonmethane Hydrocarbons in an Oceanic Atmosphere," *J. Atmos. Chem.* 2(3):257–71 (1985).

61. Rudolph, J., and A. Khedim. "Hydrocarbons in the Non-Urban Atmosphere: Analysis, Ambient Concentrations and Impact of the Chemistry of the Atmosphere," *Intern. J. Environ. Anal. Chem.* 20:265–82 (1985).

62. Dyrssen, D., and E. Fogelqvist. "Bromoform Concentrations of the Arctic Ocean in the Svalbard Area," *Oceanol. Acta* 4:313–17 (1981).
63. Berg, W., L. E. Heidt, W. Pollack, P. D. Sperry, R. J. Cicerone, and E. S. Gladney. "Brominated Organic Species in the Arctic Atmosphere," *Geophys. Res. Lett.* 5:429–32 (1984).
64. Class, T., and K. X. Ballschmiter. "Atmospheric Halocarbons: Global Budget Estimations for Tetrachloroethylene, 1,2-Dichloroethane, 1,1,1,2-Tetrachloroethane, Hexachloroethane, and Hexachlorobutadiene," *Fres. Z. Anal. Chem.* 327:198–204 (1987).
65. Penkett, S. A., B. M. R. Jones, M. J. Rycroft, and D. A. Simmons. "An Interhemispheric Comparison of the Concentrations of Bromine Compounds in the Atmosphere," *Nature* 318:550–53 (1985).
66. Singh, H. B., L. J. Salas, and R. E. Stiles. "Selected Manmade Halogenated Chemicals in the Air and Ocean Environment," *J. Geophys. Res.* 88:3675–83 (1983).
67. Cicerone, R. J., L. E. Heidt, and W. H. Pollock. "Measurements of Atmospheric Methyl Bromide and Bromoform," *J. Geophys. Res.* 93:3745–50 (1988).
68. Atlas, E., and K. Lloyd. Unpublished data.
69. Hov, O., S. A. Penkett, I. S. A. Isaksen, and A. Semb. "Organic Gases in the Norwegian Arctic," *Geophys. Res. Lett.* 11:425–28 (1984).
70. Atkinson, R. A., S. M. Aschmann, W. P. L. Carter, and A. M. Winer. "Kinetics of the Gas-Phase Reaction of OH-Radicals with Alkyl Nitrates at 299 ± 2 K," *Intern. J. Chem. Kinetics* 14:919–26 (1982).
71. Luke, W. T., and R. R. Dickerson. "Direct Measurements of the Photolysis Rate Coefficient of Ethyl Nitrate," *Geophys. Res. Lett.* 15:1181–84 (1988).

## CHAPTER 13

Distribution of Organochlorines in the Atmosphere of the South Atlantic and Antarctic Oceans

R. R. Weber and R. C. Montone

INTRODUCTION

Since organochlorine pesticides and polychlorinated biphenyls were banned, or at least severely restricted, by many of the heavily industrialized countries of the Northern Hemisphere in the mid-1970s, we should have expected a systematic decrease in their marine environmental levels. The open ocean would be the final sink of these persistent organic chemicals, as predicted by Woodwell et al.[1]

However, as Goldberg[2] correctly foretold, "the southward tilt" occurred. This phenomenon was confirmed by the work of Tanabe et al.[3] for the South Pacific and Indian Ocean. Goldberg's concern was that the Southern Hemispheric, less developed countries would continue to use these compounds in increasing levels for agricultural and public health. Tragically, Goldberg's concern was well warranted. India continues to use DDT in cotton agriculture and malaria control.[4] Brazil used hexachlorocyclohexane (HCH or BHC) and DDT extensively from 1970 to 1980 in agriculture and public health services. HCH was used specifically for treatment of coffee, soybean, and cotton cultures as well as in the control of Chagas' disease; DDT was used for malaria control. It was not until 1980 that they were banned for agricultural use in Brazil. Even now mirex, HCH, and DDT are still allowed for public health purposes.

The main route of these "recalcitrant chemicals" to the open ocean is atmospheric transport, as first shown by the pioneer work of Risebrough et al.[5] in Barbados and Prospero and Nees[6] with the Sahara dusts. This was confirmed by the subsequent works of Bidleman and Olney,[7] Harvey and Steinhauer,[8] and Giam and Atlas[9] in the Northwest Atlantic. Near the British Isles, the transport was confirmed by Dawson and Riley.[10]

Tanabe and Tatsukawa,[11] working in the South Pacific and Indian Oceans, and Bidleman and Leonard,[4] in the North Indian Ocean and adjacent seas, also found these compounds in measurable levels in the oceanic atmosphere.

Consequently, there is considerable interest currently in measuring the levels of these compounds in air all over the world's oceans.

To our knowledge there is no information on the concentrations of chlorinated pesticides in the atmosphere of the Southwest Atlantic and its Antarctic sector. This chapter is a report on the concentration and some aspects of the distribution of these compounds on a fixed sampling point on the Brazilian coast (23°49′ S, 45°25′ W) and the open ocean between 25 and 65° S, latitude transects of the South Atlantic and Antarctic Oceans.

EXPERIMENTAL

Air samples were collected at a stationary setting in São Paulo and over the open ocean on board ship. For stationary sampling, they were collected on a monthly schedule between November 1985 and November 1986 in São Sebastião (23°49′ S, 45°25′ W), São Paulo, Brazil. Sampling was done on the tower of a wind-driven propeller power generator, located on a small hill in the backyard of the Marine Biological Center of São Paulo University. Open ocean air samples were collected during the V Brazilian Antarctic Cruise in the summer of 1987 with R/V *Prof. W. Besnard* of São Paulo University (Figure 1).

Sampling was done by pumping air through an adsorption column of polyurethane foam over a 24–96 hr period (Figure 2). The dimensions of the foam plug were 19 mm i.d. × 140 mm. The polyurethane foam was previously purified according to the method of Bidleman and Burdick.[12] The pump was a low-volume unit, originally developed for hospital use (Fanem Dia Pump, Model CAL I). The final volumes were recorded with a gas flow meter (Liceu de Artes e Ofícios de São Paulo).

The volume of sampled air varied between 97 and 589 m^3. These volumes were a function of the available pumping system, a low-volume unit delivering 10–14 L/min, leading necessarily to very long sampling periods. In fact, the sampled volumes were a compromise between the minimum volumes necessary for DDT and PCB analysis and the maximum volumes allowed for HCH collection without breakthrough.

Once finished, the adsorption column was wrapped in aluminum foil and stored in a desiccator at room temperature until analysis. In the laboratory, the foam packaging was extracted twice in a Soxhlet apparatus with 125 mL *n*-hexane each time for 4–8 hr. Schematically, the procedure is shown in Figure 3. The cleanup of the hexane extract and the separation of the chlorinated biphenyls from the pesticides basically followed the procedure of Wells and Johnstone[13] and Osterroth and Schneider.[14] The hexane extracts were reduced to 2–3 mL on a rotary evaporator at 40°C, and then to 0.5 mL in a pear-shaped calibrated flask by gently blowing nitrogen stream.

The concentrated extracts were cleaned up and separated by two chromatographic-column steps. The cleanup column consisted of 3 g of alu-

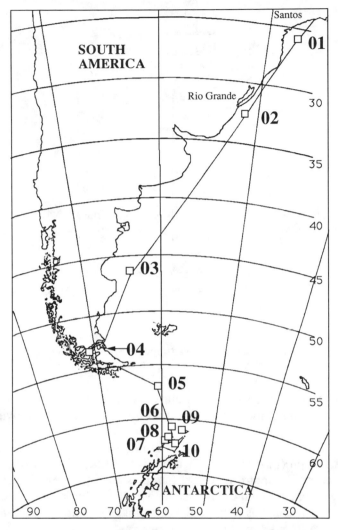

Figure 1. Survey route of R/V *Prof. W. Besnard* on the V Brazilian Antarctic Expedition. Sampling period: January 24-March 4, 1987. Air was continuously pumped between each pair of points, starting in Santos, São Paulo, Brazil.

mina (Merck 1017) packed in a long tube (4 mm i.d. × 15 cm). The packing had been previously activated at 800°C, with subsequent deactivation with 3% water. Elution was done with 12 mL of *n*-hexane, pesticide grade (Merck 4371). The extract was concentrated and then applied to the top of a similar chromatographic column containing 2 g of silica gel (Merck 7734), activated at 120°C and partially deactivated with 2% water. The first fraction, eluted with 5 mL of *n*-hexane, contained the chlorinated biphenyls and *p,p'*-DDE. The second fraction, eluted with 12 mL of 10% diethylether in hexane, contained

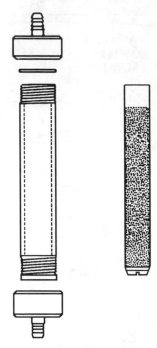

Figure 2. Cartridge for organochlorine adsorption. External cylinder is made of inox steel. Internal cylinder with the adsorbent is made of glass with a glass frit.

the other pesticides. The two fractions were each concentrated down to 0.5 mL on a rotary evaporator at 40°C and then in a calibrated flask by gentle nitrogen stream evaporation.

Analysis of the organochlorine compounds was done on a CG 35370 gas chromatograph (Instrumentos Científicos CG Ltda, SP, Brazil), fitted with a tritium electron capture detector. Samples of 4 μL were injected into two separate chromatographic columns, held at 200°C for identification and quantification. The columns were of nickel tubing, 3 mm i.d. × 1.8 m, and operated at a flow of 35 mL/min nitrogen carrier gas.

One column was filled with 1.5% OV-17 plus 1.95% QF-1, and the other with 5% OV-210, both coated on Chromosorb WHP (100–120 mesh). Standards of pesticides and chlorinated biphenyls were obtained from the U.S. Environmental Protection Agency Pesticide Repository.

Collection efficiency of organochlorines on polyurethane foam was evaluated according to the procedure of Lewis and McLeod,[15] and the results are presented in Table 1.

Special care was taken to obtain minimum blank values for solvents, glassware, and sorbents employed in the overall procedure. When necessary, double cleanup through the alumina column was performed.

Quantification of the samples was obtained by comparison of the chromatographic peak areas of the samples to those of the standards. For the chlori-

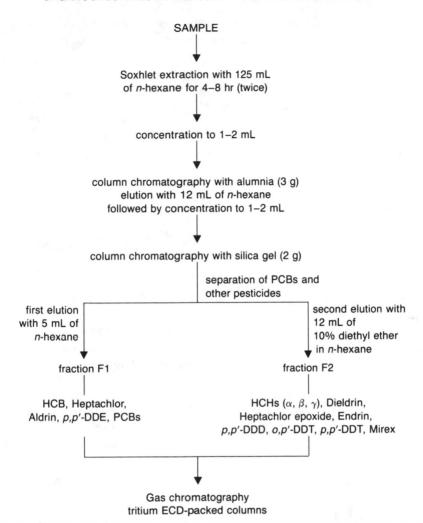

Figure 3. Scheme of the extraction, cleanup, and separation procedures for organochlorine compounds.

nated biphenyls, Arochlor 1254 and 1260, ten major peaks were used for quantification purposes. In our chromatograms, Arochlor 1254 appeared with 10 peaks and 1260 with 12 peaks. Chromatograms of blanks, standards, and samples are presented in Figures 4–8.

RESULTS AND DISCUSSION

São Sebastião Data

The organochlorines that occurred most frequently in our São Sebastião samples, as shown in Table 2, were α-HCH (nd [not detected] to 58 pg/m³), γ-

Table 1.   Collection Efficiencies for Organochlorines in Air Using PUF as Adsorbent

Compound	Quantity Added (ng)[a]	Collection Efficiency (%, n = 2)
HCB	10	98.3
α-HCH	10	97.8
γ-HCH	10	72.9
Heptachlor	10	71.2
Heptachlor epoxide	20	74.5
Aldrin	20	86.4
Dieldrin	30	77.2
Endrin	60	66.0
p,p'–DDE	30	76.0
o,p'–DDD	60	74.4
o,p'–DDT	80	65.9
p,p'–DDD	60	64.1
p,p'–DDT	100	64.1
Mirex	100	74.8
Arochlor 1254	100	121.9

[a]Air volume: 1.0 m^3.

HCH (nd to 22 pg/m^3), p,p'-DDE and p,p'-DDT (nd to 404 pg/m^3), and PCBs (nd to 1370 pg/m^3). Higher concentrations of α-HCH were always present than of γ-HCH, with the exception of the sample of 03/86 where they were of equal values. Values for these more volatile compounds in air, however, can still be questioned because our sampling system had no backup plugs and used high air flow rates. The maximum air volume recommended for our system, similar to that employed by Bidleman and Burdick,[12] for collecting these compounds efficiently was 150 m^3. As is shown in Table 2, this volume was surpassed in many instances.

The levels of HCHs and DDT analogs in the air of São Sebastião are similar

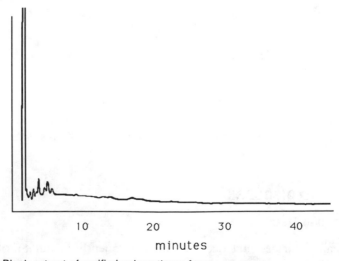

Figure 4.   Blank extract of purified polyurethane foam.

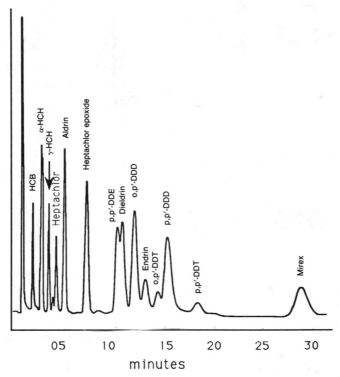

Figure 5. Chromatogram of chlorinated pesticide standards.

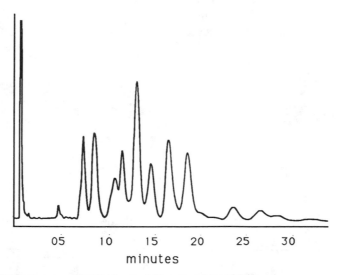

Figure 6. Chromatogram of PCB standard (2 ng Arochlor 1254).

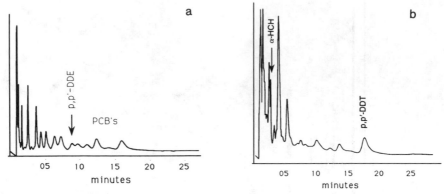

Figure 7. Chromatograms of the air sample of August 1986, São Sebastião, São Paulo, Brazil: (a) fraction F-1 and (b) fraction F-2.

to those registered by other authors in other oceanic and coastal areas of the Northern Hemisphere as reported by Atlas and Giam.[16] They are also similar to the results of Bidleman et al.[17] for the Bermuda and Barbados areas. However, the HCHs are, on the average, many times smaller than those registered for the Indian and Pacific Oceans by Tanabe et al.[13], who found very high levels, especially near India and Japan.

Although the majority of the organochlorine air concentration data found in the literature refer to Northern Hemispheric areas, some data are available for southern climes. Table 3 shows some of the data found for the Southern Hemispheric atmosphere over selected oceanic and coastal areas.

As a whole our results are similar to the mean values presented in Table 3. Our α- to γ-HCH ratios are greater than unity, which is what Atlas and

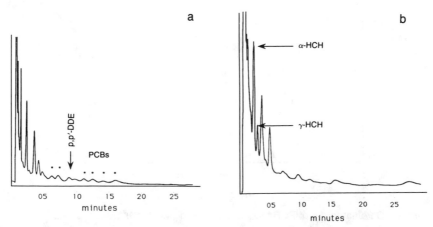

Figure 8. Chromatograms of air sample 7 (summer 1987), Antarctic (Atlantic sector): (a) fraction F-1 and (b) fraction F-2.

Table 2.   Organochlorine Pesticide and PCB Levels in the Air of São Sebastião, São Paulo, Brazil

Date	Air (m³)	α-HCH	γ-HCH	total HCH	p,p'- DDT	p,p'- DDT	Total DDT	Total PCBs
11/85	97	58	22	80	31	35	66	nd[a]
12/85	137[b]	nd	nd	nd	nd	nd	nd	132
01/86	12[c]	35	nd	35	18	50	68	1370
02/86	515	nd	nd	nd	nd	nd	nd	nd
03/86	371	20	20	40	10	nd	10	69
04/86	415	39	19	58	7	nd	7	nd
05/86	278	nd	nd	nd	11	nd	11	nd
06/86	569	nd	nd	nd	nd	146	146	38
07/86	458	35	nd	35	nd	nd	nd	nd
08/86	372	18	nd	18	12	392	404	54
10/86	152	nd	nd	nd	20	317	337	132
11/86	256	20	nd	20	nd	nd	nd	78

[a]nd: not detected < 1 pg/m³ for HCHs
                      < 3 pg/m³ for DDTs
                     < 10 pg/m³ for PCBs
[b]Sampling on board of R/V *Prof. W. Besnard.*
[c]Low volume sampled due to pumping problems.

Table 3.   Organochlorine Pesticide and PCB Concentrations in the Atmosphere over the Southern Hemispheric Oceans

Location	α-HCH	γ-HCH	p,p'-DDE	PCBs
São Sebastião, São Paulo, Southwest Atlantic (sample 4/86)[a]	39	19	7	nd[b]
Open ocean, Southwest Atlantic (sample 3)[a]	6	nd	nd	nd
Open ocean, Antarctic (Atlantic Sector) (sample 7)[a]	89	42	4	280
American Samoa South Pacific[c]	HCHs = 32		—	—
Open ocean, South Pacific[d]	45	74	34	59
Open ocean, Indian[d]	60	100	59	—
Open ocean, Antarctic (Pacific Sector)[d]	14	38	33	—

[a]This study.
[b]nd: not detected < 1 pg/m for α-HCH and γ-HCH
                      < 3 pg/m for p,p'-DDE
                     < 10 pg/m for PCBs as Arochlor 1254.
[c]Atlas and Giam.[18]
[d]Tanabe and Tatsukawa.[11]

Table 4. Organochlorine Pesticide and PCB Concentration in the Atmosphere of the Southwest Atlantic and Antarctic Oceans, Summer 1987 (January 24–March 4).

Sample	Air Vol. (m³)	α-HCH	γ-HCH	Total HCH	p,p'- DDE	p,p'- DDD	p,p'- DDT	Total DDT	PCBs
1	31	nd[a]	nd	nd	nd	nd	141	141	nd
2	67	nd	nd	nd	nd	nd	nd	nd	nd
3	40	6	nd	6	nd	nd	nd	nd	nd
4	58	51	nd	51	nd	nd	88	88	360
5	32	104	nd	104	nd	nd	nd	nd	370
6	51	87	69	156	nd	nd	86	86	150
7	83	89	42	131	4	nd	79	83	280
8	70	14	3	17	nd	nd	nd	nd	120
9	95	66	nd	66	nd	24	nd	24	140
10	51	23	nd	23	nd	nd	nd	nd	210

[a]nd: not detected < 1 pg/m³ for HCHs
                       < 3 pg/m³ for DDTs
                       < 10 pg/m³ for PCBs as Arochlor 1254.

Giam[16,18] found. On the other hand, Tanabe et al.[3] found these ratios to be less than unity.

To try to assess the sources of these compounds relative to the local land area, we made a survey of the existing wind regime from pilots' charts of 1951 to 1972 from the Hydrographical and Navigation Directory of the Brazilian Navy. This survey showed that the prevailing winds on an average yearly basis were predominantly from the east, occurring less frequently from the southwest. Therefore, the measured levels reflect a larger contribution from marine atmospheric transport, which is prevailing, than from the continental area.

South Atlantic and Antarctic Ocean area (25–65° S)

The air volumes sampled on the cruise were lower than those collected in São Sebastião, so better recoveries were expected for the HCH data. On the other hand, the volumes were relatively low for detection sensitivity for DDT analogs and PCBs. The ideal sampling strategy would have been to use a low volume sampler for the HCHs and a high volume sampler for the DDT compounds and PCBs together. Unfortunately, the high volume sampler was not available on this cruise. Despite the limitations already discussed, and approaching conservatively the few data obtained on this cruise, some conclusions on the distribution of organochlorine compounds in the atmosphere of the study area can be drawn with confidence.

We found α- and γ-HCH, DDT analogs, and PCBs in measurable quantities in the air of this region, as shown in Table 4. The levels varied between nd and 104 pg/m³ for α-HCH, and nd to 69 pg/m³ for γ-HCH. α-HCH, as in the São Sebastião area, is in greater proportion than γ-HCH. The levels of HCHs and the relative abundances of α- and γ-isomers agree well with the results of Atlas and Giam[18] for the American Samoa in the South Pacific.

Tanabe et al.,[3] however, in the only other study relevant to the Southern Hemisphere, found more γ- than α-HCH in the majority of their air samples. Only in their Northern Hemispheric samples did they observe a higher proportion of γ- over α-HCH. The occurrence of the different proportions of α- and γ-HCHs, depending on the hemisphere, deserves more investigation. As already pointed out by Tanabe et al.,[3] "It is an interesting question as to whether the predominance of γ-HCH extends all over the oceans in the Southern Hemisphere."

Technical HCH has been reported to contain 70% α-HCH, 7% β-HCH, 13% γ-HCH, 5% δ-HCH, and 5% impurities.[19] Lindane is the commercial name of the purified γ-isomer material that in recent times has been commercially made available. Indeed, some governmental units have been requiring its use. Unfortunately, little information is available on the relative use of lindane and technical HCH in South American countries that border the Southwest Atlantic. Hence, little can be determined about the sources of HCH to the South Atlantic atmosphere.

Total DDT varied between nd and 141 pg/m^3. p,p'-DDT was the more abundant isomer. Its concentration decreased slightly with increasing latitude. PCB levels varied between nd and 370 pg/m^3; there was no visible trend with the latitude. PCB peaks matched reasonably with the high molecular weight peaks of Arochlor 1254.

Quantification of the HCHs and PCBs in our samples should be regarded as tentative because we used only packed columns. Even so, we can conclude that our results do show the occurrence of chlorinated pesticide and PCB residues in the atmosphere of the study area.

The levels are of the same order of magnitude as those measured in the atmosphere of the other oceans, showing that these compounds are largely disseminated in the atmosphere as a whole. Because the interhemispheric exchange of the atmosphere is quite slow (see Chapter 6) on account of the impeding action of the Hadley cell of the tropical areas, the concentrations found in the study area reflect the use of chlorinated pesticides and PCBs in the South American countries that border the Southwest Atlantic.

Nevertheless, a much more systematic study, with proper collection techniques and capillary gas chromatography, must be made in order to provide good quality data on the levels of these compounds in the area. Such data are essential for understanding and modeling the transport of these compounds to the Southwest Atlantic and its Antarctic counterpart.

ACKNOWLEDGMENTS

The assistance of the following individuals and institutions is appreciated: the officers and crew of R/V *Prof. W. Besnard* and M.Sc. M. C. Bicego for collecting the air samples on the 1987 cruise, and the staff of the Marine Biological Center of São Paulo University for the facilities provided. We would also thank Dr. T. F. Bidleman of the University of South Carolina for all the advice and help during this work, and Dr. D. A. Kurtz of Pennsylvania State University for the invitation to present work at the ACS June 1988 Toronto meeting and his valuable comments on the manuscript. This work was supported by FAPESP, PROANTAR, and the ACS Petroleum Research Fund.

REFERENCES

1. Woodwell, G. M., P. P. Craig, and H. A. Johnson. "DDT in the Biosphere: Where Does It Go?" *Science* 174:1101–7 (1971).
2. Goldberg, E.D. "Biostimulants and Chlorinated Hydrocarbons: The Marine Pollutants of the 1980's," Sympos. Util. Coast. Ecosyst. Rio Grande-RS. Duke Univ./ Fund. Univ. Rio Grande, Brazil, 1:159–68 (1982).
3. Tanabe, S., R. Tatsukawa, M. Kawano, and H. Hidaka. "Global Distribution and Atmospheric Transport of Chlorinated Hydrocarbons: HCH (BHC) Isomers and DDT Compounds in the Western Pacific, Eastern Indian, And Antarctic Oceans," *J. Oceanogr. Soc. Japan* 38:137–48 (1982).
4. Bidleman, T. F., and R. Leonard. "Aerial Transport of Pesticides over the Northern Indian Ocean and Adjacent Seas," *Atmos. Environ.* 16(5):1099–1107 (1982).
5. Risebrough, R., R. J. Huggett, J. J. Griffin, and E. D. Goldberg. "Pesticides: Transatlantic Movements in the Northeast Trades," *Science* 159:1233–36 (1968).
6. Prospero, J. M., and R. T. Nees. "Some Additional Measurements of Pesticides in the Lower Atmosphere of the Northern Equatorial Atlantic Ocean," *Atmos. Environ.* 6:363–64 (1972).
7. Bidleman, T. F., and C. E. Olney. "Chlorinated Hydrocarbons in the Sargasso Sea Atmosphere and Surface Water," *Science* 183:516–18 (1974).
8. Harvey, G. R., and W. G. Steinhauer. "Atmospheric Transport of Polychlorobiphenyls to the North Atlantic," *Atmos. Environ.* 8:777–82 (1974).
9. Giam, C. S., and E. Atlas. "Phthalate Esters, PCBs and DDTs Residues in the Gulf of Mexico Atmosphere," *Atmos. Environ.* 14:65–69 (1980).
10. Dawson, R., and J. P. Riley. "Chlorine Containing Pesticides and Polychlorinated Biphenyls in British Coastal Waters," *Estuar. Coast. Marine Sci.* 5(1):55–71 (1977).
11. Tanabe, S., and R. Tatsukawa. "Chlorinated Hydrocarbons in the North Pacific and Indian Oceans," *J. Oceanogr. Soc. Japan* 36:217–26 (1980).
12. Bidleman, T. F., and N. F. Burdick. "Frontal Movement of Hexachlorobenzene and Polychlorinated Biphenyls Vapors through Polyurethane Foam," *Anal. Chem.* 53:1926–9 (1981).
13. Wells, D. E., and S. J. Johnstone. "Methods for Separation of Organochlorine Residues Before Gas-Liquid Chromatography Analysis," *J. Chrom.* 140:17–28 (1977).
14. Osterroth, C., and R. Schneider. "Residues of Chlorinated Hydrocarbons in Cod Livers from the Kiel Bight in Relation to Some Biological Parameters," *Meeresforsc.* 25:105–14 (1976).
15. Lewis, R. G., and K. E. McLeod. "Portable Sampler for Pesticides and Semivolatile Industrial Organic Chemicals in Air," *Anal. Chem.* 54:310–15 (1982).

16. Atlas, E. L., and C. S. Giam. "Sea-Air Exchange of High Molecular Weight Synthetic Organic Compounds," in *The Role of Air Sea Exchange in Geochemical Cycling,* P. Buat-Menard, Ed. (Dordrecht, Netherlands: Reidel, 1986), pp. 295–330.

17. Bidleman, T. F, E. J. Christensen, W. N. Billings, and R. Leonard. "Atmospheric Transport of Organochlorines in the North Atlantic Gyre," *J. Marine Res.* 39(3):443–64 (1981).

18. Atlas, E. L., and C. S. Giam. "Ambient Concentrations and Precipitation Scavenging of Atmospheric Organic Pollutants," *Water Air Soil Poll.* (in press).

19. Baumann, K., J. Augerer, R. Heinvich, and G. Lehnert. "Occupational Exposure of Hexachlorocyclohexane. 1. Body Breeder of HCH Isomers," *Int. Arch. Occup. Environ. Health* 47:119–27 (1980).

## CHAPTER 14

Regional Atmospheric Transport and Deposition of Pesticides in Maryland

D. E. Glotfelty, G. H. Williams, H. P. Freeman, and M. M. Leech

INTRODUCTION

The Chesapeake Bay is a vital natural resource that is generally conceded to be under stress from many different sources of contamination. In December 1987, after nearly a decade of federal and state research to define problems in the Bay, the Chesapeake Bay Agreement was signed, in which the governors of Maryland, Virginia, and Pennsylvania; the mayor of the District of Columbia; U.S. EPA; and the Chesapeake Bay Commission, representing the legislatures of the states involved, pledged federal, state, and local cooperation to preserve, protect, and restore the Bay. The measure adopted to judge the success of the efforts was the restoration of the living resources of the Bay to conditions as they are perceived to have existed in the 1950s.

One of the major components of the Bay ecosystem that needs to be restored is the submerged aquatic vegetation. Rooted aquatic macrophytes have undergone a dramatic decline in recent decades.[1] Questions have been raised about the potential role of agricultural chemicals in the overall decline of the Bay in general, and the decline of submerged aquatic vegetation in particular.

Two routes can be identified by which agricultural chemicals may move to the Bay:

1. by transport in water through runoff, erosion, or leaching, or by transport in air followed by rainout, fallout
2. by direct deposition from the atmosphere at the Bay surface

We conducted several studies to determine the roles of waterborne and airborne transport in the movement of various pest control chemicals to Wye River, a tributary of Chesapeake Bay. An assessment of the amounts of atrazine and simazine carried into the Wye River by surface runoff is reported elsewhere.[2] That study showed that the waterborne delivery of atrazine and simazine to Wye River was dependent upon the timing and intensity of runoff with respect to local corn planting times. They were transported almost entirely as dissolved chemicals, mixed conservatively with saline water in the Bay, and exhibited about a 30-day half-life in the estuary. The peak concentration of atrazine measured in Wye River was about 15 μg/L following a critical runoff event.

We report here the results of a two-year study of the atmospheric concentrations and deposition of atrazine, simazine, alachlor, metolachlor, and toxaphene to Wye River and compare the atmospheric transport route to surface runoff.

EXPERIMENTAL

Study Area and Sampling Locations

Wye River is a shallow, well-mixed estuary that lies just east of the Eastern Bay region of Chesapeake Bay (38°55′ N, 76°10′ W). The study focused on that portion of Wye River that lies north of Wye Narrows. This portion of the river is about 7.5 km long, averages 2 m in depth, and has a surface area of 3.8 km^2 and a volume of 9.2×10^6 m^3. Land use in the surrounding watershed is fairly typical of the Eastern Shore region of Maryland. About half the 27.5 km^2 watershed is cultivated; the remaining area is divided among forests, pastures, and small housing subdivisions.[2] The cultivated acreage is almost equally divided between corn and soybeans. Frequently an overwinter crop of barley or wheat is grown, completing a corn, small grain, soybean rotation every two years. During the study period most of the corn and about half of the soybeans were planted using conventional tillage practices. When soybeans followed small grain, the prevailing practice was no-tillage planting into the stubble residue after grain harvest. Corn is planted around May 1, conventional soybeans around June 1, and no-till soybeans around July 1.

Table 1 and Figure 1 give the sampling locations and the dates during which rain and air samples were collected. In the initial phases of the study, rain and air samples were collected exclusively in the Wye River area at stations only a few kilometers apart (Figure 1a), but near the end of the study, rain samplers were placed at other locations across central Maryland to assess the variability in pesticide concentration in rain over distances of about 100 km (Figure 1b).

One of the stations was located at the Patuxent Wildlife Center near Laurel, Maryland (A in Figure 1b). This large wildlife refuge, operated as a research facility by the U.S. Fish and Wildlife Service, is a tract of woodland and marshes that is surrounded by suburban housing developments. Consequently, there is no farming in the vicinity. The sampling station was within the wildlife center, and no pesticides would have been applied within a several-kilometer radius of the sampler.

Another station was at Sandy Point State Park (B in Figure 1b), located at the western terminus of the Chesapeake Bay Bridge, near Annapolis, Maryland. This site was on the shore of the main stem of the Bay and was also well

Table 1.   Locations and Dates of Rain and Air Sampling.

Designation	Location	Dates of Operation	
		Rain Samples	Air Samples
Wye Island			
East	38°53′07″ N 76°06′47″ W	Mar. 31, 1981– Aug. 7, 1981	—
West	38°52′51″ N 76°10′14″ W	Mar. 31, 1981– Aug. 7, 1981	—
Middle	38°53′31″ N 76°08′59″ W	Oct. 15, 1981– May 26, 1982	—
Wye Institute	38°54′41″ N 76°09′03″ W	Mar. 31, 1981– Aug. 11, 1982	Mar. 31, 1981– Mar. 26, 1982
Wye Plantation	38°55′08″ N 76°07′43″ W	May 26, 1982– Aug. 11, 1982	May 26, 1982– Aug. 11, 1982
Patuxent Wildlife Refuge	39°02′01″ N 76°48′03″ W	Feb. 24, 1982– Aug. 11, 1982	—
Sandy Point State Park	39°01′17″ N 76°23′49″ W	May 26, 1982– Aug. 11, 1982	—

isolated from any farming or pesticide applications. Table 2 gives the structures and the physical properties of the pesticides we studied.

Rain Sampling

The sampling strategy one uses to determine the deposition of pollutants from the atmosphere is dictated by the resources available and by the objectives of the experiment. Deposition from the atmosphere occurs by "wet" and "dry" processes, "wet" referring to material deposited with rain, and "dry" deposition referring to the direct accumulation of both gases and particles at the surface. "Bulk" deposition is the sum of these wet and dry processes. In order to completely understand these processes, it is necessary to develop a sampling strategy that not only samples wet and dry deposition separately but also discriminates between gaseous-and particulate-phase pollutant in the atmosphere, and between dissolved and particulate-phase species in rainwater. Distinguishing between dissolved and particulate-phase pollutant in rainwater requires that the water be filtered through a highly efficient filter as it is collected.

However, the present study had limited objectives and limited resources. We wished to compare the amounts of selected pesticides deposited in Wye River by atmospheric transport to the amounts entering the river by surface runoff. For this objective we needed to know only bulk deposition and something about its spatial variability. Our sampling strategy was designed to meet these requirements.

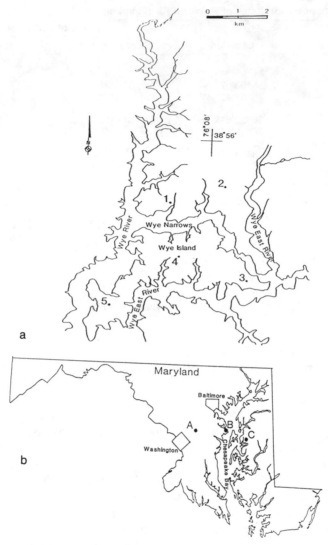

Figure 1.   (a) Air and rain sampling locations in the vicinity of Wye River: **1,** Wye Institute; **2,** Wye Plantation; **3,** Wye Island East; **4,** Wye Island Middle; **5,** Wye Island West. (b) Rain sampling locations in Maryland: **A,** Patuxent Wildlife Center; **B,** Sandy Point State Park; **C,** Wye River.

Rain was sampled using a 1-m^2 galvanized steel catch basin. Rainwater caught in the catch basin trickled through a 70-cm^3 bed of Amberlite XAD-4 resin (Mallinckrodt) packed in a stainless steel pipe nipple (2.6 cm i.d. $\times$ 15 cm). A 4-cm pad of glass wool was placed on top of the resin bed to remove coarse particulate material, but the rainwater was not filtered before it entered the resin bed. The catch basin, which was screened on top to keep out extraneous materials, provided 1 L of water to the resin cartridge for each millimeter

Table 2. Structures and Physical Properties of the Pesticides Studied.

Compound	Structure	Vapor Pressure mPa (20°C)	Solubility g/m^3 (20°C)	Air-Water Distribution Coefficient (Dimensionless)
Alachlor		3	130	2.5×10^{-6}
Metolachlor		1.7	530	3.7×10^{-7}
Atrazine		0.04	30	1.2×10^{-7}
Simazine		0.0085	5	1.4×10^{-7}
Toxaphene	polychlorinated camphene ($C_{10}H_{10}Cl_8$)	0.5	0.5	1.7×10^{-4}

Source: Metolachlor data from Worthing, C. R., and S. B. Walker, *The Pesticide Manual: A World Compendium* (Thornton Heath, UK: The British Crop Protection Council, 1987). All other data from Suntio, L. R., W. Y. Shiu, D. Mackay, J. N. Seiber, and D. E. Glotfelty, *Rev. Environ. Contam. Toxicol.* 103:1–59 (1988).

of rainfall. The volume of rain sampled was determined by placing a rain gauge beside each collector.

The XAD-4 resin was prepared by the method described by Williams.[3] Impure resin was washed thoroughly with distilled water (taking care to decant the fines), rinsed with methanol to remove excess water, then extracted consecutively with methanol, then 1:1 hexane-acetone, for 24 hr each in a Soxhlet apparatus. Clean resin was refrigerated in an amber bottle under methanol until used. The resin cartridge was packed by pipetting the methanol-resin slurry. At the end of the sampling period, the resin and glass wool pad were recovered and stored at –10°C until analyzed.

The resin was extracted four times with 80-mL volumes of peroxide-free ether. The resin was shaken with the solvent for about a minute, then allowed to stand for 20 min. The solvent was then filtered off and the process was repeated. Finally, the solvent was evaporated to a small volume and exchanged for benzene. Recovery of alachlor, metolachlor, atrazine, and simazine from 20 L of water spiked at the 0.5 ppb level averaged about 90 ± 8%.[3] Toxaphene recovery, determined by comparison to direct methylene chloride extraction of rainwater, was essentially quantitative. The glass wool pad and its associated particulate matter, if any, were analyzed separately using the resin extraction procedure.

Air Sampling

Air was sampled using a Sierra Misco High Volume Air Sampler (Model 680), equipped with a 20 cm × 24.5 cm sample head (Model 810-P). The sampler was operated on an interval timer that turned the sampler on from 0900–1500 hr each day. Recirculation of sampled air through the sampler is a potential problem when winds are calm. We judged that this would be less likely to occur during the 6-hr period around midday when atmospheric turbulence and winds tend to be greatest. Restricting sampling to this period also allowed us to set flow rates at values that could be controlled accurately, while still limiting the total volume of air sampled. The flow rate was adjusted to about 1.4 m^3/min, and the total volume sampled was about 3000 m^3. The sample media consisted of a glass-fiber filter (P810-G) and two 20 cm × 24.5 cm × 5 cm layers of porous polyurethane foam (PUF). The gray, ester-type PUF was cut from commercially available sheets. The PUF was prepared by a thorough water rinse, an acetone rinse to remove excess water, and then overnight Soxhlet extractions, first with acetone, then 1:1 acetone-hexane. The glass-fiber filters were precleaned by sealing the individual filters in aluminum foil and baking overnight at 400°C. At the end of the sampling period, the filter and PUF media were extracted for 6 hr (6 cycles/hr) in a Soxhlet apparatus, using 1:1 hexane-acetone. The extract was evaporated to a small volume and exchanged for benzene.

Cleanup and Analysis

The extracts from the resin, glass wool pad, glass-fiber filter, and PUF were split into toxaphene and herbicide fractions by column chromatography on activated (130°C overnight) Florisil PR (Floridin Co.). Ten grams of adsorbent were packed into a glass chromatography column (Kontes, K-420600), topped with a 1.5-cm layer of anhydrous Na_2SO_4, then wetted with benzene. The samples were applied to the top of the column in a small volume of benzene. Toxaphene was eluted quantitatively with 100 mL benzene; alachlor, metolachlor, atrazine, and simazine were collected together with 150 mL of 5% acetone in benzene. Both fractions were reduced to a small volume using a Kuderna-Danish Evaporative Concentrator with a three-ball Snyder Column (Kontes K503000, K570001, K570050).

The herbicide fraction was concentrated just to dryness under a gentle flow of N_2, the sample residue redissolved in 0.5 mL of 50% methanol-water, and then subjected to high-pressure liquid chromatographic fractionation using a Spectra Physics Model 3100 HPLC equipped with a 25 cm × 0.46 cm i.d. 10 μm ODS (C-18) reversed-phase column (DuPont Zorbax). The sample loop on the six-port rotary valve was replaced with a C-18 guard column (Brownlee Labs), and the solvent in and out lines were reversed. In the "load" position, the sample plus two 0.25 mL 50% methanol rinses were injected into the guard column (1.0 mL being just less than the volume required for simazine to begin to elute from the guard column). In the "inject" position, HPLC solvent backflushed the guard column to the head of the analytical column, and the chromatographic run was started. After an initial hold at 50% methanol, the solvent was programmed linearly up to 75% methanol. The herbicides eluted in the order simazine, atrazine, and alachlor plus metolachlor. The HPLC was used as a residue-polishing step: the column effluent was collected only during the time over which the compounds of interest eluted. Simazine was collected alone in fraction 1; atrazine, alachlor, and metolachlor were combined into a single fraction 2. The collected fractions were diluted with 75 mL of water, extracted in a small separatory funnel with 15 mL of benzene, then concentrated to an appropriate volume for gas chromatographic analysis.

The herbicides were analyzed by element-selective gas chromatography, using a Tracor 560 gas chromatograph (Tracor Instruments, Austin, TX) equipped with a Hall 700A electrolytic conductivity detector operated in the N mode. The herbicides were separated using a 1.83 m × 2 mm i.d. coiled glass column packed with 3% FFAP on 80–100 mesh Anakrom Q (Analabs, Inc., North Haven, CT). A number of the samples were confirmed by GC/MS, and many were confirmed by reaction with sodium methoxide.[4]

Toxaphene was isolated from PCBs and other chlorinated insecticides by silicic acid chromatography as described by Bidleman et al.[5] It was analyzed by capillary gas chromatography using an SE-54 WCOT column and a Hall electrolytic conductivity detector operated in the halogen mode. The limit of detection was about 0.005 μg/L for a 10 L rain sample.

In the analyses of both herbicides and toxaphene, the gas chromatographs were calibrated on a routine basis using an external standard calibration curve that was rechecked after every few samples to correct for any changes in instrument sensitivity. Samples with final volumes ≥ 1.0 mL were prepared in Class A volumetric glassware. The volumes of samples concentrated below 1.0 mL were estimated either by the precalibrated graduation marks on the concentrator receiver tubes or by taking the samples up in a calibrated syringe.

RESULTS AND DISCUSSION

Pesticides in Rain

Although, in the final analysis, the conclusions we reach do not depend upon the distinction between bulk and wet-only deposition, strictly speaking, we measured pesticides in bulk precipitation, that is, the sum of material deposited in the collector by rain, plus that material deposited with dust when it was not raining. Several considerations, however, lead us to conclude that dry deposition in our collectors was negligible. First, only on rare occasions could we find traces of any of the pesticides in the particulate material retained by the glass wool pad at the top of the resin column in the rain collector. Second, our measurements (described more fully below) indicate that the pesticides we studied were present in the atmosphere almost wholly in the vapor phase and would thus not be expected to deposit in the collectors. Third, subsequent studies[3] that carefully eliminated dry deposition confirmed the overwhelming predominance of wet deposition in the movement of alachlor, metolachlor, atrazine, and simazine from the atmosphere to the surface. For these reasons we will describe our results in terms of rainfall deposition.

The analytical methods were very sensitive and reliable, especially for the herbicides because of the HPLC residue-polishing step that was used. Figure 2 shows the gas chromatographic analysis of herbicides in rain that was sampled at Wye Institute during the week ending May 26, 1982. As can be seen, the chromatograms are clean and unambiguous. They correspond to 0.08 μg/L simazine, 1.1 μg/L alachlor, 0.8 μg/L metolachlor, and 0.6 μg/L atrazine. The detector responds to nitrogen and is therefore about five times more sensitive for atrazine and simazine than it is for alachlor and metolachlor. Based on a 1-cm rainfall, the limits of detection (signal:noise $\approx$ 3) are about 0.001 μg/L for atrazine and simazine, and about 0.005 μg/L for alachlor and metolachlor. The herbicides can be confirmed at the limit of detection by the methoxylation reaction. Mass spectrometric confirmation is somewhat less sensitive but never failed to confirm the presence of the herbicides at slightly higher levels.

Figure 3 shows a chromatogram of the toxaphene fraction of rain samples at Wye Plantation during the week ending June 15, 1982. The toxaphene concentration was about 0.20 μg/L. As can be seen, the pattern of peaks in the sample chromatogram closely matches the characteristic pattern of peaks obtained with an authentic toxaphene standard.

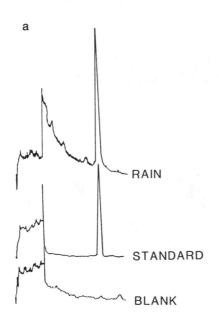
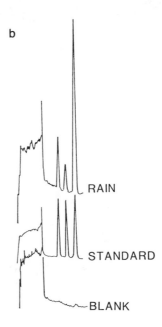

Figure 2. Gas chromatograms of the herbicide fraction of rain sampled at Wye River during the week ending May 26, 1982. (a) HPLC fraction 1: top chromatogram = 1/10 aliquot of the rain sample (12.2 mL rain); middle = 0.5 ng simazine; bottom = 1/165 resin blank (74 mL rain); (b) HPLC fraction 2: top chromatogram = 1/10 aliquot of the rain sample (1.2 mL rain); middle = 2.5 ng alachlor, 2.5 ng metolachlor, and 0.5 ng atrazine, eluting in that order; bottom = 1/175 resin blank (70 mL rain). Gas chromatographic conditions given in the text.

Rain samples were collected at weekly intervals, except during the winter months when samples were collected at 2-week intervals. Fortunately throughout the study there were only two sample intervals that had no rain. The average concentration of atrazine in rain measured with the collectors stationed in the vicinity of the Wye River is shown in Figure 4. As can be seen, the concentration of atrazine in rain is highly seasonal, exhibiting a sharp peak coinciding with local corn-planting time in early May. This seasonality was noted by Richards et al.[6] for a number of compounds but contrasts markedly with the lack of seasonality in atrazine concentration in rain as reported by Wu,[7] who found that winter rainfall often contained higher concentrations of atrazine than summer rain. In 1981 the peak concentration of atrazine was 1.3 μg/L, which occurred in the week ending May 4. In 1982 the peak concentration of 3.3 μg/L occurred during the week ending May 13. Simazine mirrored

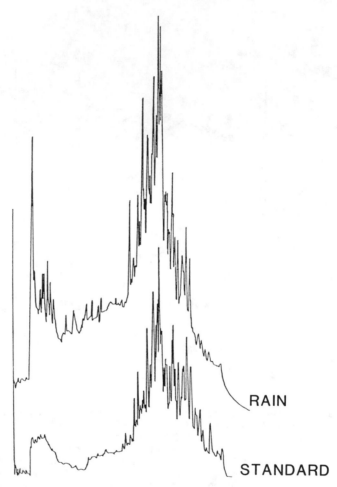

Figure 3. Gas chromatogram of the toxaphene fraction of rain sampled at Wye River during the week ending June 15, 1982. Top chromatogram = rain sample; bottom = toxaphene standard. Gas chromatographic conditions given in the text.

the atrazine behavior for the most part, except at lower concentration. For example, in 1982 the simazine:atrazine ratio in rain, based on 51 analyses, was 0.28 ± 0.04.

Although the concentrations of these two herbicides decline to low values in the period October to December, they were above the limit of detection and significantly above blank values in all samples. They are thus present at very low levels in rain throughout the year, even in winter months. The major uses of atrazine and simazine are as preplant and preemergent herbicides for the control of broadleaf weeds in corn. They are labeled for other purposes, such as weed control on rights-of-way and industrial sites, and on various crops such as alfalfa and asparagus, but the total usage for these purposes is negligi-

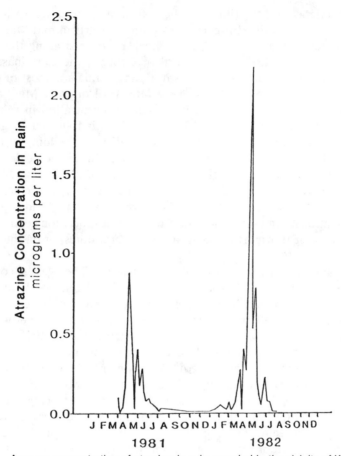

Figure 4.   Average concentration of atrazine in rain sampled in the vicinity of Wye River.

ble compared to corn applications. Other sources of these pesticides in the atmosphere might also include losses from storage facilities or losses during shipment and distribution.

A more likely wintertime source is persistent volatile releases from previously treated fields. The strength of this source is unknown. Atrazine and simazine dissipate from treated soil by a first-order process with a half-life on the order of a few weeks.[2] In Maryland soils about 5% of both herbicides will be carried over through the winter months until the next growing season. It is conceivable that small volatile losses could continue through the winter from these carryover residues.

We believe that the annual pattern of atrazine and simazine in rain gives evidence of regional-scale, and perhaps long-range, transport in the atmosphere, and that the atrazine and simazine measured in early spring precipitation come from regions to the south with earlier corn planting dates. January and February are winter months, and March and April are typically cold and rainy in Maryland, so little tilling of the soil occurs, and no corn is planted until at least late in April. However, corn planting starts in mid-March in the Gulf Coast states, and in a very narrow strip northward along the Atlantic seaboard. Corn is planted in early April in Georgia, the Carolinas, and in much of the southeast. In Maryland, as in Pennsylvania and most of the corn belt states, corn is not planted until very late April or early May. Figure 4 shows a small, but significant, rise in atrazine concentration in rain over a several-week period in January and February. This is followed by gradually increasing concentrations throughout March and April, building to the early May peak. This period of increasing concentrations in rain coincides with the progress of corn planting from the Gulf Coast to states nearer to Maryland. The rise in concentration measured in January and February coincides with the planting and application of herbicides to sweet corn in Florida. The data suggest that atrazine and simazine persist in the atmosphere long enough for at least regional-scale transport to occur, across several states, or on the order of 10^3 km.

Alachlor and metolachlor show a different pattern. The average concentration of alachlor in rain collectors stationed in the vicinity of Wye River is shown in Figure 5. Metolachlor mirrors alachlor, except at lower concentration. For example, for 30 samples collected in 1982 the ratio of metolachlor to alachlor was 0.47 ± 0.05. In contrast to atrazine and simazine, the data suggests that alachlor and metolachlor are very transient in the atmosphere. Figure 5 shows that the alachlor concentration in rain went through very pronounced, short-lived peaks and valleys. The peak concentrations appeared to occur about May 1, June 1, and July 1, with the last being the highest. Before April and after mid-July, alachlor and metolachlor were undetectable (< 0.005 μg/L) in rain. These peaks occurred at times of the year when alachlor and metolachlor are typically applied to local crops: corn about May 1, conventionally tilled soybeans about June 1, and no-till soybeans in small grain stubble about July 1. It is likely that this pattern will be influenced by local management practices and local weather.

Alachlor and metolachlor are only present in the atmosphere for a short time around local planting times; their concentrations in rain drop to undetectable levels within 1 to 2 weeks of the end of local applications. We see no March/April rise in alachlor concentrations analogous to the behavior of atrazine, corresponding to transport into Maryland from more southerly states. We conclude that alachlor and metolachlor are very transient in the atmosphere, persisting no more than a day or two; otherwise regional-scale transport would occur. This transient behavior suggests that these two herbicides are susceptible to fairly rapid vapor-phase degradation in the atmosphere.

The average concentration of toxaphene in rain for the collectors located in the vicinity of Wye River is shown in Figure 6. We were able to analyze only a selected set of samples for toxaphene, and therefore we have fewer data than we have for the herbicides and some periods are missing. However, most of the

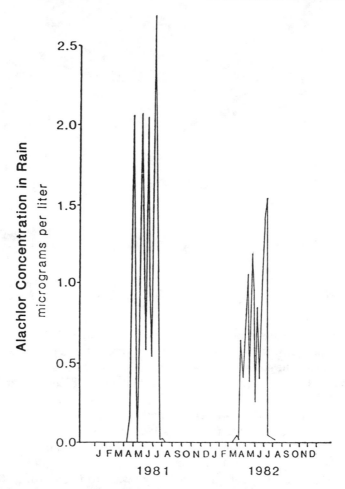

Figure 5. Average concentration of alachlor in rain sampled in the vicinity of Wye River.

data in Figure 6 are an average of 2 or 3 samples. All of the summer samples were above the limit of detection of 0.005 μg/L.

The data show that the concentration of toxaphene in rain also depended strongly upon the season, with concentrations peaking during local corn-planting time. This is to be expected since the predominant use of toxaphene in Maryland was to control corn rootworm, and it usually was applied as a tank mixture with atrazine. There is evidence also of at least regional-scale transport of toxaphene; concentrations in rain begin to rise in early April in a manner analogous to atrazine. Actually, there is abundant literature giving evidence that toxaphene is susceptible to very long distance transport in the atmosphere.[8-10]

The highest measured toxaphene concentration in rain was 1.4 μg/L measured in mid-May 1982. This high concentration coincided with a sharp rise in

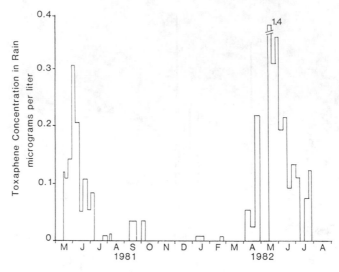

Figure 6. Average concentration of toxaphene in rain sampled in the vicinity of Wye River. Missing data means samples were not analyzed.

the levels of toxaphene in air, up to values in the range of 200 to 400 ng/m³ (discussed more fully below). The mean concentration of toxaphene in rain at Wye River during the summer of 1981 was 0.12 ± 0.10 μg/L (n = 22), and the range was 0.021 to 0.34 μg/L. During the summer of 1982, the mean was 0.23 ± 0.30 μg/L (n = 21), and the range was 0.021 to 1.4 μg/L. In contrast to the summer values, late summer/winter rain had a mean concentration of 0.018 $\pm$ 0.013 μg/L (n = 11), and the range was from below detection to 0.04 μg/L.

These values are very similar to those reported for measurements at North Inlet estuary (South Carolina): during 1977, the mean toxaphene concentration was 0.159 ± 0.133 μg/L, and the range was from ≤0.016 to 0.497 μg/ L;[11,12] during 1981–1983, the arithmetic mean concentration was 0.018 μg/L, and the range was from < 0.002 to 0.115 μg/L.[13] They are also similar to data for eight rain samples obtained near Baltimore, Maryland, in 1974, which had a mean of 0.10 μg/L and ranged from below detection to 0.28 μg/L.[14]

Pesticides in Air

A summary of the measured concentrations of the five pesticides in summer air are given in Table 3; winter values are recorded in Table 4. These data are average concentrations in noontime air on a weekly basis. The concentrations of the pesticides in air show the same seasonal pattern as their concentrations in rain, that is, they peak rather sharply at times of the year when they are used locally. Atrazine and simazine could be measured in the atmosphere at all times of the year, even in the winter. Winter values for atrazine in air were about 1% of summer values; winter values for simazine were about 10% of summer values. Alachlor and metolachlor could not be detected (< 0.005 ng/

Table 3.   Concentrations and Filter-Retained Fractions of Selected Pesticides in Air Near Wye River, May–June, 1982.

| Compound | Concentration (ng/m^3) | | | Fraction Retained by Glass-Fiber Filter (Mean, %) |
	Range	Mean	Standard Deviation	
Atrazine	0.1–20	3.7	6.4	5
Simazine	0.02–0.35	0.15	0.12	26
Alachlor	0.06–7.3	2.4	2.5	1.5
Metolachlor	0.07–1.7	0.8	0.6	nd[a]
Toxaphene	11–376	98	130	nd[a]

[a]nd = none detected.

m^3) in the atmosphere from about mid-August until mid-April. No winter samples were analyzed for toxaphene. The toxaphene concentrations in summer air measured at Wye River were about an order of magnitude greater than those reported by Bidleman et al.[11] for 1977 data in South Carolina and by Rice et al.[10] for Mississippi in 1981. We were unable to find literature values for ambient air concentrations of atrazine, simazine, alachlor, and metolachlor.

In summer all of the pesticides except simazine (which has a very low vapor pressure) were found ≥95% in the polyurethane foam vapor trap. The glass-fiber filter retained 26% of the simazine. In the colder winter months, the filter-retained fractions of both atrazine and simazine were significant.

Glotfelty et al.[15] recently reported the field-measured volatilization rates of alachlor, atrazine, simazine, and toxaphene. Since at the time of the study these compounds were all applied essentially simultaneously to corn during planting time, it is illuminating to compare the measured concentrations in air to what one would expect to find based upon their relative volatilities and amounts applied. This comparison, normalized to atrazine, is given in Table 5. The concentrations of atrazine and toxaphene in summer air are very close to their "expected" proportions based on amounts applied[16] and volatilities. Simazine concentration is about 30% of the "expected" value. In contrast, alachlor concentrations were only about 6% of the "expected" value. This suggests that alachlor is far less persistent than atrazine in the atmosphere,

Table 4.   Concentrations and Filter-Retained Fractions of Selected Pesticides in Air Near Wye River, December 1981–February 1982.

| Compound | Concentration (ng/m^3) | | | Fraction Retained by Glass-Fiber Filter (Mean, %) |
	Range	Mean	Standard Deviation	
Atrazine	0.008–0.04	0.026	0.019	32
Simazine	0.003–0.04	0.013	0.015	39
Alachlor	nd[a]	—	—	nd
Metolachlor	nd	—	—	nd
Toxaphene	na[b]	—	—	na

[a]nd = none detected.
[b]na = not analyzed.

Table 5.   Comparison of Expected and Measured Relative Concentrations in Air for Atrazine, Simazine, Alachlor, and Toxaphene (Data Normalized to Atrazine).

Compound	Relative Volatility[a]	Relative Amount Applied[b]	Relative Concentration in Air		Ratio, Measured to Expected
			Expected[c]	Measured[d]	
Simazine	0.5	0.26	0.13	0.04	0.3
Alachlor	8	1.4	11	0.65	0.06
Toxaphene	14	1.8	25	26	1.0

[a]Relative volatility = $k_{pesticide}/k_{atrazine}$, where k is the field-measured volatilization rate constant.[15]
[b]Pesticide usage in Queen Annes County, MD, in 1982 (metric tons): atrazine, 16.0; simazine, 4.2; alachlor, 22.6; and toxaphene, 29.0.[16]
[c]Relative volatility × relative amount applied.
[d]Ratio of mean concentrations in air (Table 3).

which agrees with earlier conclusions drawn from the seasonality of alachlor concentrations in rain.

Cumulative Pesticide Deposition in Wye River

Our primary rain data were the amounts of the various pesticides retained by the resin cartridge at the end of each collection period. This directly yielded the amount deposited, in $\mu g/m^2$, for each week. Summing up the weekly increments yields the cumulative amount of pesticide deposited, from which the total amounts of the various pesticides delivered to Wye River by rainfall can be estimated. Figure 7 shows the cumulative deposition of atrazine during 1982. Although atrazine deposition began in January, more than 90% of the annual deposition occurred between April and July.

Table 6 gives the amounts of atrazine, simazine, alachlor, and toxaphene deposited by summer rains in the vicinity of Wye River for 1981 and 1982 and at Beltsville, Maryland, in 1984.[3] Compared to the amounts applied for pest control in agricultural fields, the amounts deposited by rain are very small: except for alachlor in 1984, all the deposited amounts were well less than 0.1% of a nominal application (range, 0.006 to 0.08%). In 1984 the measured deposition of alachlor at Beltsville was approximately 0.3% of the amount typically recommended for control of weeds in corn and soybeans.

The 1981 and 1982 data in Table 6 are based upon averages of the measurements made with the several collectors stationed near Wye River and are thus representative of the amounts of the various pesticides delivered to Wye River in rain. In 1982 summer rainfall (March 30 to July 27) deposited 0.4 kg of atrazine and toxaphene, 0.04 kg of simazine, and 0.8 kg of alachlor directly into Wye River.

Glotfelty et al.[2] measured the concentrations and total amounts of atrazine and simazine in Wye River in 1980, 1981, and 1982. They concluded, based on the relationship between atrazine concentration and salinity and from the coincidence of increases of herbicide amounts in the estuary with major runoff events, that atrazine and simazine moved to the estuary primarily in surface

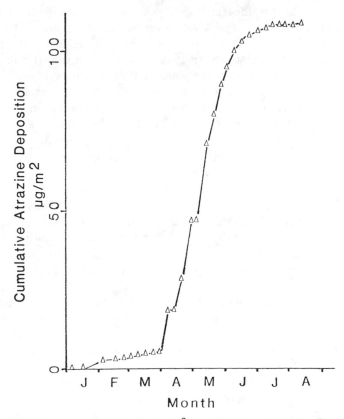

Figure 7. Cumulative amount of atrazine (μg/m^2) deposited by rain in Wye River in 1982.

Table 6.   Amounts of Selected Pesticides Deposited by Summer Rainfall in the Vicinity of Wye River and at Beltsville, Maryland.

Period	Total Wet Deposition (μg/m^2)			
	Atrazine	Simazine	Alachlor	Toxaphene
March 31– July 30, 1981	74	11	310	45
March 30– July 27, 1982	103	9.5	205	95
May 1– Sept. 1, 1984 (Beltsville)[a]	54	12	820	—

[a]Williams.[3]

runoff. The first year of the study, 1980, was very dry, and the total amount of atrazine and simazine in Wye River never exceeded 1.5 and 0.9 kg, respectively. A single, "critical" runoff event[17] in 1981 delivered 23 kg of atrazine and 1.4 kg of simazine to the estuary. In 1982 the first major runoff event occurred on June 1, about one month after planting time. Before this event there was about 2.1 kg of atrazine and 0.7 kg of simazine in the river, of which only about 15 and 3%, respectively, had been deposited by rain in the preceding six weeks. The June 1 runoff delivered 13 kg of atrazine and 1.1 kg of simazine, further reducing the role of rainfall deposition to less than 3 and 2%, respectively. Clearly, the major source of atrazine and simazine in Wye River was surface runoff from treated fields.

Alachlor residues in Wye River in 1981 were in the range of 10 to 30% of the atrazine levels.[18] Following the "critical" runoff event in May of that year, there was approximately 4.4 kg of alachlor in Wye River. Comparable data for 1982 are not available. As in the case of atrazine, runoff is clearly the predominant mode of alachlor movement to Wye River, but the fraction delivered by rainfall may be more significant, perhaps on the order of 20% or more of the total load.

The 95 $\mu g/m^2$ of toxaphene deposited in Wye River in 1982 (Table 6) was two to three times as large as reported deposition of toxaphene to North Inlet estuary (South Carolina) during the summers of 1977 and 1982[12,13] and was many times larger than the atmospheric deposition of toxaphene in remote areas. Recently Rapaport and Eisenreich[19,20] estimated the historical inputs of toxaphene and other chlorinated compounds from the atmosphere into peat bogs at seven locations across the midlatitudes of eastern North America (Minnesota to Nova Scotia). Total atmospheric toxaphene deposition between 1947 and about 1984 ranged from approximately 30 to 200 $\mu g/m^2$ at the various locations. In the peak year, 1978, depositional fluxes to these sites ranged from 1.8 to 9 $\mu g/m^2/year$. In recent years (after 1980), the fluxes ranged from 0.8 to 5.5 $\mu g/m^2/year$. Thus, on an annual basis, deposition of toxaphene to Wye River in 1982 was 10 to 100 times greater than deposition in nonagricultural areas across the midlatitudes of eastern North America.

When diluted by the total volume of Wye River, the toxaphene deposited by rain in 1982 corresponds to a concentration in the river of about 0.050 $\mu g/L$. We were not able to conduct a complete survey for toxaphene in the river, but 30 samples analyzed in 1982 had a range of 0.034 to 0.44 $\mu g/L$, with a mean value of 0.17 $\mu g/L$. Wet deposition could thus have contributed a major part of the toxaphene in Wye River.

The concentrations of toxaphene in Wye River in 1982 greatly exceeded the levels found to cause problems in juvenile fish. For example, the growth of brook trout fry was significantly depressed at levels as low as 0.039 $\mu g/L$,[21] and fathead minnows suffered growth inhibition and bone deformation when exposed to toxaphene concentrations as low as 0.055 $\mu g/L$.[22] In November 1982, the U.S. EPA canceled most registrations of toxaphene because of its toxicity to aquatic life and its carcinogenic hazard to humans. Toxaphene

Table 7.   Concentrations of Selected Pesticides in Rain (μg/L) at Four Separate
Locations in Maryland, June–July 1982.

Date	Wye Institute	Wye Plantation	Patuxent Wildlife Center	Sandy Point S.P.
		Atrazine		
June 22	0.11	0.21	0.17	0.086
June 30	0.10	0.057	0.20	0.11
July 7	0.074	0.48	0.075	0.092
July 15	0.031	0.12	0.033	0.068
		Simazine		
June 22	0.045	0.057	0.069	0.049
June 30	0.022	0.040	0.018	0.021
July 7	0.015	0.015	0.007	0.011
July 15	0.005	0.029	0.013	0.010
		Alachlor		
June 22	0.32	0.48	0.84	0.40
June 30	0.35	1.63	0.43	0.42
July 7	1.89	0.94	0.14	0.35
July 15	0.54	2.57	0.23	0.50
		Metolachlor		
June 22	0.089	0.12	na[a]	na
June 30	0.093	0.23	0.081	0.046
July 7	0.097	0.29	0.064	0.16
July 15	na	na	na	0.11

[a]na = not analyzed.

input to the environment continued, however, since "existing stocks" could be used through 1986 for various purposes, including control of armyworm and cutworm in corn. In 1982, before the ban, just over 160 metric tons of toxaphene were used in Maryland, largely in counties bordering the Chesapeake Bay.[16] In 1985, toxaphene use in Maryland was still over 94 metric tons.[23]

Atmospheric Deposition of Pesticides to Chesapeake Bay

Our study of the Wye River shows that runoff produced locally high concentrations, while atmospheric transport and deposition lead to low-level but widespread contamination. With the possible exception of toxaphene, the relative contribution of pesticides to Wye River by rainfall was minor. Wye River has a high loading function, about 3 square meters of watershed per cubic meter of river volume. It is therefore appropriate to consider whether rainfall deposition would be a more significant source of pesticides to other tributaries with smaller loading functions or to the Chesapeake Bay proper.

Because of the rapid and widespread dispersion that occurs with atmospheric transport, and because of the widespread nature of the agricultural sources, the measurements of atmospheric concentration and rain deposition taken at the Wye River and the other sampling locations are probably valid for larger areas. Table 7 gives data for four sampling locations that were in simul-

Table 8.   Projected Amounts of Pesticides Entering Chesapeake Bay via Summer (May-August) Rainfall.

Pesticide	Wet Deposition (metric tons)[a]		
	1981	1982	1984
Atrazine	0.88	1.2	0.64
Simazine	0.13	0.11	0.14
Alachlor	3.7	2.4	9.8
Metolachlor[b]	1.7	1.1	4.6
Toxaphene	0.54	1.1	na[c]

[a]Pesticide data from Table 6. Area of Chesapeake Bay and tributaries is 11.9 × 10⁹ m², from Cronin, W. B., "Volumetric, Areal, and Tidal Characteristics of the Chesapeake Bay Estuary and Its Tributaries," Spec. Report 20, Johns Hopkins University, Baltimore, MD (Ref No. 71–2) (1971).
[b]Only a partial set of measurements is available for metolachlor; the projected amounts of metolachlor entering Chesapeake Bay are based upon the relative concentrations of metolachlor and alachlor in the samples for which data are available.
[c]na = not analyzed.

taneous operation for a portion of the summer of 1982. Together they spanned a transect of about 100 km across central Maryland, at about the midpoint of the north-south axis of Chesapeake Bay (see Figure 1b). Table 7 shows that the concentrations of atrazine, simazine, alachlor, and metolachlor in rain were fairly uniform across the transect for a four-week period in June-July 1982. The variability in this four-station transect was no greater than the variability between the several samplers stationed much closer together near Wye River during the two-year study.

If we assume, based on the evidence presented, that pesticide concentrations in air are uniform over a wide area, and also that there are no systematic differences in the distribution and amount of rainfall received across the Bay, then the data in Table 6 can be used to obtain a reasonable first approximation of the amount of the pesticides studied that rained out over the entire Bay. Table 8 gives the projected amounts of atrazine, simazine, alachlor, and toxaphene that entered Chesapeake Bay with rainfall during the summers of 1981, 1982, and 1984. The quantities deposited in the Bay ranged from a low of 110 kg of simazine in 1982, to almost 10 metric tons of alachlor in 1984.

There has been no detailed monitoring of the input of pesticides by tributaries to the Bay. In 1979–1980 the U.S. Geological Survey measured atrazine concentrations in the three main tributaries of the Chesapeake Bay: the Susquehanna, Potomac, and James rivers.[24] Atrazine was rarely above the limit of detection in the James River, although occasional traces were found. Atrazine was present in the Potomac and Susquehanna rivers, generally from mid-April through mid-October. Using the data given in Figure 4 of the report by Lang,[24] the total atrazine discharged to the Chesapeake Bay by these tributaries in 1979 and 1980 was estimated as the product of average atrazine concentration, average water flux, and time (May through September). These estimates are given in Table 9. In 1979, the Potomac and Susquehanna discharged about 3 metric tons of atrazine to the Bay; in 1980, about 12.5 metric tons.

Table 9.   Atrazine Flux to the Chesapeake Bay via the Susquehanna and Potomac Rivers, 1979-80.

	1979		1980	
	Potomac	Susquehanna	Potomac	Susquehanna
Average flow (10^6 L/sec)[a]	0.34	0.62	0.45	1.4
Average Atrazine concentration (μg/L)[a]	0.2	0.25	0.25	0.6
Total atrazine flux (metric tons)	0.9	2.1	1.5	11

[a]Estimated from Figure 4 of Lang[24] for the period May through September.

The water discharge rate averaged over the two years was close to the long-term average, implying that the average atrazine flux to the Bay probably lies somewhere between the estimated values. Wet deposition of atrazine by rainfall is therefore probably on the order of 10% of the total atrazine input into the Bay. It is interesting to note that when all the minor tributary sources are added in, the proportion of atrazine added by rainfall to the Bay as a whole is probably close to the proportions added by rainfall to Wye River.

One major unknown is the amount of pesticide entering the Bay water by direct adsorption of vapor at the air-water interface. This may be substantial.[25] During the summer atrazine is present in the atmosphere almost entirely in the vapor phase, so its adsorption will reflect the partitioning between the air and water. The air-water distribution coefficient for atrazine is about 1.2×10^{-7} (Table 2), so if the average summertime concentration in the air is 3.7×10^{-3} ng/L, the water concentration at equilibrium should be close to 3.1×10^4 ng/L. This is on the order of 30 times higher than the observed value of about 10^3 ng/L. This indicates that although temperature and salinity effects may reduce the difference to some degree, the water phase was undersaturated with respect to the air and the net flux would have been from the atmosphere into the bay. The magnitude of this input from the atmosphere is not known.

SUMMARY

In a two-year study the concentrations of atrazine, simazine, alachlor, metolachlor, and toxaphene in the atmosphere and rain, and the amount of these chemicals deposited by rain, were monitored at several locations near Wye River and at various points across central Maryland. The concentrations of the pesticides in air and rain were strongly dependent upon the season, with maximum concentrations during the period when local crops are planted. Although they declined to low levels by late summer, the triazine herbicides, atrazine and simazine, were present at measurable levels in air and rain the entire year. For these chemicals, there was circumstantial evidence of at least regional-scale transport in the atmosphere, on the order of 10^3 km. Toxaphene is known to undergo long-range transport. In contrast, alachlor and metolachlor show no evidence of even regional transport and are apparently rapidly degraded in the atmosphere.

Surface runoff from treated fields causes locally high concentrations in the receiving water body; atmospheric transport and redeposition results in widespread, low-level contamination. Major runoff events produced the greatest amounts of pesticides in Wye River. Rain deposited about 3% of the total atrazine in Wye River. The rain-deposited proportion of alachlor was about 20% of the total, while rain may have deposited a much larger fraction of the total toxaphene.

Pesticides are apparently distributed uniformly in air across central Maryland. By assuming widespread, uniform concentrations and no systematic differences in rainfall amount or distribution across the Bay, the study data were used to estimate the amounts of the pesticides studied that were deposited by rainfall into Chesapeake Bay. In the years 1981, 1982, and 1984, the quantities deposited in the Bay ranged 0.6 to 1.2 metric tons of atrazine, 0.11 to 0.14 metric ton of simazine, 2.4 to 9.8 metric tons of alachlor, and 0.54 to 1.1 metric tons of toxaphene. By comparison, the two major tributaries, the Potomac and Susquehanna Rivers, brought 3 metric tons of atrazine into the Bay in 1979, and 12.5 metric tons in 1980. It thus appears that rain deposited about the same proportion of atrazine in the open Bay as it did in Wye River. One major uncertainty in this analysis is the loading of pesticides into the Bay by direct absorption of particles and gases. Henry's law predicts that for atrazine, the water in the Bay is greatly undersaturated with respect to the vapor-phase concentration in the atmosphere during the summer. Therefore, the net flux will be from the atmosphere to the Bay. The magnitude of this flux, however, is not known.

REFERENCES

1. Orth, R. J., and K. A. Moore. *Estuaries* 7(4B):531–40 (1984).
2. Glotfelty, D. E., A. W. Taylor, A. R. Isensee, J. Jersey, and S. Glenn. *J. Environ. Qual.* 13:115–21 (1984).
3. Williams, G. H. "Field Measurement of Pesticide Washout in Rain Near Beltsville, Maryland," MS Thesis, University of Maryland, College Park, MD (1986).
4. Lawrence, J. F. *J. Agric. Food Chem.* 22:936–38 (1974).
5. Bidleman, T. F., J. R. Matthews, C. E. Olney, and C. P. Rice. *J. Assoc. Office Analyst. Chem.* 61:820–28 (1978).
6. Richards, R. P., J. W. Kramer, D. B. Baker, and K. A. Krieger. *Nature* 327:129–31 (1987).
7. Wu, T. L. *Water Air Soil Poll.* 15:173–84 (1981).
8. Bidleman, T. F., and C. E. Olney. *Nature* 257:475–76 (1975).
9. Bidleman, T. F., E. J. Christensen, W. N. Billings, and R. Leonard. *J. Mar. Res.* 39:443–64 (1981).
10. Rice, C. P., P. J. Samson, and G. E. Noguchi. *Environ. Sci. Technol.* 20:1109–16 (1986).

11. Bidleman, T. F., E. J. Christensen, and H. W. Harder. In *Atmospheric Pollutants in Natural Waters*, S. J. Eisenreich, Ed. (Ann Arbor, MI: Ann Arbor Press, 1981), pp. 481–508.
12. Harder, H. W., E. J. Christensen, J. R. Matthews, and T. F. Bidleman. *Estuaries* 3:142–47 (1980).
13. Bidleman, T. F., M. T. Zaranski, and M. D. Walla. In *Toxic Contamination Large Lakes. Volume I. Chronic Effects of Toxic Contaminations in Large Lakes*, N. W. Schmidtke, Ed. (Chelsea, MI: Lewis Publishers, 1988), pp. 257–84.
14. Munson, T. O. *Bull. Environ. Contam. Toxicol.* 16:491–94 (1976).
15. Glotfelty, D. E., M. M. Leech, J. Jersey, and A. W. Taylor, *J. Agric. Food Chem.* 37:546–51.
16. "Pesticide Usage in 1982," Maryland Department of Agriculture, Publication No. 215–85, Annapolis, MD.
17. Wauchope, R. D. *J. Environ. Qual.* 7:459–72 (1978).
18. Glotfelty, D. E. Unpublished data.
19. Rapaport, R. A., and S. J. Eisenreich. *Environ. Sci. Technol.* 22:931–41 (1988).
20. Rapaport, R. A., and S. J. Eisenreich. *Atmos. Environ.* 20:1367–2379 (1986).
21. Mehrle, P. M., and F. L. Mayer. *J. Fish. Res. Board Can.* 32:609–613 (1975).
22. Mehrle, P. M., and F. L. Mayer. *J. Fish. Res. Board Can.* 32:593–98 (1975).
23. "Maryland Pesticide Statistics — 1985," Maryland Department of Agriculture, Publication No. 227–87, Annapolis, MD.
24. Lang, D. L. *Water Quality of the Three Major Tributaries to the Chesapeake Bay: The Susquehanna, Potomac and James Rivers*, January 1979-April 1981, Water-Resources Investigations 82–32 (Towson, MD: U.S. Geological Survey, 1982).
25. Eisenreich, S. J., B. B. Looney, and D. J. Thornton. *Environ. Sci. Technol.* 15:30–38 (1981).

## CHAPTER 15

Long Range Atmospheric Transport and Deposition of Toxaphene

E. C. Voldner and W. H. Schroeder

INTRODUCTION

The presence of toxic chemicals in the Great Lakes ecosystem is of concern under the Canada-United States Great Lakes Water Quality Agreement. Available measurements indicate that the atmosphere may be an important, and perhaps a dominant, source of some of these chemicals. As an example, toxaphene, which has not been used extensively in the Great Lakes area, has been found in fish and in lake water as well as in air and precipitation in this region.

Currently there are few estimates of atmospheric input of toxic chemicals to the Great Lakes.[1,2] Since these estimates are generally based on environmental measurements, they are limited to compounds that have a reasonable database. Here we will describe the use of mathematical simulation techniques to estimate the atmospheric transport and deposition of toxaphene to the Great Lakes and their basins.

Toxaphene was introduced in the United States during the late 1940s. It became the most heavily used insecticide between 1960 and the mid-1970s. It was applied primarily to cotton in the southern states. In 1980 the usage had fallen to 7.3×10^6 kg. Application had widened to other crops, and usage by then extended somewhat west and north.[3] In December 1982, due to health concerns, the U.S. Environmental Protection Agency canceled most usages of toxaphene.[4] Figure 1 summarizes the recorded usage of toxaphene for the period 1964 to 1983.[5] Only a limited amount of toxaphene was used in Canada for scabies treatment on livestock.[6]

To simulate the atmospheric transport and deposition of toxaphene to the Great Lakes region, we need to follow this pesticide from its sources to its sinks. First we will present some of the major features of the mathematical simulation model, then we will describe how the emissions inventory was compiled, and finally we will show a few of the results derived from 1980. Details are provided in Voldner and Schroeder.[7] The year 1980 was selected for

k metric tons

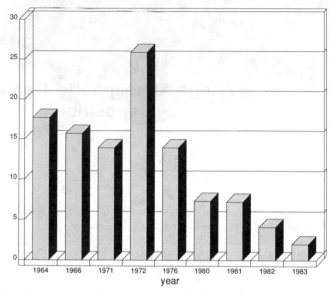

Figure 1. Recorded toxaphene usage in the United States, 1964–1983.

the modeling exercise due to the availability of use pattern data for toxaphene as well as of some environmental measurements.

THE MODEL AND INPUT DATA

Since toxaphene has primarily been applied to crops in the southern, southwestern, and southeastern United States, regional-scale atmospheric transport must be considered. The Advanced Statistical Trajectory Regional Air Pollution (ASTRAP) model[8] was chosen and modified for simulation of toxaphene transport and deposition to the Great Lakes region. ASTRAP, which is Lagrangian in approach, has been used extensively in acid rain research.[9,10]

The model simulates monthly, seasonal, or multiyear average air concentrations, as well as cumulative wet and dry deposition over spatial scales from about 50 km to continental. Trajectories are initiated every 6 hr from a site and followed up to 7 days. Horizontal dispersion is estimated by the mean position and spread of an ensemble of simulated tracer trajectories as a function of plume age for each virtual source. Vertical diffusion is calculated through numerical integration of the standard diffusion equation. Incorporation of nine vertical layers allows a description of slow leakage of pollutants into the free troposphere and permits treatment of emissions from surface sources as well as from tall stacks. Dry deposition fluxes, including air-surface exchange of vapor (directed upward or downward), are modeled through the concept of dry deposition velocity, i.e., as a product of a velocity and a concentration. Physical and/or chemical transformation rates are assumed to be linear. Dry deposition velocities, vertical stability profiles, and transformation rates vary diurnally and seasonally. Wet removal is assumed to be a function of rain intensity and duration and takes into account the solubility of the pollutant. It is approximated by a power function of the 6-hr precipitation rates.

The meteorological data, required for model simulation, were obtained from the Canadian Meteorological Centre (CMC). Continental wind fields, at 1000 and 850 mb levels, were obtained every 6 hr with a spatial grid resolution of 127×127 km from the CMC spectral model. Objectively analyzed 24-hr precipitation fields were obtained twice daily with a grid resolution of 127×127 km. These were allocated to 6-hr intervals.

PRIMARY AND SECONDARY EMISSIONS OF TOXAPHENE

Toxaphene is lost to the atmosphere through both primary and secondary emissions. Primary emissions are the result of dispersion of application spray, volatilization and subsequent wind dispersion from the plant canopy and soil surface, and wind entrainment of pesticides shortly after application. Secondary emissions include volatilization and reentrainment of deposited material, and volatilization from soil. These processes are congener-dependent and should, ideally, be simulated in the model. However, due to insufficient information, emissions were provided as input to the model. Hence, an emissions inventory was compiled from information on use patterns, methods and time of application, predictions from spray models, experimental studies on volatilization rates, as well as from information on residence time and fate in soil.[7]

Vlier Zygadlo[3] provided a breakdown of toxaphene use in 1980 as a function of crop type for six major regions in the United States. These use patterns were augmented with (1) information on usage in individual states, and (2) information on cropland distribution, as a function of type of crop and state, from the 1978 Census of Agriculture.[11] From these combined sources, toxaphene usage was allocated either directly to a particular state or indirectly through the assumption of direct proportionality between toxaphene usage on a particular crop and the amount of acreage of that crop in a state to that in the region to which the state belongs. The use pattern is shown in Figure 2.

We have assumed that 50% of the toxaphene applied to crops in 1980 was emitted into the atmosphere during the summer months, and 10% was emitted in the early spring. These assumptions are discussed in Voldner and Schroeder.[7] The following calculations are based on the single year 1980 and do not include emissions of toxaphene accumulated in the soil from application in previous years.

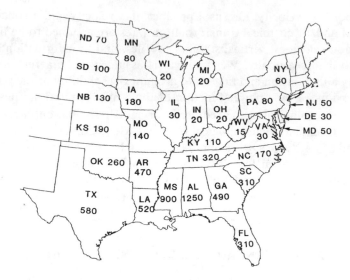

Figure 2.   1980 toxaphene usage.

MODELING RESULTS AND DISCUSSION

Before attempting to simulate the atmospheric pathways of toxaphene to the Great Lakes region, it is of interest to determine if the meteorological conditions were conducive to its transport from major source regions. Thus, using air parcel trajectories originating every 6 hr from a given source region, seasonal mean trajectories and standard deviations of individual trajectories from the mean trajectories were computed. The mean trajectory showed the "preferred" path of flow; the standard deviation indicated the dispersion potential of an inert tracer. Figure 3 shows the 1980 summer (defined as July-September) mean trajectory originating in Texas, a major source region. The trajectory has a duration of 5 days. The ellipses illustrate the horizontal standard deviations about the mean of individual trajectory endpoints at different times after release.

The figure clearly illustrates the potential exposure of the Great Lakes ecosystem to pollutants originating in the southern United States. The mean trajectory from Texas passes over the lower Great Lakes, with individual trajectories frequently passing over the upper lakes. Similarly, trajectories originating in other states show a significant potential for exposure.

Of the 3600 metric tons of toxaphene emitted in the United States in 1980, approximately 60% was deposited within North America. The remainder left the continent primarily over the eastern coast. Transport and deposition also extended into northern Manitoba, the James Bay region, and northern Quebec. Transport to the North Atlantic is supported by the measurements of Bidleman and Olney[12] and Bidleman et al.[13] Muir[14] has detected toxaphene in fish in northern Canada.

Figure 3.   Summer mean trajectory and spread, originating in Texas.

Table 1 summarizes the estimated annual deposition to the Great Lakes region for 1980. The results are provided for each individual lake drainage basin as well as for each lake directly and are broken down into wet and dry components. Atmospheric deposition to the basins as well as the lakes deserves consideration since the former constitute a secondary source of toxaphene to the lakes through, for example, revolatilization, resuspension, and runoff. Wet deposition is estimated to contribute 70–80% of the total deposition, depending on which receptor region and time period are considered.

As can be seen from Table 1, it is estimated that Lake Michigan received the largest deposition, about 10 metric tons in 1980; deposition to Lakes Superior and Huron was essentially the same (8 metric tons). The deposition fluxes over

Table 1.   Estimated Wet, Dry, and Total Deposition (metric tons) of Toxaphene to the Great Lakes and Their Basin for 1980.

	Wet	Dry	Total
Lakes			
Superior	6	2	8
Michigan	7	3	10
Huron	6	2	8
Erie	4	1	5
Ontario	2	1	3
Basins			
Superior	9	5	14
Michigan	14	5	19
Huron	12	3	15
Erie	14	4	18
Ontario	8	2	10

Lake Erie were relatively high, considering that the surface area of the lake is small. For Lake Superior the converse is true. Average air concentrations for toxaphene over the Great Lakes were generally below 1 ng/m³ during early spring and 1–2 ng/m³ during summer, resulting in an annual average concentration below 0.5 ng/m³.

Our estimated deposition of 10 metric tons to Lake Michigan in 1980 compares favorably with estimates of 3–7 metric tons for 1981 by Rice and Samson.[15] There is a great deal of uncertainty in the latter estimate since it is based on a few measurements of ambient air concentrations obtained in summer and fall of 1981 at two sites near the northern and southern ends of Lake Michigan.

Estimated deposition along a transect from north central Minnesota to Nova Scotia by Rapaport and Eisenreich[16,17] is substantially lower than that predicted by the model. Rapaport and Eisenreich determined recent deposition to peat bogs to be in the range 0.5–9 μg/m²/year; our estimates for 1980 fall in the range 50–130 μg/m²/year. The peat core samples indicate little degradation of toxaphene. Nevertheless, the composition of the environmental samples differ from standard toxaphene. The model predictions are influenced by the assumed source distribution around the Lakes and along the United States-Canada border (Figure 2). Reduction in the strength of these sources would result in lower model predictions.

The predicted average air concentration over Lake Michigan during the period July–December 1980 is about 0.7 ng/m³; Rice and Samson[15] and Rice et al.[18] reported measured average ambient air concentrations of 0.3 ng/m³ for August–November 1981. Rice[19] contends there is a great deal of uncertainty in these measurements. The mean percent of matched peaks against selected peaks in toxaphene "standard" was only 37%. Rice and Samson also report measured air concentrations closer to the source regions, i.e., Greenville, Mississippi, and St. Louis, Missouri. Average air concentrations during August-November 1981 were 7.4 and 1.2 ng/m³, with 51 and 38% matched peaks, respectively. Predicted average air concentrations over Mississippi and Missouri during July–December 1980 are 8 and 3 ng/m³. Considering the uncertainty in the measurements and the differences in time period, the agreement between measured and predicted quantities is good.

The model estimates pertain to 1980. Recently Hoff[20] detected toxaphene in air samples taken in southern Canada. Since toxaphene is banned in North America, its presence in the atmosphere is likely due to secondary emissions from soils in North America and from applications in Mexico.[7,21]

As with ambient air concentrations, there is a scarcity of measurements of toxaphene in rain in the Great Lakes region. Rice and Samson[15] report two measurements at Beaver Island, Michigan: one in April 1981 and one in September-October of the same year. The former contained 70 ng/L (5 peaks matched out of 37); the latter, 32 ng/L (14 peaks out of 37). Based on 17 collectors located around Lakes Michigan and Huron, Swain (as reported by Environmental Protection Agency[22]) found an average concentration in precipitation around these lakes in the range of 5–108 ng/L. However, no information on the period of sampling was provided. Our predicted annual average concentrations are comparable. Thus, the predicted annual average concentrations for Lakes Michigan and Huron are estimated at 130 ng/L and 100 ng/L, respectively.

SUMMARY AND CONCLUSIONS

It has been suggested that the atmosphere constitutes a primary transport route of toxaphene to the Great Lakes from the major source regions in the southern United States. Environmental measurements are too few to estimate the total input of toxaphene to the Great Lakes and their basins.

The ASTRAP model, extensively used in acid rain research, has been modified for simulation of the atmospheric pathway of toxaphene. Based on use patterns in North America during 1980, ambient air concentrations and deposition of toxaphene to the Great Lakes and their basins have been estimated for this period. The results confirm suggestions based on measurements that the atmosphere is a major transport route of toxaphene to the Great Lakes region. They also show that toxaphene can be transported out of the North American continent, over the North Atlantic.

Total estimated deposition to each of the five Great Lakes was in the range of 3–10 metric tons in 1980, and annual average air concentrations over this region were about 0.5 ng/m^3. Lake Michigan received the highest input; Lake Ontario, the lowest. Although the information on physical and chemical properties, as well as emissions inventories, is incomplete and air quality and precipitation chemistry measurements of toxaphene are few and highly uncertain, model predictions show good agreement with the measurements.

Even though toxaphene was restricted in 1982 and banned in 1986 in North America, it is still detected in air samples in the Great Lakes region. Its presence is likely due to secondary emissions from soils in North America and from applications in Mexico.

To improve model-derived estimates of deposition to the Great Lakes for priority toxic chemicals in general, and toxaphene in particular, it is essential that suitable emission inventories are compiled and information on physicochemical properties improved. To verify the model predictions, high quality air, precipitation, and lake water chemistry measurements are required.

ACKNOWLEDGMENTS

The authors wish to acknowledge the support of this work by Environment Canada and to thank Dr. J. D. Shannon for the loan of the basic ASTRAP model and for aid in making modifications, Dr. T. Bidleman for helpful discussions, Mr. G. West for computational support, and Mrs. M. Perbaud and Ms. E. Mathis for typing the manuscript.

REFERENCES

1. Strachan, W. M. J., and S. J. Eisenreich. "Appendix 1 from the Workshop on Great Lakes Atmospheric Deposition," International Joint Commission SAB/WQB/IAQAB, Scarborough, Ont., 1986, (1988), p. 113.
2. Mackay, D., S. Paterson, and W. H. Schroeder. "Model Describing the Rates of Transfer Processes of Organic Chemicals Between Atmosphere and Water," *Environ. Sci. Technol.* 20:810–16 (1986).
3. Vlier Zygadlo, L. "Domestic Usage of Toxaphene," Environmental Protection Agency, Office of Pesticide Programs, Economic Analysis Branch (1982), p. 9.
4. "Toxaphene," IRPTC Bulletin 5, No. 3 (1983), p. 27.
5. Voldner, E. C., and L. Smith. "Appendix 2 from the Workshop on Great Lakes Atmospheric Deposition," International Joint Commission SAB/WQB/IAQAB, Scarborough, Ont., 1986, (in press).
6. "Toxaphene–Environmental Health Criteria Document," Health and Welfare Canada, Health Protection Branch, Ottawa, Report 77-EHD-11 (1977), pp. 41–43.
7. Voldner, E. C., and W. H. Schroeder. "Modelling of Atmospheric Transport and Deposition of Toxaphene into the Great Lakes Ecosystem," *Atmos. Environ.* 23(9):1949–61 (1989).
8. Shannon, J. D. "A Model of Regional Long-Term Average Sulfur Atmospheric Pollution, Surface Removal and Net Horizontal Flux," *Atmos. Environ.* 15:689–701 (1981).
9. Shannon, J. D., and E. C. Voldner. "Estimation of Wet and Dry Deposition of Pollutant Sulfur in Eastern Canada as a Function of Major Source Regions," *Water Air Soil Pollut.* 18:101–4 (1982).
10. Shannon, J. D., and B. M. Lesht. "Estimation of Source-Receptor Matrices for Deposition of NO_x-N," *Water Air Soil Pollut.* 30:815–24 (1986).
11. Menzie, C. M., and D. F. Walsh. "Toxaphene: An Assessment," U.S. Fish and Wildlife Service (1982), p. 44.
12. Bidleman, T. F., and C. E. Olney. "Long Range Transport of Toxaphene Insecticide in the Atmosphere of the Western North Atlantic," *Nature* 257:475–77 (1975).
13. Bidleman, T. F., E. J. Christensen, W. N. Billings, and R. Leonard. "Atmospheric Transport of Organochlorines in the North Atlantic Gyre," *J. Marine Research* 39:443–63 (1981).
14. Muir, D., N. P. Griff, C. Ford, A. Reigner, M. Hendzel, and W. L. Lockhart. "Evidence for Long Range Transport of Toxaphene to Remote Arctic and Subarctic Waters from Monitoring Fish Tissue," Third Chemical Congress of North America, Toronto, Ont. (June 1988).
15. Rice, C. P., and P. J. Samson. "Atmospheric Transport of Toxaphene to Lake Michigan," report prepared for EPA under grant no. 808849 (1983).
16. Rapaport, R. A., and S. J. Eisenreich. "Atmospheric Deposition of Toxaphene to Eastern North America Derived from Peat Accumulation," *Atmos. Environ.* 20:2367–79 (1986).
17. Rapaport, R. A., and S. J. Eisenreich. "Historical Atmospheric Inputs of High Molecular Weight Chlorinated Hydrocarbons to Eastern North America," *Environ. Sci. Technol.* 22:931–41 (1988).
18. Rice, C. P., P. J. Samson, and G. E. Noguchi. "Atmospheric Transport of Toxaphene to Lake Michigan," *Environ. Sci. Technol.* 20:1109–16 (1986).

19. Rice, C.P. Personal communication.
20. Hoff, R. M. "Measurements of Ambient Air Concentrations of PCBs and OCs in Southern Ontario," presented at the 32nd Conference on Great Lakes Research, Madison, WI (May 1989).
21. Voldner, E. C., and L. Smith. "Emission of 14 Priority Toxic Chemicals," presented at the 32nd Conference on Great Lakes Research, Madison, WI (May 1989).
22. "Toxaphene: Decision Document," Environmental Protection Agency, Office of Pesticide Programs (1982).

## CHAPTER 16

Atmospheric Deposition of Selected Organochlorine Compounds in Canada

William M. J. Strachan

INTRODUCTION

Toxic chemicals from the atmosphere have been noted in environmental samples for many years, dating back to the mid-1960s.[1-3] Most of the reports pertain to persistent organochlorine pesticides in common use up to the late 1970s and early 1980s. Even now, however, compounds that were banned in Canada and the United States continue to be observed in rain and other atmospherically related samples.

In Canada, official interest in toxic organic compounds in the atmosphere started about 1975, following the observation of 10–100 ng PCBs/L in the rainfall of southwestern Ontario.[4] Sampling of rain and snow undertaken by the author in the Great Lakes basin found similar levels of PCBs, as well as a number of other organochlorine compounds, during the period 1976–77.[5] Investigators on the U.S. side of the lakes also found PCBs in rain and snow and reported these at the 1976 conference of the International Association for Great Lakes Research and in subsequent publications.[6,7]

Following these earlier activities, Environment Canada modified an existing sampler design and situated several of the modified versions at approximately 20 locations across Canada. The results from this network were only semi-quantitative since the collection area of the sampler did not generally permit samples of sufficient size to be obtained for the analytical detection limits of the time. Where sufficient samples were obtained, however, they tended to confirm the pattern and levels of contaminants observed earlier in the Great Lakes region. Results were reported at an American Chemical Society meeting and published as a government report.[8] The results reported here are a continuation of this concern about atmospherically transported and deposited organic chemicals in the Canadian environment.

EXPERIMENTAL

A consequence of the earlier efforts was the development of a wetfall-only rain sampler specifically designed for collection of organics. A description of its construction and the associated analytical procedures and recoveries has been reported elsewhere.[9] Briefly, the collection funnel area is 0.2 m² (1 cm of rainfall produces 2 L of sample), and the lid opening is controlled by a sensor responding to wetfall precipitation. The sample passes through an all-Teflon column containing XAD-2 resin and exits via a U-tube, which ensures that the resin remains wet at all times. All funnel surfaces in contact with the sample are Teflon-coated. The analytes are extracted from the resin with diethyl ether, dried over Na_2SO_4, cleaned on a silica gel column, and analyzed on a SE54 capillary column gas chromatograph.

Because of earlier experiences with rain sampling and the natural variability of environmental samples of this sort, it was decided to place these samplers as replicate sets of three at selected locations. During the period 1983–86, they were deployed at a number of sites across Canada. The locations are indicated in Figure 1; not all stations were operated during all years. During 1985, Environment Canada also established a network around the Great Lakes with single samplers to enable it to provide better geographical coverage and to respond to the requirements of the Great Lakes Water Quality Agreement (1978). These locations are indicated in Figure 2.

Figure 1.   Canadian replicate rain sampling sites: 1983–86.

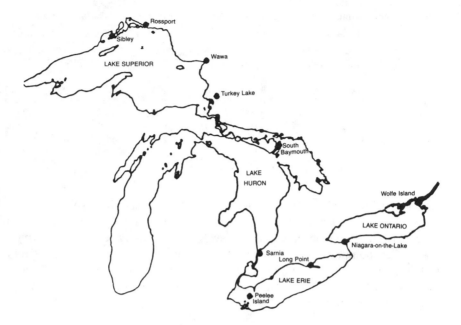

Figure 2.   Great Lakes Organic Rain Sampling Network: 1985–86.

The samplers were installed early during each wetfall season (usually in May) at locations remote from any probable sources of the analytes or evidence of urban activity; they were operated until near freeze-up (variously from the end of September to November). The sites of the replicate sampling stations were mainly in provincial or national parks, but two were at research stations of the Atmospheric Environment Service and a third was on a Coast Guard lighthouse station. Samples for these locations corresponded to approximately 100 mm rain (20 L), resulting in 3–5 replicated samples for the wetfall period. At the Great Lakes network sites, the samplers were approximately 50% at federal or provincial agency locations (and manned by their technical staff), and 50% on private property usually well removed from any industrial or vehicular activities or other obvious sources of the analytes or interfering substances. These latter samples were collected on a monthly basis during the rain season.

When the columns were removed from the samplers, they were immediately replaced with clean, conditioned columns; the exposed ones were sealed with Teflon caps and mailed to the author. The resin was extracted and analyzed for a range of 15 organochlorine pesticides and PCBs; results are corrected for blank values determined for each batch of resin. The compounds investigated included those reported here and, in addition, heptachlor, heptachlor epoxide, aldrin, the chlordanes, other forms of DDT and its transformation products, methoxychlor, the endosulfans, endrin, and mirex. The quantification limits for all of these were approximately 0.02 ng/L. A number of these compounds

Table 1.   Mean Annual Concentrations (ng/L) of Contaminants in Canadian Rain.

	L. Superior		Kouch.	Cree Lake		Suffield	Kanaka Cr.	
	1983	1984	1984	1984	1985	1985	1985	1986
# Samples	8	9	4	3	3	3	3	3
α-HCH	26.0	6.6	13.0	6.5	22.0	14.0	29.0	14.0
Lindane	6.4	3.0	6.7	1.2	6.5	5.9	5.0	2.9
Dieldrin	0.43	0.79	0.27	0.38	0.04	0.10	0.27	0.02
p,p'-DDE	0.13	0.13	0.02	0.07	0.04	0.03	nd	0.02
PCBs	6.0	2.9	1.1	3.1	3.5	5.5	nd	0.62
HCB	0.07	0.06	0.07	tr	tr	0.84	nd	0.03

nd = not detected; tr = trace (below "quantification").

were found in each column of individual replicate sets but were not found consistently at the same or other sample locations, nor throughout the sampling period. When observed, their levels were usually less than 0.5 ng/L. These other compounds are not reported on in this chapter. Mention is made of the pesticide methoxychlor since it was found in most Lake Superior samples at concentrations exceeding 1 ng/L; it was not found, however, in samples from elsewhere.

RESULTS AND DISCUSSION

Mean annual concentrations of contaminants found at the various replicate station locations over the four years are presented in Table 1. These means were determined by estimating the total loadings for the wetfall season using the several replicate sample means from each site and dividing by the total amount of rain falling during the season. The six compounds reported are the most commonly observed ones; they were found in at least 90% of all samples and were consistent (RSD approximately 30%) within a replicate sample set.

The relative concentrations of the six compounds are similar at all of the locations and for all of the years reported in Table 1. The most prominent ones are always the hexachlorocyclohexanes (α-HCH and lindane, the γ-HCH isomer), which together had an arithmetic average of the annual mean concentrations of 21 ng/L. PCBs were found at the second highest concentration levels of the six contaminants—the average of the results for the several years and sites was 3.6 ng/L. This class of compounds, as well as the remaining pesticides, accumulate in fish and other aquatic biota[10] and are responsible, in large part, for much of the concern over the whole family of organochlorine pesticides. The remaining compounds from Table 1 and their mean annual concentration levels were dieldrin, 0.29 ng/L; DDE, 0.09 ng/L; and hexachlorobenzene (HCB), 0.12 ng/L.

It is interesting to examine what the significance of these concentrations might be with regard to loadings at the different locations investigated. The use of loadings has the advantage of normalizing any comparisons among locations, although neither it nor the concentration estimations can give any

Table 2.   Contaminant Loading Rates from Rainfall (μg/m^2/year).

	L. Superior		Kouch.	Cree Lake		Suffield	Kanaka Cr.	
	1983	1984	1984	1984	1985	1985	1985	1986
Rain (mm)	607	618	1050	307	274	208	1241	1651
Snow (mm)	214	263	377	176	207	98	56	34
α-HCH	16.0	4.3	14.0	2.1	6.5	3.3	36.0	22.0
Lindane	4.0	1.9	7.3	0.39	1.9	1.3	6.2	4.8
Dieldrin	0.27	0.51	0.30	0.12	0.01	0.022	0.34	0.03
p,p'-DDE	0.083	0.084	0.02	0.02	0.01	0.007	—	0.03
PCBs	4.9	2.6	1.6	1.5	1.7	1.2	—	1.0
HCB	0.04	0.04	0.08	—	—	0.18	—	0.05

indication of the influence of intensity and number of events. These latter factors may be important determinants of the quantities of chemicals removed from the atmosphere via the rain. Table 2 presents the loadings for the replicate stations. In making the estimations, the mean rain concentrations from Table 1 were used with total rainfall to estimate the contribution from rain. For snow, the concentrations used were 10% of those in rain except for PCBs, which were taken to be the same as in rain;[5] the snowfall precipitation data used were from Environment Canada and collected in their regular weather reporting program. The means for Lake Superior for 1983 and 1984 are derived by averaging the annual means found for Isle Royale and Caribou Island (1983) and Caribou Island and Agawa (1984).

A somewhat higher loading of the HCH isomers was observed on the western coast (Kanaka Creek) compared to the other sites. This may have been the consequence of high usage of lindane in parts of Asia[11] coupled with long-range atmospheric transport; this sort of transport has been indicated by deposition of these materials in Japan[12] and Hawaii.[13] Studies on the global movement of the Chernobyl radionuclides in 1986,[14] of DDT by Peakall,[15] and of other pesticides by Cohen and Pinkerton[16] have clearly demonstrated this mode of redistribution, even at intercontinental distances.

The deposition of the persistent organochlorine compounds mentioned in this report has been a particular concern in the Great Lakes region. There, the governments of Canada and the United States, through their Great Lakes Water Quality Agreement and the International Joint Commission that administers it, have identified the same compounds accumulated in the adipose tissues of resident biota,[10] including humans.[17] The major source of these compounds in this system has been shown to be atmospheric precipitation (see Chapter 19 and references cited therein). The rain concentrations for 1985 and 1986 from a Great Lakes network are presented in Tables 3 and 4, respectively. These show no significant differences from those obtained at other locations in Canada, as reported in Tables 1 and 2. As with the replicate sampling for these and other years, the pattern of compounds is qualitatively the same; α-HCH and lindane are found at greatest concentrations, followed by the PCBs and the other three pesticides. The results from these samples were not included in

Table 3.   Mean Contaminant Concentrations (ng/L) in Great Lakes Rain: 1985.

	Superior	Huron	Erie	Ontario
# Sites	4	2	2	2
# Samples	15	7	6	8
α-HCH	25.0	6.5	16.0	11.0
Lindane	4.6	2.0	5.5	3.7
Dieldrin	0.36	0.71	0.86	0.48
p,p'-DDE	tr	0.03	nd	0.06
PCBs	0.51	1.6	2.3	2.5
HCB	tr	0.06	nd	0.04

nd = not detected; tr = trace (below "quantification").

Table 1 since they were not replicated and were gathered on a monthly rather than a volume basis. The values (in ng/L), weighted for the number of samples from each lake for 1985 and 1986, respectively, were α-HCH, 17.0 and 9.5; lindane, 4.0 and 4.9; dieldrin, 0.54 and 0.14; DDE, 0.02 and 0.16; PCBs, 1.5 and 0.76; and HCB, 0.02 and 0.07.

Only the Lake Superior data were collected for a sufficient period to consider possible trend determinations. The period 1983–86 does not show any consistent trend, although the differences between the concentrations from this period and those from earlier 1975–76 samples[5] indicate a substantial drop in PCBs and DDT residues. This observation, however, is speculative since changes in analytical procedures between the two periods make comparisons dubious. The apparent lack of a trend in the rain data is in contrast with that found in some biota from the Great Lakes.[18] Data for herring gull eggs, collected annually since 1974, show a steady downward trend, with a half-life of clearance from this subcompartment of approximately six years; fish data are more ambiguous over the same period[19] but are also generally decreasing.

The compounds reported here represent a group for which sensitive analytical techniques are available, using a gas chromatograph with an electron capture detector. There are many other compounds potentially present in the atmosphere and potentially hazardous because of their known toxic effects and persistence. They should be the subject of method development, followed

Table 4.   Mean Contaminant Concentrations (ng/L) in Great Lakes Rain: 1986.

	Superior	Huron	Erie	Ontario
# Sites	4	2	2	2
# Samples	16	9	10	10
α-HCH	12.0	5.9	7.1	11.0
Lindane	4.9	3.9	4.7	5.9
Dieldrin	0.06	0.18	0.27	0.08
p,p'-DDE	0.02	0.15	0.28	0.26
PCBs[a]	0.82	0.34	0.41	1.4
HCB	0.03	0.22	0.04	tr

tr = trace (below "quantification").
[a]High blank values.

by application in the field. In addition, research is sorely needed on several processes affecting the levels in the atmosphere, including volatilization from the water compartment, washout by rain and snow, partitioning between vapor and particle-adsorbed states, and deposition of particle-bound contaminants. All of these affect the concentrations and ultimately the loadings to both the terrestrial and the aquatic environments, and all are poorly understood.

REFERENCES

1. Wheatley, G. A., and J. A. Hardman. "Indications of the Presence of Organochlorine Insecticides in Rainwater in Central England," *Nature* 207:486–87 (1965).
2. Risebrough, R. W., R. J. Huggett, J. J. Griffin, and E. D. Goldberg. "Pesticides: Transatlantic Movements in the Northeast Trades," *Science* 159:233–36 (1968).
3. Peterle, T. J. "DDT in Antarctic Snow," *Nature* 224:620 (1969).
4. Sanderson, M., and R. Frank. "Report of the Task Force on PCB to the Joint Department of Environment and National Health and Welfare Committee on Environmental Contaminants," Environ. Canada Report EPS-CCB-77-1 (1977).
5. Strachan, W. M. J., H. Huneault, W. M. Schertzer, and F. C. Elder. "Organochlorines in Precipitation in the Great Lakes Region," in *Hydrocarbons and Halogenated Hydrocarbons in the Aquatic Environment*, B. K. Afghan and D. Mackay, Eds. (New York, NY: Plenum Press, 1980), pp. 387–96.
6. Murphy, T. J., and C. P. Rzeszutko. "Precipitation Inputs of PCBs to Lake Michigan," *J. Great Lakes Res.* 3:305–12 (1977).
7. Swain, W. R. "Chlorinated Residues in Fish, Water and Precipitation from the Vicinity of Isle Royale, Lake Superior," *J. Great Lakes Res.* 4:398–407 (1978).
8. Brooksbank, P. L. "The Canadian Network for Sampling Organic Compounds in Precipitation," Environment Canada, Inland Waters Technical Bulletin 129 (1983).
9. Strachan, W. M. J., and H. Huneault. "Automated Rain Sampler for Trace Organic Substances," *Environ. Sci. and Technol.* 18:27–30 (1984).
10. "Report on Great Lakes Water Quality," International Joint Commission (1983).
11. Kaushik, C. P., M. K. K. Pillai, A. Raman, and H. C. Agarwal. "Organochloride Insecticide Residues in Air in Delhi, India," *Water Air Soil Poll.* 32:63–76 (1987).
12. Marahiro, O., and H. Takahisa. "Alpha-and Gamma-BHC in Tokyo Rainwater (December 1968 to November 1969)," *Environ. Pollut.* 9:283–88 (1975).
13. Benvenue, A., J. N. Ogata, and J. W. Hylin. "Organochlorine Pesticides in Rainwater, Oahu, Hawaii," *Bull. Environ. Contam. Toxicol.* 8:238–41 (1972).
14. Joshi, S. R. "The Fallout of Chernobyl Radioactivity in Central Ontario, Canada," *J. Environ. Radioactivity* 6:203–11 (1988).
15. Peakall, D. B. "DDT in Rainwater in New York Following Applications in the Pacific Northwest," *Atmos. Environ.* 10:899–900 (1976).
16. Cohen, J. M., and C. Pinkerton. "Widespread Translocation of Pesticides by Air Transport and Rain-Out," *Organic Pesticides in the Environment*, R. F. Gould, Ed., ACS Advances in Chemistry Ser. 60 (Washington, DC: American Chemical Society, 1966), pp. 163–76.
17. Mes, J., D. J. Davies, and D. Turton. "Polychlorinated Biphenyls and Other

Hydrocarbon Residues in Adipose Tissue of Canadians," *Bull. Environ. Contam. Toxicol.* 28:97–104 (1982).

18. Mineau, P., G. A. Fox, R. J. Norstrom, D. V. Weseloh, D. J. Hallett, and J. A. Ellenton. "Using the Herring Gull to Monitor Levels and Effects of Organochlorine Contamination in the Canadian Great Lakes," in *Toxic Contaminants in the Great Lakes*, J. O. Nriagu and M. S. Simmons, Eds. (New York, NY: John Wiley and Sons, 1984), pp. 425–52.

19. "1987 Report on Great Lakes Water Quality," International Joint Commission, Great Lakes Water Quality Board, Windsor, Ont. (1987).

## CHAPTER 17

Transport of Soluble Pesticides Through Drainage Networks in Large Agricultural River Basins

David B. Baker and R. Peter Richards

INTRODUCTION

Most studies on long-range and atmospheric transport of synthetic organic chemicals and pesticides have focused on compounds that are persistent and that tend to bioaccumulate, such as PCBs, DDTs, PAHs, and toxaphene. Such a focus is certainly justified since these two characteristics combine to confer upon such compounds the potential for significant and long-term ecological damage, as well as concern regarding possible human health impacts. Awareness of such damage and concern has led to restrictions and even banning of the use and manufacture of many of these compounds.

Although some pesticides have been banned, total pesticide use in the United States increased by about 170% between 1964 and 1985.[1] Fortunately, most of the newer generation pesticides are much less persistent and have a much lower tendency to bioaccumulate. Much of the increase in pesticide use has been comprised of large increases in the use of herbicides. In the corn and soybean production areas of the Ohio drainage to Lake Erie, herbicides comprised 92% by weight of the total pesticides used in 1986.[2] For many of these compounds, volatilization is a significant pathway leading to their dissipation from application sites.[3] The axiom "what goes up must come down" has again been confirmed in the recent reports of relatively "high" concentrations of many current generation pesticides in dew[4] and in rainfall.[5]

Although the concentrations of current generation pesticides in rainfall are generally high in comparison to concentrations of the persistent organochlorine insecticides in rainfall, by far the highest widely occurring offsite concentrations of current generation pesticides are associated with surface water runoff from edges of fields directly into streams and rivers.[6] Accidental spills or leaks can result in extremely high, localized concentrations, but such incidents are infrequent and localized in comparison with exposures associated with runoff events from fields receiving normal pesticide applications.

Many of the currently used compounds have low affinities for sediments and are transported primarily in the dissolved state rather than through attachment to sediments.[7] Where surface waters containing these soluble compounds are used for public water supplies, conventional water treatment does little to lower pesticide concentrations,[8,9] leading to pesticide exposures through drinking water.

Both the human health risks posed by pesticides in drinking water supplies and the ecological risks posed by pesticides in surface waters are dependent on the interaction of two major sets of factors, the exposure to the pesticides and the toxicity of the pesticides (Figure 1). Given long-standing concerns regarding human health and ecological effects of pesticides, it is rather surprising that there is very little exposure data for current generation pesticides in streams and rivers. In a 1978 review of edge-of-field losses of pesticides, Wauchope noted a paucity of information regarding the fates and effects of pesticides after they left the edge of the field.[10]

This lack of exposure data for current generation pesticides is a consequence of several factors. Most federal and state water quality monitoring programs involve fixed station approaches with sampling frequencies ranging from monthly to annually, depending largely on the cost of the analyses.[11] Such programs provide reasonable information regarding the impacts of point sources of pollution, but they are inappropriate for non-point source pollution studies. A nationwide pesticide monitoring program in rivers, conducted between 1975 and 1980 by the U.S. Geological Survey, involved collection of four samples per year at 160 stations and analysis for 18 insecticides and 4 herbicides.[12] The program focused on studies of the disappearance of banned pesticides rather than the occurrence of current generation compounds. The 22 pesticides together made up less than 33%, by weight, of the pesticides used during that period. Fewer than 10% of the river samples contained reportable

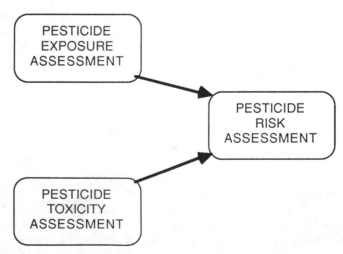

Figure 1. Relationships between exposure assessment, toxicity assessment, and risk assessment for pesticides.

pesticide concentrations. More recently, federally mandated pesticide monitoring has been restricted to annual tests in municipal water supplies for six pesticides (endrin, lindane, methoxychlor, toxaphene, 2,4-D, and 2,4,5-TP). Recommended maximum contaminant levels (RMCLs) were set for these six compounds through implementation of the Safe Drinking Water Act, triggering required monitoring for these compounds. These compounds comprise a very small portion of current pesticide use. In Ohio, they make up approximately 1% of the total amount of pesticides applied. They are rarely detected in public water supplies and apparently have never been observed to exceed the RMCLs. In Ohio, as well as in many other states, no mandated pesticide monitoring was in force through 1987 for 99% of the pesticides used in today's agriculture. In 1988, the Ohio EPA required water treatment plants utilizing surface waters to analyze one sample, collected during the May through July period, for alachlor and metolachlor.

Extensive studies of current generation pesticides have been conducted for research plots and individual fields at land grant universities and agricultural research centers.[10] The U.S. EPA's pesticide monitoring strategy is largely based on the use of models to predict in-stream concentrations.[13] Edge-of-field pesticide runoff data are used as input for the models. Although these models have been used to predict pesticide concentrations in streams and rivers, with few exceptions,[14] these predictions have rarely been verified with actual concentration data from detailed monitoring programs.

The streams and rivers draining into Lake Erie have been the focus of detailed studies of nutrient and sediment runoff, beginning in the early 1970s.[15,16] In 1981, pesticide runoff studies were added to the program, in part to assess whether conservation tillage would aggravate pesticide runoff problems. Conservation tillage is viewed as a fundamental component of agricultural nonpoint pollution control programs aimed at reducing phosphorus loading to Lake Erie. It is likely that the resulting pesticide runoff studies in streams and rivers of the Lake Erie Basin are the most detailed of their type in the United States. In the absence of comparable studies in other regions, it is unclear whether the results observed in this region reflect unusually high, normal, or low degrees of pesticide exposure. It is likely, however, that the exposure patterns and characteristics that we have observed in this region will also occur in other streams and rivers draining watersheds with intensive row crop production. In this chapter, we will summarize those exposure patterns and characteristics, using data from the streams and rivers of the Lake Erie Basin as a case study.

METHODS

To characterize pesticide exposure patterns in streams draining agricultural landscapes, we have focused our efforts at a relatively small number of locations; at those locations, however, we collect samples at frequent intervals, especially during runoff events. All of our sampling locations are located at U.S. Geological Survey stream gauging stations in the Lake Erie Drainage Basin (Figure 2). These stations provide continuous discharge data so that both pesticide loadings and concentration patterns can be determined. The drainage areas upstream from the sampling stations range from 11.3 to 16,395 km². Most of the soils have a relatively heavy texture (clays, silty clay loams, and silty loams) and use of subsurface tile drainage is common. The drainage areas and land use upstream from each station are summarized in Table 1.

Corn and soybeans are the major crops grown in the region. Together, they account for 99.7%, by weight, of the herbicides and 90.1%, by weight, of the insecticides used in 1986 in the Ohio portion of the Lake Erie Basin.[2] The 20 pesticides used in the largest quantities in this region are listed in Table 2. The 15 herbicides on the list accounted for 97.3% of the total herbicide use and the 5 listed insecticides accounted for 87.9% of the total insecticide use. The bulk of the pesticide use occurs within the northwestern and north central parts of Ohio, which are drained primarily by the Maumee and Sandusky rivers, respectively. Table 2 also includes a 1982 ranking of pesticides, by amount used, for the entire state of Ohio.[17] Although there have been some changes in

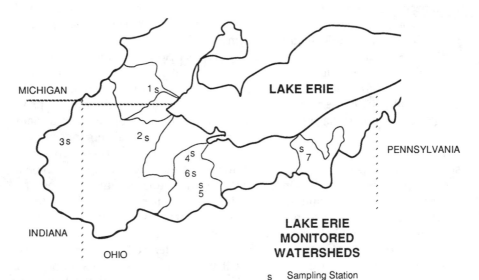

Figure 2. Locations of pesticide monitoring stations for the Lake Erie Basin agricultural runoff studies. The United States Geological Survey stream gauging station identification numbers are listed alongside the name for each of the tributary sampling stations.

Table 1. Watershed Areas and Land Use for the Pesticide Monitoring Stations in the Lake Erie Basin

Watershed	Watershed Area (km^2)	Cropland (%)	Pasture (%)	Forest (%)	Water (%)	Other (%)
Maumee River at Waterville, OH	16,395	75.6	3.2	8.4	3.5	9.4
Sandusky River at Fremont, OH	3,240	79.9	2.3	8.9	2.0	6.8
River Raisin near Monroe, MI	2,699	67.1	6.8	9.0	3.0	14.1
Cuyahoga River at Independence, OH	1,831	4.2	43.1	29.1	3.0	20.6
Honey Creek at Melmore, OH	386	82.6	0.6	10.0	0.5	6.3
Rock Creek at Tiffin, OH	88.0	80.9	2.3	11.8	0.9	4.2
Lost Creek Trib. near Farmer, OH	11.3	83.0	0	10.6	1.4	5.0

Source: Baker.[16]

the rankings of pesticide use during the course of these studies, the general pattern of pesticide use has changed very little.

For most stations, automatic samplers (ISCO Model 2700 or equivalent) are used to collect from two to four 1-L samples per day during the period from mid-April through mid-August, which encompasses most of the period of high ambient pesticide concentrations and loadings. The automatic samplers are housed in the U.S. Geological Survey stream gauging station shelters. Submersible pumps in the streams deliver water to sampling wells inside the shelters, and the sampler pumps withdraw water from the wells. During periods of runoff events, all of the samples from the automatic samplers are analyzed, while during non-event periods, two samples per week are analyzed. From mid-August through the winter to mid-April, grab samples are collected twice per month at each station.

Since the frequency of sampling varies in relation to the expected pesticide concentrations, with samples collected more frequently during periods of high concentrations, calculation of average exposures involves weighting of individual samples in relation to the time they are used to characterize the stream systems. These time-weighted mean concentrations (TWMCs) are calculated as follows:

$$\text{TWMC} = \frac{\Sigma c_i t_i}{\Sigma t_i}$$

where c_i are the observed concentrations and t_i are the times represented by each sample.

Table 2.   Pesticide Use in the Lake Erie Basin for 1986 and Ranking by Use in the State of Ohio for 1982

Pesticide	Brand	Type[a]	Quantity Used (metric tons)	1986 Rank, by Use	1982 Rank by Ohio Use[b]
Alachlor	Lasso	H	1319	1	1
Metolachlor	Dual	H	897.2	2	3
Atrazine	Aatrex	H	783.9	3	2
Cyanazine	Bladex	H	273.4	4	4
Metribuzin	Lexone, Sencor	H	255.5	5	7
Chloramben	Amiben	H	155.5	6	6
Linuron	Lorox, Linurex	H	133.0	7	8
Terbufos	Counter	I	74.67	8	10
Trifluralin	Treflan	H	65.7	9	12
Butylate	Sutan, Genate plus	H	58.19	10	5
Dicamba	Banvel	H	54.39	11	19
Pendimethalin	Prowl	H	49.49	12	17
Bentazon	Basagran	H	49.36	13	15
Carbofuran	Furadan	I	48.23	14	11
2,4-D	2,4-D	H	44.44	15	14
Chlorpyrifos	Dursban	I	36.33	16	NR
EPTC	Eradicane, Eptam	H	34.67	17	13
Phorate	Thimet	I	31.51	18	NR
Fonofos	Dyfonate	I	25.10	19	9
Simazine	Princep	H	24.82	20	16

Total herbicide use in the Lake Erie Basin The 15 herbicides listed above make up 97.3% of the total herbicide use in the Lake Erie Basin.		4315.5
Total insecticide use in Lake Erie Basin The 5 insecticides listed above make up 87.9% of the total insecticide use in the Lake Erie Basin.		245.6

Sources: Waldron.[2,17]
[a]H, Herbicide; I, Insecticide.
[b]NR: not in top 20 in 1982.

   A multiresidue scan, using gas chromatography, capillary columns, temperature programming, and nitrogen-phosphorus detectors, is used for pesticide analysis. Samples are first extracted with methylene chloride, concentrated with Kuderna-Danish apparatus, and transferred to 2-propanol for subsequent analysis. Simultaneous injection into DB1 and DB5 capillary columns allows separation, identification, and quantification of most of the major pesticides used in this region. A partial list of the pesticides included in the scan, along with detection limits and spike recoveries, is shown in Table 3. All of the data presented in this chapter have been corrected for recoveries less than 100%, using either the average recovery for each compound each year or, where data from several years are combined, using the overall average recovery. The scan provides data for 89% by weight of the herbicides used in this region and for 75% by weight of the insecticides.

   Quality control procedures have included the analysis of blanks, replicates, reference standards, and spikes, as well as interlaboratory exchanges with

Table 3. Approximate Detection Limits and Lower Limits of Linear Response Ranges

Pesticide	Detection Limit (ng/L)	Lower Limit of Linear Response[a] (ng/L)	Mean Percent Recovery					Used (ACS)
			1984	1985	1986	1987	1988	
Alachlor (Lasso)	100	500	71	64	76	74	70	71
Metolachlor (Dual)	250	250	87	67	69	77	81	76
Atrazine (Aatrex)	50	500	86	69	76	76	70	75
Cyanazine (Bladex)	250	500	98	79	85	62	51	75
Metribuzin (Lexone, Sencor)	100	2500	54	65	73	70	74	67
Linuron (Lorox, Linurex)	1500	5000	—	80	86	—		83
Terbufos (Counter)	100	nd	76	54	73	73		69
Butylate (Sutan, Genate plus)	50	200	73	70	—	61		68
Pendimethalin (Prowl)	50	nd	80	71	71	71		73
Carbofuran (Furadan)	200	500	89	77	92	100		90
EPTC (Eradicane, Eptam)	50	nd	76	66	66	61		67
Fonofos (Dyfonate)	50	150	60	57	43	62		56
Simazine (Princep)	250	2500	88	74	77	89	87	83

Note: Based on analysis of dilution series of mixed standards; mean percent recoveries of spikes, by year; and mean recoveries used for calculations using the entire data set.

[a]nd: not determined.

Table 4. Comparison of Time-Weighted Mean Concentrations (TWMC) of Five Major Herbicides During the "Winter" and "Summer" Periods

Pesticide	Summer		Winter		Ratio
	N	TWMC (μg/L)	N	TWMC (μg/L)	(Summer/Winter)
Maumee River					
Alachlor	267	1.84	78	0.05	40.1
Metolachlor	267	2.80	78	0.22	12.7
Atrazine	267	4.05	78	0.79	5.1
Cyanazine	267	1.17	78	0.07	15.9
Metribuzin	267	0.87	78	0.11	7.6
Sandusky River					
Alachlor	307	2.48	69	0.06	44.1
Metolachlor	307	4.79	69	0.33	14.5
Atrazine	307	5.06	69	0.83	5.9
Cyanazine	307	4.79	69	0.04	31.1
Metribuzin	307	1.07	69	0.10	10.9
Honey Creek					
Alachlor	443	3.32	96	0.22	15.4
Metolachlor	443	5.38	96	0.86	6.3
Atrazine	443	6.72	96	1.55	4.3
Cyanazine	443	1.15	96	0.16	7.3
Metribuzin	443	0.88	96	0.10	4.3

Note: The summer period is defined as April 15 to August 15, the winter period as August 16 to April 14 of the following year. The data represent the period from April 1983 to November 1987.

pesticide manufacturers. Additional details on the sampling procedures have been described by Baker,[16] and the analytical procedures have been described by Kramer and Baker.[18] The analytical method is very similar to the U.S. EPA's Draft Method 507 (nitrogen- and phosphorus-containing pesticides in water by gas chromatography with a nitrogen-phosphorus detector) as revised in 1988 and recommended for use in evaluation study WS023 by U.S. EPA, Office of Research and Development, Environmental Monitoring and Support Laboratory, Cincinnati, Ohio.

RESULTS AND DISCUSSION

Seasonal Characteristics of Pesticide Exposure

The concentrations of currently used pesticides in rivers of this region are much higher in the late spring through midsummer season, following the major periods of pesticide application, than they are during other times of the year. In Table 4, TWMCs of five major herbicides are shown for three monitoring stations for the April 15-August 15 period ("summer") and the August 16-April 14 period ("winter"). These data were collected between April 1983 and November 1987. At these stations, the ratios of the summer to winter TWMCs ranged from 4.3 to 44. The magnitude of the ratio varied among the five herbicides, apparently in relationship to the persistence of the compound. At each site, the ratio of summer to winter concentrations was highest for alachlor, the least persistent of the five herbicides.[19] The ratio of summer to winter concentrations was lowest for atrazine, the most persistent of these herbicides. In the remainder of this chapter, will be referred to as the pesticide runoff season for this region.

Pesticide Concentrations in Relation to Stream Flow

Comparisons of plots of pesticide concentrations as a function of time (chemographs) with plots of stream discharge as a function of time (hydrographs) clearly shows that during the pesticide runoff season, pesticide concentrations increase in association with runoff events.[16] Pesticide chemographs and the discharge hydrograph for Honey Creek at Melmore, Ohio, during the 1985 pesticide runoff season are shown in Figure 3. These data also illustrate that there is no clear relationship between the magnitude of a storm's peak discharge and magnitude of the associated pesticide concentrations. Three successive storm events with greatly differing peak discharges all had similar

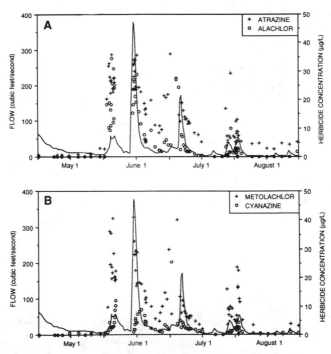

Figure 3. Chemographs for atrazine, alachlor, metolachlor, and cyanazine in relation to the streamflow hydrograph for the Honey Creek sampling station during the April 15-August 15, 1985 pesticide runoff season.

peak pesticide concentrations. However, by mid-July, seasonal aspects of pesticide runoff begin to appear. For example, the mid-July runoff events in Honey Creek (Figure 3) had lower peak concentrations than the May and June runoff events, especially for alachlor. Several factors could account for the lower peak herbicide concentrations of the July storms, including breakdown of the herbicides, depletion of available herbicides at the soil surface by previous runoff, herbicide dissipation through volatilization, herbicide uptake by plants, or herbicide movement into deeper soil layers via leaching.[6]

Plots of herbicide concentration in relation to stream discharge for Honey Creek during the April 15 to August 15 periods illustrate the large amount of variability in the relationships between instantaneous discharge and stream flow (Figure 4). The scattering in Figure 4 occurs because runoff events of greatly differing magnitude can be accompanied by similar pesticide concentrations and runoff events of the same magnitude can have greatly differing

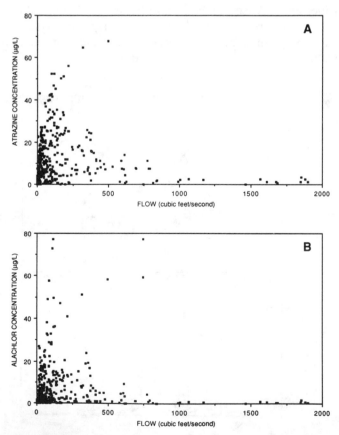

Figure 4.   Relationships between herbicide concentrations and stream flow at the Honey Creek sampling stations for samples collected during the April 15-August 15 period for 1983-1987. (A) atrazine; (B) alachlor.

pesticide concentrations. Consequently, stream flow is, by itself, not a good predictor of pesticide concentration, even during the pesticide runoff season.

The relationships between pesticide concentrations and runoff events evident in Honey Creek also occur in other watersheds, both larger and smaller.[16] Often storm events are superimposed on one another, so that streams do not return to base flow between storm events. These circumstances lead to complex pesticide concentration patterns in streams and rivers.

Relationships Between Pesticide, Sediment, and Nitrate Transport

The same rainstorms that move pesticides off fields into streams and rivers also move sediments and nitrates into the streams and rivers. During individual storm runoff events, the concentration patterns for pesticides differ from those of sediments and nitrates in a systematic fashion. The timing of the peak concentrations and the chemograph shapes can be used to infer characteristics of the pathways of material movement from fields into streams. A storm in Honey Creek during May 1986 illustrates the typical pattern (Figure 5).

Sediment concentrations usually peak during the rising portion of the hydrograph and are already declining by the time discharges reach their peak values (Figure 5a). Two explanations of the advanced sediment peaks are often proposed. One of these notes that even in runoff studies of individual fields or

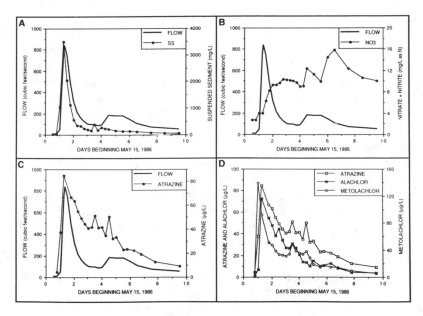

Figure 5. Pollutant chemographs and the storm hydrograph for a runoff event at the Honey Creek station beginning May 15, 1986. (A) suspended solids concentrations in relation to the hydrograph; (B) nitrate concentrations in relation to the hydrograph; (C) atrazine concentration in relation to the hydrograph; (D) comparison between the atrazine, alachlor, and metolachlor chromographs.

plots, the initial surface runoff water has higher sediment concentrations than water leaving the field later during the event.[20] This occurs even in rainfall simulator studies where the rainfall intensity is held constant. Apparently there is a "pool" of readily erodible sediment on the soil surface that comes off with the early runoff water at the beginning of the storm. Advanced sediment peaks in rivers are then thought to be accounted for by the routing of the water from individual fields into and through the stream networks.[20]

An alternative explanation for the advanced sediment peaks involves the role of resuspension and deposition of sediments between the water column and the stream bottom.[21] This explanation considers stream bottom sediment as the major source of sediment observed in a stream during runoff events. A flood creates a wave which propagates down the river, just as a stone thrown into a pond creates ripples (small waves) which propagate across the pond's surface. The time trace of the wave as it passes a given point is what we call the hydrograph. The wave moves independently of the water molecules, which move only in small circles in the pond, but move downstream in the river, more slowly than the wave. As the flood wave moves downstream, it resuspends sediment from the stream bottom into the water column, yielding high sediment concentrations on the rising limb of the hydrograph. Since the flood wave moves downstream more quickly than the water flows downstream, water comprising the rising limb of the hydrograph at an upstream site will comprise part of the falling limb of the hydrograph at a downstream site. Under the channel conditions at the downstream site, much of the sediment originally incorporated into the water column at the upstream site would have settled out of the water column. However, at the downstream site, as the flood wave moved past, it would have picked up sediment during the rising stage of the hydrograph, yielding an advanced sediment peak at the downstream site. According to this hypothesis, the advanced sediment concentration peaks in rivers are a consequence of the movement of the flood front, as a kinematic wave, resuspending sediment as it moves through the drainage network, and redepositing it at downstream sites.

In contrast to sediment, the highest concentrations of nitrate occur during the falling limb of the hydrograph (Figure 5b). Large proportions of the cropland in these watersheds have been systematically tiled. These fields use an array of clay tiles or plastic drainage pipes placed 3–6 ft below the surface, to facilitate drainage. The tiles drain into the tributaries, either directly or via drainage ditches. Plot and field studies indicate that most of the nitrate export from fields in this region occurs through the tile systems.[22] At river sampling stations, the proportion of tile drainage water to surface runoff water increases during the falling limb of the hydrographs, accounting for the observed peak nitrate concentrations during the later portion of the runoff event.

The concentration patterns of those pesticides that are transported primarily in the dissolved state are similar to, but do not coincide with, either the sediment or the nitrate concentration patterns. For atrazine, during storm events the peak concentrations occur near the time of peak discharge, but the concentrations do not decrease nearly as rapidly as the sediment concentrations (Figure 5c). In contrast with sediment, the atrazine apparently is carried into streams throughout the time of surface runoff from the fields. Thus, the atrazine chemograph is much broader than the sediment graph. Also, the peak atrazine concentrations precede the peak nitrate concentrations. Since nitrate serves as a marker for tile effluent, the peak atrazine concentrations cannot be attributed to tile flow. Studies of atrazine concentrations in tile effluent from this region show much lower atrazine concentrations than we observe in the stream systems.[23]

The concentration patterns of other herbicides largely parallel the patterns for atrazine (Figures 3 and 5d). Since peak exposures for multiple herbicides closely coincide, the possibilities of synergistic interactions among pesticides need to be evaluated.

Effects of Pesticide Use Rates on Pesticide Concentrations

In general, the concentration of a particular pesticide in river systems is closely related to the quantity of that pesticide used in watersheds upstream from the monitoring site. In Figure 6, the TWMCs of various pesticides at the Maumee and Sandusky river monitoring stations are plotted in relation to their 1986 use in the Lake Erie Basin. Alachlor, metolachlor, and atrazine are used in the largest quantities in the Lake Erie Basin (Table 2). These three herbicides have the highest TWMCs.

These data also illustrate the importance of factors other than the amount of

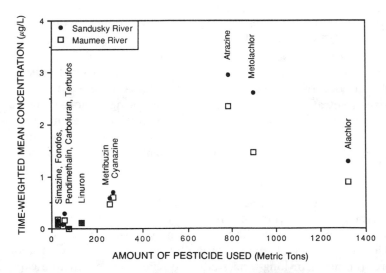

Figure 6.   Relationships between the TWMCs for several pesticides at the Maumee and Sandusky river stations and the quantities of pesticide used in the Lake Erie Basin in 1986. The TWMCs were based on all samples collected between April 1983 and November 1987. *Source*: Quantities of pesticide used from Waldron.[2]

use in affecting environmental exposures. Within this group of three herbicides, the time-weighted mean concentrations are inversely related to the amount of herbicide used. Of the three herbicides, alachlor is the least persistent, and atrazine is the most persistent.[19] One of the pesticides with a very low apparent TWMC relative to the amount applied is terbufos. Terbufos not only breaks down rather quickly in the soil and water, but its application involves incorporation into the soil, thereby greatly reducing its movement in surface runoff water. Furthermore, the sample storage conditions used in these studies (refrigeration for up to three weeks before extraction) are unsatisfactory for terbufos—in the analytical methods for use in the EPA's National Pesticide Survey, terbufos is noted as failing the 14-day sample storage tests.[24] Consequently, we could be underestimating its concentrations.

Efforts are currently underway to develop watershed-specific pesticide use data. Small watersheds are more likely to have significant deviations from regional pesticide use patterns than are larger watersheds, such as those of the Maumee and Sandusky rivers. Such data will allow assessment of the effects of differences in pesticide use patterns among the smaller watersheds on their pesticide concentration patterns. Watershed-specific pesticide use data for the smaller watersheds will also increase the value of these data sets for model development and calibration. It should also be noted that small watersheds are more likely to differ from one another in their deviations from normal regional rainfall patterns.

Annual Variability in Pesticide Runoff

Both the TWMCs and observed peak concentrations of individual pesticides vary considerably from year to year at each station. Examples of annual variability in TWMCs for three major herbicides are shown in Table 5 for the Maumee and Sandusky rivers and for Honey Creek. Two- to fourfold variations in average concentrations during the pesticide runoff season for these herbicides occurred within the six-year period. Annual variability in observed peak concentrations at these same three monitoring stations is shown in Table 6. The variability in observed peak concentrations is even larger than the variability in TWMCs. It should be noted that the actual peak concentrations occurring in the stream systems would probably be higher than the peak concentrations observed from the sampling program. Pesticide concentrations change rapidly during storm events and it is unlikely that the sampler would happen to collect a sample at the precise time when a particular pesticide reaches its peak concentration. It is also unlikely that the peak concentrations of different pesticides would occur precisely at the same time.

The large extent of annual variability in pesticide concentrations underscores the need for long-term studies to characterize pesticide exposure patterns in river systems. Such variability extends to pesticide loading, as well as to concentration patterns. Extensive annual variability is also characteristic of other agricultural contaminants, such as sediments, nitrates, and phos-

Table 5. Annual Variations in Time-Weighted Mean Concentrations of Atrazine, Alachlor, and Metolachlor During the Pesticide Runoff Season for the Maumee and Sandusky Rivers and for Honey Creek

Year	Alachlor (μg/L)			Metolachlor (μg/L)			Atrazine (μg/L)		
	Maumee	Sandusky	Honey Cr.	Maumee	Sandusky	Honey Cr.	Maumee	Sandusky	Honey Cr.
1983	1.36	0.69	1.99	1.38	2.79	3.87	2.35	2.22	4.07
1984	2.71	1.90	2.65	2.15	3.44	3.31	3.49	3.01	4.91
1985	0.75	2.63	3.25	1.91	6.65	7.27	2.79	5.98	7.49
1986	2.01	4.10	4.44	3.68	5.92	7.89	4.86	7.81	9.32
1987	1.93	2.20	3.50	4.76	4.64	4.62	5.98	4.74	6.38
1988	0.16	0.09	0.33	0.40	0.36	1.25	1.23	0.90	2.63

Table 6. Annual Variations in Peak Observed Concentrations of Atrazine, Alachlor, and Metolachlor During the Pesticide Runoff Season for the Maumee and Sandusky Rivers and for Honey Creek

Year	Alachlor (μg/L)			Metolachlor (μg/L)			Atrazine (μg/L)		
	Maumee	Sandusky	Honey Cr.	Maumee	Sandusky	Honey Cr.	Maumee	Sandusky	Honey Cr.
1983	10.54	6.94	12.49	9.25	21.97	30.82	7.22	10.63	23.30
1984	17.64	8.75	22.01	13.73	19.45	35.41	13.62	10.15	37.46
1985	5.63	26.52	27.06	8.52	42.36	35.36	9.00	28.20	29.33
1986	8.70	33.52	72.17	31.57	39.00	138.76	13.16	32.38	66.96
1987	7.66	13.18	74.14	13.71	26.12	30.86	13.05	21.64	46.20
1988	0.76	0.40	7.07	1.72	0.73	7.96	3.00	2.19	8.84

phorus.[16] This variability greatly complicates the task of assessing the effectiveness of certain types of management practices in reducing agricultural nonpoint pollution.

Watershed Scale Effects

As storm runoff water moves through a watershed's drainage network, it is continually mixing with water from other parts of the watershed. As streams merge, even when they are of similar stream order, they are seldom in the same phase of their hydrographs and chemographs. Thus, they will have differing discharge rates and differing pesticide concentrations. The resulting pesticide concentrations will depend on the discharge rates and the pesticide concentrations of each of the tributaries that merged and will always fall between the concentrations of the two parent streams. The continuous operation of this process within drainage networks gives rise to systematic changes in pesticide concentration patterns in relation to position in the drainage network, even when soil types, land uses, and management practices are similar throughout the entire watershed. These systematic changes in pesticide concentration patterns reflect the operation of what we refer to as "scale effects" within the watershed. Two important aspects of these scale effects are that peak concentrations decrease as drainage area increases and that intermediate concentrations persist for longer durations as drainage area increases.

Concentration exceedency curves provide a convenient way to compare pesticide exposure patterns among various sites. Concentration exceedency curves are constructed by ranking the concentrations in decreasing order and plotting them as a function of the cumulative time the samples represent. The curves allow one to determine the proportion of time any specified concentration is exceeded, or the proportion of time characterized by a specified concentration interval. In Figure 7, atrazine concentration exceedency curves are shown for Lost Creek (11.3 km²) and the Maumee River (16,395 km²). Peak concentrations are much higher in Lost Creek than in the Maumee River. However, the concentration exceedency curves cross so that intermediate concentrations persist for a much longer time in the Maumee River than in Lost Creek. Median pesticide concentrations are generally higher in large watersheds than in small watersheds.[16]

In Table 7, peak observed concentrations of several herbicides are shown for various watersheds in relation to watershed size. As watershed size decreases, peak herbicide concentrations increase. This relationship is also evident in the data of Table 6. This trend apparently continues to watersheds much smaller than the smallest we have observed. The peak pesticide concentrations that have been reported for edge-of-field studies[10] are much higher than those that we have observed in our smallest watersheds. Edge-of-field studies can give rise to particularly high concentrations of individual compounds since individual compounds are generally used over the entire field. As watershed size

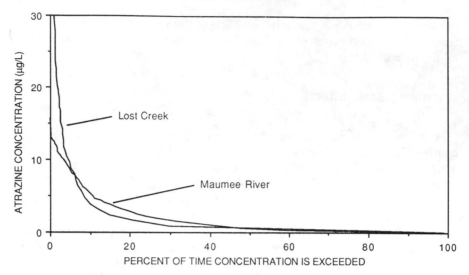

Figure 7.   Atrazine concentration exceedency curves at the Maumee River and Lost Creek sampling stations. Curves include the entire period between April 1983 and November 1987.

increases, the percent of area covered by single compounds drops off to the regional averages.

Although scale effects are evident for peak concentrations and median concentrations, they are not readily evident in terms of TWMCs (Table 8). For small watersheds, the short durations of high concentrations coupled with the long durations of low concentrations tend to balance the long durations of intermediate concentrations and the lower peak concentrations of the larger watersheds. The TWMCs of the major herbicides are very similar for the Lost Creek station, the smallest of the study watersheds, and the Maumee River station, which has the largest drainage area.

Effects of Land Use and Soil Texture

In Figure 8, the concentration exceedency pattern for atrazine is compared for the Sandusky River, the River Raisin, and the Cuyahoga River. As a group these rivers are similar in size. The watershed of the Cuyahoga River is dominated by forests and urban/suburban land uses, while the River Raisin and the Sandusky River have similar proportions of cropland. The soils of the River Raisin Basin have a much coarser texture (70% loams and sandy loams) than those of the Sandusky River Basin.[25] The concentration exceedency patterns (Figure 8) show that atrazine concentrations are much higher in the Sandusky River than in the River Raisin, and that the concentrations in the Raisin are much higher than in the Cuyahoga. Concentrations of sediment, nitrate, and phosphorus are also much higher in the Sandusky River than the River Raisin,[16] even though average gross erosion rates are higher in the River Raisin

Table 7. Peak Observed Concentrations of Major Herbicides at the Monitoring Stations During the Interval from April 1983 to October 1987

River	N	Alachlor	Metolachlor	Atrazine	Cyanazine	Metribuzin	Linuron
Maumee River	340	25.85	28.66	15.61	13.27	8.61	8.79
Sandusky River	375	35.88	37.34	32.81	26.49	13.82	8.27
Honey Creek	534	77.28	125.98	67.85	21.02	13.31	18.67
Rock Creek	477	32.96	127.53	57.70	33.03	23.81	13.65
Lost Creek	410	91.47	83.74	281.41	30.17	37.54	16.19
Cuyahoga River	94	1.64	7.09	3.80	1.81	1.57	6.08
River Raisin	134	10.59	7.77	16.62	5.00	3.67	2.32

Note: All concentrations are given in $\mu g/L$. The first five stations are listed in the order of decreasing watershed size.

Table 8. Time-Weighted Mean Concentrations of Major Herbicides at the Monitoring Stations During the Interval from April 1983 to October 1987

River	N	Alachlor	Metolachlor	Atrazine	Cyanazine	Metribuzin	Linuron
Maumee River	340	0.94	1.51	2.43	0.63	0.50	0.12
Sandusky River	375	1.30	2.61	2.97	0.70	0.60	0.12
Honey Creek	534	1.74	3.05	4.07	0.65	0.49	0.39
Rock Creek	477	0.66	2.29	2.00	0.27	0.37	0.23
Lost Creek	410	1.04	0.96	2.97	0.73	0.28	0.08
Cuyahoga River	94	0.13	0.33	0.55	0.15	0.12	0.23
River Raisin	134	0.70	0.46	1.30	0.38	0.19	0.08

Note: All concentrations are given in $\mu g/L$. The first five stations are listed in order of decreasing watershed size (see Table 1).

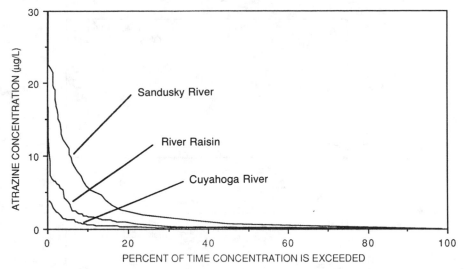

Figure 8. Atrazine concentration exceedency curves for the Sandusky River (agricultural land use, fine textured soils), the River Raisin (agricultural land use, coarser textured soils), and the Cuyahoga River (urban and forested watersheds). Curves include the entire period between April 1983 and November 1987.

Basin (9.75 tons/ha/year) than in the Sandusky River Basin (8.25 tons/ha/year).[26]

Herbicide Exposures in Relation to Proposed Lifetime Health Guidance Levels

The Maumee and Sandusky rivers, as well as Honey Creek, serve directly as raw water sources for 11 community water intakes supplying public drinking water to 124,000 residents of northwestern Ohio.[27] Most of the current generation pesticides pass directly through conventional water treatment plants, giving rise to pesticide exposures through drinking water.[8,9] The occurrence of such exposures generates frequent questions concerning accompanying human health risks. These risks have been discussed in some detail in Chapter 25.

The U.S. EPA has recently published lifetime health guidance levels for several major herbicides.[28] Concentration exceedency curves provide a convenient means of summarizing exposure patterns for comparison with lifetime health guidance levels. In Figure 9, concentration exceedency curves for alachlor, atrazine, and metolachlor at the Sandusky River station are shown. The curves are based on data collected between April 1983 and October 1987. For each herbicide, the proposed lifetime health guidance level, the percentage of time the lifetime health guidance level is exceeded, and the TWMC are indicated. As is evident in Figure 9, the proposed health guidance levels are exceeded about 10% of the time for alachlor, 17% of the time for atrazine, and 5% of the time for metolachlor. The TWMCs for all three herbicides were

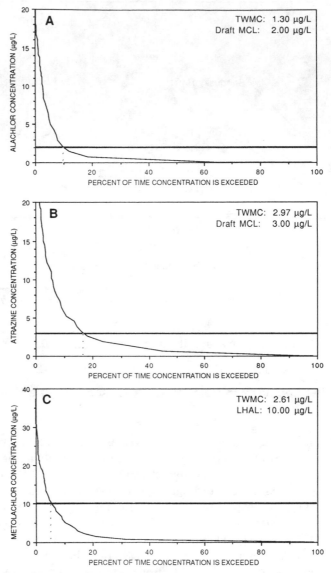

Figure 9. Relationships between lifetime health advisory levels (LHAL) or draft maximum contaminant levels (Draft MCL), and concentration exceedency curves for alachlor (A), atrazine (B), and metolachlor (C) at the Sandusky River station. Curves include the entire period between April 1983 and November 1987.

below the lifetime health advisory levels, although in the case of atrazine, the TWMC and the lifetime health advisory level were very similar.

A very important characteristic of pesticide exposure patterns in rivers is that high concentrations are present for relatively short durations of time. These periods of high concentrations not only can exceed lifetime health guidance levels, but they also contribute greatly to the TWMCs of these compounds. Because of these exposure patterns, treatment to remove pesticides at drinking water treatment plants for relatively short periods during high concentrations can simultaneously reduce or prevent the occurrence of concentrations in excess of the standards and efficiently lower the TWMCs. One effective option for pesticide removal would be the use of powdered activated carbon (PAC) during storm runoff events during the late spring and early summer periods.[8] Most treatment plants utilizing rivers for water supplies have experience with and facilities for the addition of PAC, since such treatment is frequently used to deal with periodic taste and odor problems. Other treatment options include granular activated carbon,[8] reverse osmosis, or ozone oxidation.[29]

Ecological Significance of Current Generation Pesticides

Within the Great Lakes Region, there has been great concern over toxic chemicals in aquatic systems, both in terms of human health effects through eating contaminated fish or drinking contaminated water, and in terms of direct impacts on aquatic communities. Although several current generation pesticides have been observed in waters of the Great Lakes, they are notably absent from the lists of toxic compounds of concern in the Great Lakes Region.[30] Persistent organics, such as earlier generation chlorinated insecticides, industrial organics, combustion by-products, and various metals, comprise the lists of toxic substances. Fish consumption advisories in the Great Lakes Region, such as those provided by the Ontario Ministry of the Environment,[31] are based on metals and persistent organics. Current generation pesticides, if detected at all, are not deemed to constitute risks to those eating fish from this region. There is considerable uncertainty regarding the water quality impacts of current generation pesticides and the extent of water quality benefits that would accompany reduced exposures to these compounds.[32]

The major herbicides that occur in rivers of northwestern Ohio have very low acute toxicities to fish.[33] Some representative LC_{50}s for major current generation herbicides and insecticides are shown in Table 9. The peak herbicide concentrations we have observed in small streams are at least an order of magnitude below these acute toxicity values. The acute toxicities of some of the currently used insecticides are much greater (i.e., they have lower LC_{50}s) than those of the herbicides.[33] If stream concentration patterns were strictly proportional to pesticide use, such that the ratios of insecticide concentrations to herbicide concentrations were proportional to their use rates, one could expect rather frequent fish kills in streams due to insecticide runoff. However, the rather rapid breakdown of many of the insecticides, coupled with application techniques that usually involve incorporation into the soil rather than broadcasting on the soil surface, result in ambient insecticide concentrations in streams that are much lower than herbicide concentrations, relative to the quantities applied. We have rarely, if ever, observed insecticide concentrations

Table 9.   Acute Toxicities to Fish of Major Herbicides and Insecticides as Represented by Their LC_{50}s in 24 hr and 96 hr Static Bioassays.

Pesticide	Rainbow Trout		Bluegill		Channel Catfish		Fathead Minnows	
	24 hr	96 hr	24 hr	96 hr	24 hr	96 hr	24 hr	96 hr
Herbicides								
Alachlor								
Technical material	—	2,400	11,500	4,300	—	—	—	—
Emulsifiable conc.	4,300	1,400	7,800	3,200	—	—	—	—
Metolachlor								
Technical material	—	—	—	—	—	—	—	8,000
Emulsifiable conc.	—	—	—	—	—	—	—	8,400
Ciba Geigy data	—	2,000	—	15,000	—	4,900	—	11,000
Atrazine 4L								
43% liquid	—	24,000	48,000	42,000	—	—	—	—
Ciba Geigy data	—	4,900	—	6,700	—	—	—	15,000
Cyanazide								
80% wettable powder	12,000	9,000	22,500	20,300	13,400	10,400	19,700	19,400
Metribuzin								
Technical material	—	42,000	—	92,000	>100,000	—	—	—
Linuron								
Technical material	—	—	—	—	—	2,900	—	—
Wettable powder	—	—	—	—	3,200	1,800	—	—
Trifluralin								
Technical material	167	92	120	47	400	210	360	150
Chloramben								
Wettable powder	—	>10,000	—	>10,000	—	—	—	—
Butylate								
Technical material	4,000	2,100	900	470	—	—	—	—

Table 9, continued

Pesticide	Rainbow Trout		Bluegill		Channel Catfish		Fathead Minnows	
	24 hr	96 hr	24 hr	96 hr	24 hr	96 hr	24 hr	96 hr
Insecticides								
Terbufos								
Technical material	24	10	4.8	1.7	—	—	390	390
Granular formation	23	8.8	3.1	1.7	1,800	1,800	210	150
Carbofuran								
Technical material	680	380	101	88	372	248	883	872
Wettable powder	—	—	370	240	—	—	—	—
Chlorpyrifos								
Technical material	53	7.1	—	2.5	410	280	—	—
Phorate								
Technical material	25	13	7.6	2.0	500	280	—	—
Fonofos								
Technical material	109	20	45	6.8	—	—	—	—

Note: All data are from studies at the Columbia National Fisheries Research Laboratory, Columbia, MO,[33] except for additional data for atrazine and metolachlor provided by Ciba Geigy Corporation, Greensboro, NC. Concentrations given in $\mu g/L$.

in the range of their LC_{50}s. The lower concentrations of the insecticides, coupled with their less frequent occurrence and their lower stabilities during sample storage, combine to make detailed exposure studies for insecticides much more difficult than comparable studies of herbicide exposures.

Although fish kills in farm ponds are often attributed to runoff of current generation insecticides from adjacent fields, actual documentation of the role of insecticides in such incidents is much less common. Reports of fish kills in stream systems, due to runoff of current generation insecticides from fields, are not common. It should be noted that fish kills in farm ponds are much more likely to be observed than fish kills in small streams during runoff events. Impacts of pesticides on fish reproduction in streams and rivers could also be very difficult to observe.

Algal and rooted aquatic plant communities are much more likely to be directly impacted by herbicides than are the fish and invertebrate communities. The herbicide concentrations we have observed in streams and rivers do reach levels that have shown inhibitory or toxic effects on plant communities in both microcosm and mesocosm studies.[34,35] However, within these stream systems, the turbidity associated with sediment transport that co-occurs with the herbicide exposures would likely have a greater effect on the productivity of the plant communities than would the herbicides. Often the effects of herbicide exposures in experimental studies are transitory and the communities quickly recover following the exposures, although the species composition at lower trophic levels may be shifted.[35,36]

The Ohio EPA has made extensive use of both fish indices and macroinvertebrate indices in their assessments of water quality in the streams and rivers of Ohio.[37] For each ecoregion within Ohio, the indices for stream segments are compared to the indices for the "best" streams within that ecoregion. This approach indicates that many of the stream segments within the Huron/Erie Lake Plain ecoregion and the Eastern Cornbelt ecoregion, which encompass the agricultural watersheds discussed in this chapter, have impaired aquatic communities.[37] The five major sets of factors that are thought to interact in determining biological community performance in streams are chemical variables, flow regimes, habitat structure, energy sources, and biotic interactions. Intensive agricultural land use directly affects at least four of the sets of factors (all except for the biological interactions). In addition to pesticide exposures and nutrient enrichment, current agricultural land use in this region results in habitat modification through sedimentation and channelization, extremes in discharge at both the high and low flow ranges, and high light intensities in low-order streams due to a lack of streamside vegetation. While there is no doubt that agricultural land use has had a major impact on the streams and rivers of this region, there is considerable doubt as to whether currently used pesticides have directly resulted in the associated ecological impairments.

SUMMARY OF PESTICIDE EXPOSURE PATTERNS IN STREAMS AND RIVERS

The pesticide data sets for Lake Erie tributaries illustrate many characteristics of pesticide transport in river systems. Although many of these characteristics would be expected based on extrapolations from the numerous edge-of-field studies of pesticide behavior or based on modeling programs, actual data sets confirming these expectations and providing quantitative illustrations are rare. These general characteristics are the following:

1. Pesticide concentrations in rivers are much higher during the three months immediately following major spring pesticide applications (May, June, and July) than they are during the remainder of the year.
2. During the three months of "high" concentrations, the pesticide concentrations are highest during periods of storm runoff events and drop to lower concentrations between runoff events.
3. The concentration patterns of soluble pesticides in streams during individual storm runoff events are distinct from both the suspended sediment concentration patterns and the nitrate concentration patterns.
4. The concentration patterns of individual pesticides parallel one another so that the peak concentrations of individual pesticides often coincide.
5. In general, the concentrations of individual pesticides in stream systems are proportional to the amount of their use within the watersheds. Factors such as persistence and mode of application also strongly affect the peak and average concentrations.
6. There is very large year-to-year variability in peak and average pesticide concentrations, and in pesticide loadings, depending on the frequency, duration, and intensity of runoff-generating rainfall events in relation to the timing of pesticide applications.
7. The patterns of pesticide concentrations in streams are greatly affected by watershed size. As watershed size increases, peak pesticide concentrations decrease but the length of time that intermediate concentrations are present becomes extended.
8. For short periods of time, concentrations of major herbicides in streams and associated municipal water supplies do reach levels in excess of the U.S. EPA's proposed lifetime health advisories related to possible chronic effects. However, the TWMCs for herbicide concentrations in streams and public water supplies are usually less than the lifetime health advisory levels, even for those herbicides used in the largest quantities.
9. In public water supplies withdrawn from rivers, TWMCs for pesticides can be efficiently lowered by removal treatment during the rather short durations when high concentrations are present.
10. Although impairment of biological communities is evident in the streams and rivers of this region and is associated with intensive row crop agriculture, it is not clear that pesticides, which are present during runoff events following periods of application, contribute significantly to this impairment.

REFERENCES

1. "Pesticide Industry Sales and Usage: 1985 Market Estimates," U.S. Environmental Protection Agency, Office of Pesticide Programs, September 1986, Table 8.
2. Waldron, A. C. "Surveying Application of Potential Agricultural Pollutants in the Lake Erie Basin of Ohio: Pesticide Use on Major Crops," Ohio Cooperative Extension Service, Bulletin 787 (Columbus, OH: Ohio State University, 1988).
3. Taylor, A. W., and D. E. Glotfelty. "Evaporation from Soils and Crops," in *Environmental Chemistry of Herbicides,* R. Grover, Ed., Vol. 1 (Boca Raton, FL: CRC Press, 1988), Chapter 4.
4. Glotfelty, D. E., J. N. Seiber, and L. A. Liljedahl. "Pesticides in Fog," *Nature* 325:602–5 (1987).
5. Richards, R. P., J. W. Kramer, D. B. Baker, and K. A. Krieger. "Pesticides in Rainwater in the Northeastern United States," *Nature* 327:129–31 (1987).
6. Leonard, R. A. "Herbicides in Surface Waters," in *Environmental Chemistry of Herbicides,* R. Grover, Ed., Vol. 1 (Boca Raton, FL: CRC Press, 1988), Chapter 3.
7. Wagenet, R. J. "Processes Influencing Pesticide Loss with Water under Conservation Tillage," in *Effects of Conservation Tillage on Groundwater Quality: Nitrates and Pesticides,* T. J. Logan, J. M. Davidson, J. L. Baker, and M. R. Overcash, Eds. (Chelsea, MI: Lewis Publishers, Inc., 1987), Chapter 11.
8. Miltner, R. J., D. B. Baker, T. F. Speth, and C. A. Fronk. "Treatment of Seasonal Pesticides in Surface Waters," *Journal AWWA* 81:43–52 (1989).
9. Baker, D. B. "Regional Water Quality Impacts of Intensive Row-Crop Agriculture: A Lake Erie Basin Case Study," *J. Soil Water Cons.* 40:125–32 (1985).
10. Wauchope, R. D. "The Pesticide Content of Surface Water Draining from Agricultural Fields—A Review," *J. Environ. Qual.* 7:459–72 (1978).
11. "Better Monitoring Techniques Are Needed to Assess the Quality of Rivers and Streams," Vol. 1, U.S. General Accounting Office, Report to the Congress CED-81-30 (1981).
12. Gilliom, R. J., R. B. Alexander, and R. A. Smith. "Pesticides in the Nation's Rivers, 1975–80, and Implications for Future Monitoring," U.S. Geological Survey Water Supply Paper 2271 (1984).
13. "National Pesticide Monitoring Plan," U.S. Environmental Protection Agency, Office of Pesticide and Toxic Substances, Washington, DC (1985).
14. Donigian, A. S., J. C. Imhoff, and B. R. Bicknell. "Predicting Water Quality Resulting from Agricultural Nonpoint Source Pollution via Simulation-HSPF," in *Agricultural Management and Water Quality,* F. W. Schaller and G. W. Bailey, Eds. (Ames, IA: Iowa State University Press, 1983), Chapter 12.
15. Baker, D. B. "Fluvial Transport and Processing of Sediments and Nutrients in Large Agricultural River Basins," EPA-600/3-83-054 (1984).
16. Baker, D. B. "Sediment, Nutrient, and Pesticide Transport in Selected Lower Great Lakes Tributaries," Great Lakes National Program Office, EPA-905/4-88-001 (1988).
17. Waldron, A. C. "Pesticide Use on Major Field Crops in Ohio—1982," Cooperative Extension Service, Ohio State University, Bulletin 715 (1984).
18. Kramer, J. W., and D. B. Baker. "An Analytical Method and Quality Control Program for Studies of Currently Used Pesticides in Surface Waters," in *Quality Assurance for Environmental Measurements,* J. K. Taylor and T. W. Stanley, Eds.,

ASTM STP 867 (Philadelphia, PA: American Society for Testing Materials, 1985), pp. 116–32.

19. Nash, R. G. "Dissipation From Soil," in *Environmental Chemistry of Herbicides, R. Grover, Ed., Vol. 1* (Boca Raton, FL: CRC Press, 1988), Chapter 5.

20. Guy, H. P. "Fluvial Sediment Concepts," in *Techniques of Water-Resources Investigations of the United States Geological Survey,* U.S. Dept. of the Interior, U.S. Government Printing Office, (1970).

21. Melfi, D., and F. Verhoff. "Material Transport in River Systems During Storm Events by Water Routing," Lake Erie Wastewater Management Study, U.S. Army Corps of Engineers, Buffalo, NY (1979).

22. Logan, T. J. "Maumee River Basin Pilot Watershed Study: Summary Pilot Watershed Report," International Reference Group on Great Lakes Pollution from Land Use Activities, International Joint Commission, Windsor, Ont. (1978).

23. Keim, A. M. "A Field Study of Selected Agricultural Herbicides in Shallow Ground Water and Tile Drainage in Lucas and Ottawa Counties, Ohio," Masters Thesis, Department of Geology, University of Toledo, Toledo, OH (1989).

24. "National Pesticide Survey," U.S. Environmental Protection Agency, DU-87-A194, Attachment D.

25. "Application of the Universal Soil Loss Equation in the Lake Erie Drainage Basin, Appendix I," Lake Erie Wastewater Management Study, U.S. Army Corps of Engineers, Buffalo, NY (1978).

26. Logan, T. J., D. R. Urban, J. R. Adams, and S. M. Yaksich. "Erosion Control Potential with Conservation Tillage in the Lake Erie Basin: Estimates Using the Universal Soil Loss Equation and the Land Resource Information System (LRIS)," *J. Soil Water Cons.* 37:50–55 (1982).

27. "The Northwest Ohio Water Plan: Public Water Supply," Ohio Department of Natural Resources, Division of Water, Columbus, OH (1986).

28. "Draft Health Advisories on Pesticides," U.S. Environmental Protection Agency, Office of Drinking Water (1987). Health Advisory Communications based on these Health Advisories are available from the U.S. EPA, Office of Water, Washington, DC.

29. Fronk, C. A., and D. B. Baker. "Elimination of Pesticides from Drinking Water Using Ozone, Reverse Osmosis, Aeration, and Powdered Activated Carbon at a Water Treatment Plant," U.S. Environmental Protection Agency, unpublished draft report, Cincinnati, OH (1988).

30. Fitchko, J. "Literature Review of the Effects of Persistent Toxic Substances on Great Lakes Biota," International Joint Commission, Great Lakes Regional Office, Windsor, Ont. (1986).

31. "Guide to Eating Ontario Sport Fish," Ontario Ministry of the Environment, Ministry of Natural Resources, Toronto, Ont. (1988).

32. Crosson, P. R., and J. E. Ostrov. "Alternative Agriculture: Sorting Out Its Environmental Benefits," *Resources* No. 92 (Washington, DC: Resources for the Future, 1988).

33. Mayer, F. L., Jr., and M. R. Ellersieck. "Manual of Acute Toxicity: Interpretation and Data Base for 410 Chemicals and 66 Species of Freshwater Animals," U.S. Department of the Interior, Fish and Wildlife Service, Resource Publication 160 (1986).

34. Krieger, K. A., D. B. Baker, and J. W. Kramer. "Effects of Herbicides on Stream

Aufwuchs Productivity and Nutrient Uptake," *Arch. Environ. Contam. Toxicol.* 17:299–306 (1988).

35. Brockway, D. L., P. D. Smith, and F. E. Stancil. "Fate and Effects of Atrazine in Small Aquatic Microcosms," *Bull. Environ. Contam. Toxicol.* 32:345–53 (1984).

36. DeNoyelles, F., W. D. Kettle, and D. E. Sinn. "The Responses of Plankton Communities in Experimental Ponds to Atrazine, the Most Heavily Used Pesticide in the United States," *Ecology* 63:1285–93 (1982).

37. "Ohio Water Quality Inventory: 1988 305(b) Report, Volume I, Executive Summary," Ohio Environmental Protection Agency, Columbus, OH (1988).

## CHAPTER 18

Studies on the Transport and Fate of Chlordane in the Environment

Ravi K. Puri, Carl E. Orazio, S. Kapila, T. E. Clevenger, A. F. Yanders, Kathleen
E. McGrath, Alan C. Buchanan, James Czarnezki, and Jane Bush

INTRODUCTION

Chlordane is a chlorinated hydrocarbon insecticide, which has proven to be quite persistent in the environment. The commonly marketed form of chlordane is a mixture of variously chlorinated products of "chlordene," a Diels-Alder condensate of cyclopentadiene and hexachlorocyclopentadiene.[1] The bulk of technical-grade chlordane is formed by 10 constituents with six to nine chlorine atoms. Due to the nonselective chlorination process and the presence of impurities such as pentachlorocyclopentadiene and tetrachlorocyclopentadiene in hexachlorocyclopentadiene, a larger number of other chlorinated compounds have also been observed in the technical formulation and detected in the environmental samples. A number of studies dealing with the analysis and structure elucidation of chlordane have been published, and typical formulations of technical-grade chlordane have been shown to contain more than 40 different constituents.[2-9]

Chlordane has been used in the United States for approximately 40 years. It is estimated that over ten million kilograms of chlordane were produced and applied for control of a variety of agricultural pests. Chlordane has also been used extensively for control of termites and ants in and around domiciles. However, due to its carcinogenic activity in laboratory animals, its use was suspended as of April 1987 by the U.S. Environmental Protection Agency.

The extensive use and persistent nature of the chlordane constituents have led to the widespread presence of these chemicals in various phases of the environment worldwide. Some of the highly chlorinated constituents and stable metabolites, such as oxychlordane, have been detected by 0.1–19.0 picograms/m³ levels in air from such remote regions as the eastern Indian and Antarctic oceans as well as the arctic areas of Canada.[10,11] These determinations point towards an airborne long-range transport of the pesticide similar to that observed for other chlorinated pesticides. Chlordane constituents have been detected in sediments, water, and aquatic organisms.[12-14] These chemicals have also been monitored in human blood and milk samples from both occupationally exposed individuals and the nonoccupationally exposed general population, and a direct correlation between the frequency of occupational exposure of blood residue levels has been reported. The chlordane uptake in the general population is linked to consumption of contaminated foods and direct routes of exposure, such as dermal absorption and inhalation, especially in individuals living in houses improperly treated with chlordane.[12-17]

During the past few years, chlordane constituents have been found to be the major chlorinated hydrocarbon contaminants in the fish from many lakes and rivers in Missouri. Residue levels as high as 3–4 ppm have been detected; these levels are well in excess of the Food and Drug Administration (FDA) "action level" of 0.3 ppm.[18] These residues are attributed to the past agricultural usage and the more recent use of the chemical for subterranean termite control. Although the most recent surveys indicate a moderate decrease in concentration over the last five years, the decrease in concentration levels is not as dramatic as that reported from Okinawa Bay, Japan, where a decrease of more than an order of magnitude was observed over a four-year period.[19]

Wide variability in the rate of chlordane loss from the application site has been observed, and half-life estimates of chlordane have ranged from a few days to several years.[20-22] The loss of pesticides with low water solubility has been linked to volatilization, photodecomposition, and chemical/biochemical transformation.

Substantial volatilization losses of surface-applied pesticides, including chlordane, have been observed. The rate of volatilization and subsequent vapor-phase loss was found to be dependent on soil moisture content and increased with an increase in soil moisture.[23] In general, little or no downward movement of chlordane constituents has been reported, and it is postulated that volatilization and photodegradation are the major routes of chlordane loss following agricultural application. However, volatilization and photodegradation decrease appreciably with an increase in soil layer thickness, and these losses should be quite small from typical subsurface termiticide application.

Chemical/biochemical transformations and erosion of particle-bound chlordane are the most probable routes of transport and loss of the pesticide from subsurface application; erosion is the major contributor to transport of these chemicals into the aquatic system. However, a number of questions related to high levels of chlordane residue in Missouri fish and other aquatic organisms still remain unanswered. These questions pertain specifically to the relative contribution of agricultural and termiticide applications, the persistence of residue levels in aquatic and terrestrial environments, and the effect of copollutants on the partition and bioavailability of chlordane in an aquatic environment.

A cooperative study was initiated by the Missouri Department of Conservation and the Environmental Trace Substances Research Center, University of

Missouri, to answer some of these questions. The specific objectives of the study were

- to delineate the source of chlordane contamination
- to determine the persistence of soil/sediment-bound chlordane constituents in the terrestrial and aquatic environment
- to investigate the partition behavior of chlordane constituents in varied soil/water, including the effect of such cocontaminants as xenobiotic surfactants

The initial results of this multitiered study are presented in this report.

CONTAMINATION SURVEY STUDIES AND LABORATORY EXPERIMENTS

Fish Contamination Survey

A systematic fish contamination survey of rivers and streams was initiated by the Department of Conservation in 1983. As part of the survey, fish were collected from various locations along the two major rivers and from certain lakes in Missouri. These included five sites around urban centers and rural areas along the Missouri River. Gill nets and electrofishing gear were used to collect at least 20 specimens of each of the following species: shovelnose sturgeon, carp, river carpsucker, channel catfish, and flathead catfish. These species were selected because of their trophic positions as well as their importance to sport fishing and commercial fisheries. A record of length, weight, sex, date, and location of capture was maintained for each fish. The fish were skinned, and two fillets were removed from each fish; these were wrapped individually in aluminum foil, labeled, and kept frozen until analysis. Generally, one fillet sample was analyzed for pesticide residues, and the other was archived for later analysis if and when required.

Sediment Load Analysis Survey

These analyses were undertaken as part of a study to identify the sources of chlordane in the lower Meramec River, a tributary of the Missouri River. The study included determination of chlordane residue levels in sediments from watersheds in the area and correlation of the loadings with land use information. Two sediment samples were collected from each site. These samples were taken from stream sediment deposits and consisted of six composited subsamples.

Chlordane-Treated Soil Sampling

This part of the study was undertaken to monitor the effects of weathering on the concentration of chlordane constituents in soils treated for termite control. Soil samples were taken from 12 sites around the University of Mis-

Table 1. Soil Characteristics.

Soil Sample	% Sand	% Silt	% Clay	% Organic	pH
Times Beach, MO	11.4	52.7	35.9	2.4	6.9
Eglin, FL	91.7	6.3	2.0	1.6	4.7
Visalia, CA	44.1	34.5	21.2	1.7	8.1

souri–Columbia campus with controlled and documented application. The time period between application and sampling varied from 2 to 10 years. The samples were taken with a split-spoon sampler to a depth of 30 cm. The sample cores were split into two halves. Correlations between changes in the concentration of chlordane constituents and physicochemical and biochemical transformations were made.

Leaching and Transformation Studies with Simulated Termite Control Operations

These studies were designed to measure the transformation and loss of chlordane constituents under simulated termite control operations. The studies were carried out with three types of soils. The soils were obtained from Times Beach, Missouri, Eglin Air Force Base, Florida, and Visalia, California. The soils were characterized for typical soil parameters to their use. The values obtained are given in Table 1.

A single batch of thoroughly homogenized soil was used in all cases. The soil samples were air dried and passed through a 10 mesh sieve. The soils were then packed into 25 cm × 8 cm i.d. aluminum columns. Careful attention was paid to ensure uniformity of the column packing. A separate representative aliquot of each soil was taken and spiked with 500 ppm of technical chlordane. In one set of experiments, this contaminated soil was placed on top of the soil-packed column in a 5 mm thick layer; in another set of experiments, the contaminated soil was packed between layers of uncontaminated soils (Figure 1). The soil columns were placed in a controlled environment chamber described earlier.[24] Water containing 0, 10, or 100 ppm sodium lauryl sulfate was added periodically to the columns. Upon removal from the chamber, the columns were sectioned with a device capable of slicing the soil in 2 mm increments.[24] The column sections were analyzed for chlordane constituents. A summary of the protocol for column studies is given in Figure 2.

Soil/Water Partition Studies

Studies on the partition behavior of chlordane constituents in a soil/water system were conducted with the three types of soils described earlier. The experiments were carried out with soils ranging in chlordane concentration from 1 to 500 ppm. The spiked soils were transferred to 4-L amber glass bottles, and 3 L of organic free water or water containing 0.01% sodium lauryl sulfate was added to each bottle. The partition behavior was determined under three pH regimes. The pH values were monitored and maintained at 4.5, 7.0,

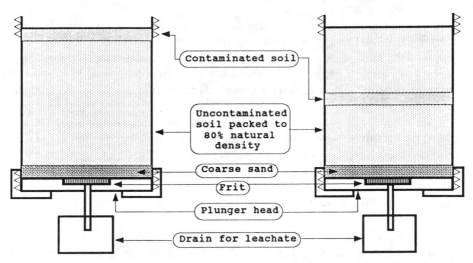

Figure 1. Schematic of soil column used for simulating surface and subsurface application of chlordane.

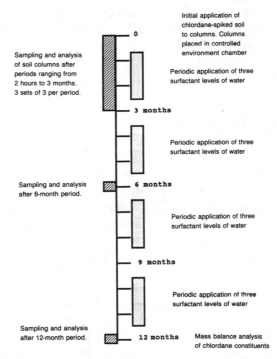

Figure 2. Protocol for soil column experiments.

and 8.5 by periodic addition of acid or base during the course of the study. A 150-mL aliquot of water was removed from each container after 24 hours, 30, 60, 90 and 180 days. A sampling tube with a 0.5-μ stainless steel filter was used for water sample collection. The aliquots were centrifuged at 4500 rpm for 30 min to remove any remaining suspended particles and were analyzed for chlordanes.

The experiments were terminated after a 10-month period and the concentration of residual chlordanes in soil and water was determined. The concentration of lauryl sulfate was monitored at regular intervals during the course of the study.

Analysis of Water, Soil/Sediment, and Fish Tissue Samples

All samples were analyzed with a gas chromatograph equipped with an electron capture detector. The overall flow diagram for the extraction and cleanup procedures employed during the study is given in Figure 3. The final determinations of chlordane in cleaned extracts were carried out with a 30 m × 0.25 mm i.d. fused silica capillary column and an electron capture detector. Confirmatory analyses for selected samples were carried out with a gas chromatograph interfaced to a quadrupole mass spectrometer. In addition, the presence of low levels of chlordene and heptachlor in field sediment samples was confirmed with a dual column reaction gas chromatographic system developed in our laboratory.[25,26]

RESULTS AND DISCUSSION

Chlordane constituents, *trans*-chlordane (1-exo, 2-endo, 4,5,6,7,8,8-octachloro-3a,4,7,7a-tetrahydro-4,7-methanoindane) — I, *cis*-chlordane (1-exo, 2-exo, 4,5,6,7,8,8-octachloro-3a,4,7,7a-tetrahydro-4,7-methano-indane) — II, *trans*-nonachlor (1-exo, 2-endo, 3-exo-4,5,6,7,8,8-nonachloro-3a,4,7,7a-tetrahydro-4,7-methanoindane) — III, and *cis*-nonachlor (1-exo, 2-exo, 3-exo, 4,5,6,7,8,8-nonchloro-3a,4,7,7a-tetrahydro-4,7-methano-indane) — IV, were invariably found to be the major chlorinated hydrocarbon contaminants in the fish tissue samples from various lakes, rivers, and streams in Missouri. This was true for fish from both the urban and rural areas. The chromatographic profile for a representative contaminated fish tissue sample is shown in Figure 4. In general, the residue levels in fish from urban areas were appreciably higher than the levels in fish from rural areas. A summary of typical levels from urban and rural areas is given in Table 2. A correlation between chlordane concentration and urban development is clearly observed.

A direct comparison of the chlordane residue data obtained during this study with the historical residue data obtained under the National Pesticide Monitoring Program (NPMP) could not be made due to the lack of chlordane-related data in the NPMP and the variability in the analysis- and residue-

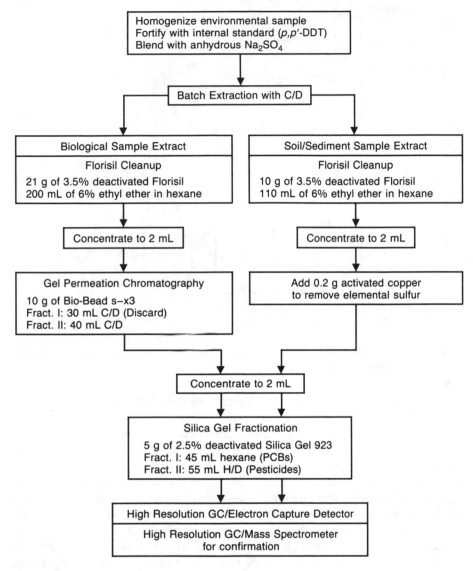

Figure 3. Sample extraction and cleanup procedure used for determination of chlordane residue in environmental samples.

reporting methodologies. However, a decrease in the concentrations of other chlorinated pesticide and polychlorinated biphenyls (PCBs) residue was observed. The results obtained during the past five years indicate only a moderate, if any, decrease in chlordane levels at most sites across the state. A typical set of levels obtained over a four-year period is summarized in Table 3.

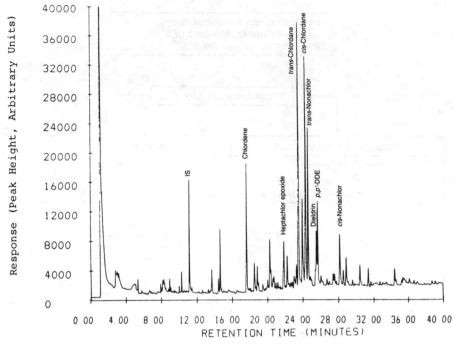

Figure 4. Chlorinated hydrocarbon residue profile of a fish tissue sample obtained from the Missouri River.

These results are representative of the trends observed elsewhere in the state. It is clear that the levels have decreased little during the past four-year period. This is in contrast to the sharp decrease in the sediment concentration observed in Japan and points towards a steady source, presumably the recently banned termiticide application.[19] A rough correlation between chlordane residue profiles in fish tissues and the technical chlordane formulations used for termiticide application could be observed. The concentrations of various constituents in the formulations were generally quite similar, except for nonachlors, whose concentration varied from 3 to 15% of the total. The residue patterns, i.e., the relative distribution of chlordane constituents, are dependent on the partition behavior and transformation rates of the constituents in the environment. Generally, a higher bioconcentration factor is observed for the persistent constituents with greater numbers of chlorines.[12] The relative concentrations of major chlordane constituents in technical chlordane formulations and the fish tissue are summarized in Table 4. The concentration of readily metabolized, transformed constituents such as chlordene-C, heptachlor, and other dienes in fish tissue is considerably less than in the technical chlordane. No significant differences in the relative concentration of residual constituents could be detected among different fish species.

The distribution of chemicals in the soil environment is expressed by the following general equation:

Table 2. Chlorinated Hydrocarbon Residue Levels (ppb) in Fish Tissue From Urban and Rural Areas (1986).

Fish Species	Locality	Aldrin	Chlordane (Tech)	p,p'-DDD	p,p'-DDE	p,p'-DDT	Dieldrin	Hept. Epox.	PCBs[a]
Channel Catfish	St. Louis Urban	3.0	2,420	27.0	40.0	<5.0	17.0	11.0	240
Carp	St. Louis Urban	<2.0	560	6.0	13.0	<5.0	7.0	5.0	206
Largemouth Bass	St. Louis Urban	<2.0	490	<5.0	13.0	<5.0	8.0	6.0	410
Carp	St. Louis Urban	<2.0	450	8.0	49.0	<5.0	3.0	<2.0	80
Channel Catfish	St. Louis Urban	<2.0	350	<5.0	15.0	<5.0	3.0	<2.0	100
White Bass	Jefferson County Urban	<2.0	310	<5.0	6.0	<5.0	8.0	<2.0	250
Carp	Johnson County Rural	<2.0	30.0	<5.0	<5.0	<5.0	<2.0	<2.0	<50
Channel Catfish	Johnson County Rural	<2.0	80.0	<5.0	<5.0	<5.0	<2.0	<2.0	60
Largemouth Bass	Knox County Rural	<2.0	<15.0	<5.0	<5.0	<5.0	<2.0	<2.0	<50
White Bass	Johnson County Rural	<2.0	<15.0	<5.0	<5.0	<5.0	<2.0	<2.0	<50
Channel Catfish	Knox County Rural	<2.0	60.0	<5.0	<5.0	<5.0	<2.0	<2.0	<50

Note: Each sample was a composite of five fillets from different fish of comparable size and maturity.
[a]Residue levels araclor 1260, 1254, and 1242.

Table 3. Chlorinated Hydrocarbon Residue Levels (ppb) in Fish Tissue from Missouri River.

Fish Species	Aldrin	BHC*	Chlordane (Tech)	p,p'-DDD	p,p'-DDE	p,p'-DDT	Dieldrin	Hept. Epox.	PCBs[a]
1984									
Shovelnose Sturgeon	<2.0	5.0	610	11.0	39.0	5.0	126	37.0	310
Carp	<2.0	<5.0	160	<5.0	12.0	<5.0	80	22.0	80
Channel Catfish	<2.0	<5.0	310	10.0	56.0	<5.0	93	14.0	56
River Carpsucker	<2.0	<5.0	160	<5.0	20.0	<5.0	80	14.0	80
1986									
Shovelnose Sturgeon	<2.0	<5.0	860	15.0	67.0	5.0	62	17.0	560
Carp	<2.0	<5.0	120	<5.0	9.0	<5.0	3.5	<2.0	<50
Channel Catfish	<2.0	<5.0	520	<5.0	20.0	<5.0	<2.0	17.0	75
River Carpsucker	<2.0	<5.0	140	<5.0	16.0	<5.0	33	11.0	70
1987									
Shovelnose Sturgeon	<2.0	<5.0	749	6.0	47.0	<5.0	100	10.0	303
Carp	<2.0	<5.0	180	<5.0	11.0	<5.0	41	6.5	70
Channel Catfish	<2.0	<5.0	295	8.0	22.0	<5.0	85	13.0	110
River Carpsucker	<2.0	<5.0	234	<5.0	16.0	<5.0	14	7.0	<50

Note: Collection site was Easley, MO. All samples were composites of fillets from five fish and were of comparable size and maturity.
*:   Total isomers.
a: Residue levels araclor 1260, 1254, 1242.

Table 4.   Percent Composition of Chlordane Constituents in Technical Chlordane Formulations and Fish Tissue Samples.

	Hepta-chlor	trans-Chlor-dane	cis-Chlor-dane	trans-Nona-Chlor	cis-Nona-Chlor	Compound K	Chlor-dene-C	α-Chlor-dene	β-Chlor-dene	γ-Chlor-dene	Hepta-chlor Isomer	Hepta-chlor Epoxide	oxy-chlor-dane
Formulation													
CTC, MFA Oil	10.1	18.7	15.1	14.7	3.8	2.2	9.0	1.4	5.5	5.0	2.8	<5.0	<5.0
Green Up Curry Cartwright Corp.	14.5	22.2	14.2	17.0	4.1	2.6	7.3	1.7	7.8	4.7	3.7	<5.0	<5.0
Chlordane, Tech, U.S. EPA	14.2	20.1	20.5	7.1	2.2	3.0	13.7	1.9	6.5	7.4	2.7	<5.0	<5.0
Goldcrest Velsicol, Chicago	11.2	21.6	16.1	14.9	4.4	4.4	8.9	1.8	7.1	6.8	3.8	<0.5	<0.5
Fish Tissue													
Channel Catfish	<0.2	14.4	19.9	27.7	11.1	<0.5	8.6	<6.5	<0.5	6.5	<0.5	3.6	7.2
Carp	<0.2	18.5	23.7	26.4	13.7	<0.5	8.6	<0.5	<0.5	<0.5	<0.5	2.8	5.6
Shovelnose Sturgeon	<0.2	18.0	17.3	36.7	18.0	<0.5	2.9	<0.5	<0.5	<0.5	<0.5	2.4	4.8
Buffalo Fish	<0.2	12.6	16.5	29.5	7.9	<0.5	14.2	<0.5	<0.5	<0.5	<0.5	0.9	1.8
Flathead Catfish	<0.2	18.6	24.2	33.6	12.3	<0.5	11.2	<0.5	<0.5	<0.5	<0.5	<0.5	<0.5
River Carpsucker	<0.2	17.0	35.4	34.6	23.6	<0.5	16.5	<0.5	<0.5	<0.5	<0.5	2.0	4.0

$$C_t = \alpha C_s + \beta C_l + \gamma C_g$$

where C_t = Total concentration
         C_s = Concentration in the solid phase
         C_l = Concentration in the liquid phase
         C_g = Concentration in the gas phase
              and $\alpha, \beta,$ and γ are proportional constants

In the case of compounds with low water solubility such as chlordane, the vapor-phase losses from soil surfaces can be quite significant. This is especially true for chlordane constituents with lower chlorine (5–7) substitutions. The rate of volatilization loss is, however, dependent on the depth of penetration of the solute in the soil. While vapor-phase transport is significant from the soil surface and the leaves of plants, as in the case of typical agricultural applications, the volatilization losses of persistent polychlorinated organics decrease significantly with an increase in the soil layer thickness,[27,28] and as a result, chemical/biochemical transformation and erosion of the particle-bound pesticide are likely to play the most dominant roles in determining the mode of transport and fate of chlordane used for subterranean applications.

These observations are supported by our survey of chlordane-treated sites around the University of Missouri–Columbia campus. The results indicate a minimal loss of chlordane constituents at sites that are protected from water deposition and erosion. At such sites, more than 70% of the originally applied chlordane could be accounted for seven years after application. Furthermore, in these cases the chromatographic profiles of the residues were almost identical to those of the original formulation, indicating that little or no chemical/biochemical transformation had occurred. Transformation products such as 1-hydroxychlordene and heptachlor epoxide were not detected. A sharp decrease in the concentration of heptachlor and chlordenes was, however, observed from sites with higher moisture content and more organic debris, pointing toward greater chemical transformation under these conditions.

The gas chromatographic profiles for residues from termite control application sites are shown in Figure 5. Chromatogram A represents residue from a dry site where all constituents remained largely intact. Chromatogram B represents residue from a site with higher organic debris and moisture content. In this site, less than 5% of the heptachlor remained intact. Rapid conversion of heptachlor to 1-hydroxychlordene in terrestrial and aquatic environments has been reported.[29,30] The lower partition coefficient (k_d) of 1-hydroxychlordene and its rapid biochemical transformation presumably account for the low levels of hydroxychlordene observed at the site.

These results indicate that routes other than volatilization play a dominant role in determining the transport and fate of chlordane constituents from subterranean application sites. These observations are supported by the results

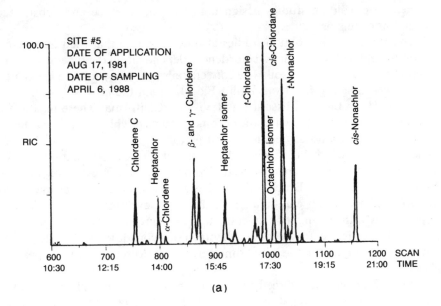

(a)

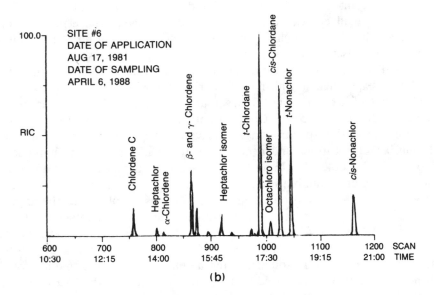

(b)

Figure 5. Chromatograms of chlordane residues in two soil samples. A: Soil sample from site with low (3.5%) moisture content. Residual technical chlordane concentration, 2300 ppm (approximately 75% of the originally applied concentration); B: Soil sample from site with high (8.5%) moisture content. Residual technical-chlordane concentration, 450 ppm (approximately 15% of the originally applied concentration).

obtained from column studies designed to simulate surface and subsurface application of this pesticide.

The results from the column studies also demonstrate that the rate of conversion of heptachlor to hydroxychlordene is dependent on the type of soil and the moisture content of the soil. The conversion rates were found to be significantly lower in sandy soil from Eglin, Florida, than the clay, silt, and loam soils from Times Beach, Missouri, and Visalia, California. These results are shown in Figure 6, which depicts the chromatograms of chlordane constituents in these soils after a 30-day equilibration period.

Clear differences in the downward movement and transformation rate of chlordane constituents were observed. Despite the higher clay and organic matter content of Times Beach soil, the degree of translocation was highest in this soil. This deviation can be attributed to the development of cracks and pores in soils with the higher clay content. The degree of translocation in the sandy soil was highest for hexa- and heptachloro constituents. The degree of movement was highest in the initial time period and decreased with the passage of time.

The initial movement is directly related to the presence of surfactants in the technical chlordane formulations. The technical chlordane formulation used in this study contained 8% anionic/cationic surfactants by weight. An enhanced leachability of lipophilic solutes has been demonstrated with the use of surfactant solutions,[31] and our soil column experiments show that the presence of surfactants can lead to a slight, but discernible, increase in the depth of penetration of all chlordane constituents. As expected, the effect is more pronounced in the case of chlordane constituents with lower numbers of chlorines and decreases with an increase in chlorine substitutions.

The soil column results were found to agree with the results from partition coefficient experiments, which were carried out with the same three soils. Analysis performed after one- to seven-day intervals showed no measurable change in the relative concentration of major constituents in the soil. The partitioning into the aqueous phase was found to be dependent on the number of chlorine substitutions and decreased with an increase in the number of chlorines in the molecule. Rapid transformation of heptachlor was observed, and after a 30-day equilibration period almost no heptachlor was observed in the aqueous phase. The decrease in heptachlor was coupled with the appearance of 1-hydroxychlordene. The transformation of heptachlor to 1-hydroxychlordene was found to be dependent on soil type. The transformation rate was highest in the clay soil, followed by the silt loam soil and then the sandy soil. Analysis of the soil after a ten-month period showed that almost all of the heptachlor was converted to 1-hydroxychlordene in the first two types of soil, but heptachlor could still be detected in the Eglin sands (Figure 7). The residual concentration of chlordane constituents in the predominantly anaerobic soil/water systems employed during the study are summarized in Table 5. The results show that, even though the rate of transformation was lowest in the Eglin sand, the overall losses were highest. These losses can be attributed to

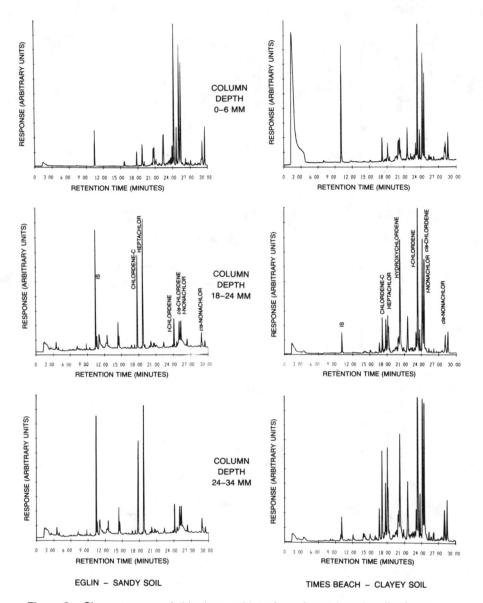

Figure 6.   Chromatograms of chlordane residues from clay and sandy soil columns.

the lack of strong binding sites on the sandy soil, which would lead to increased partition into the aqueous phase and subsequent volatilization losses from the aqueous phase.

Heptachlor was found to be the least stable constituent in the aquatic environment, followed by other dienes. Similar results have been obtained by Eickelberger and Lichtenberg.[30] The saturated and highly substituted constitu-

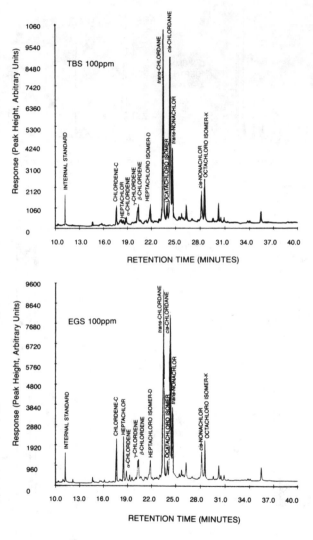

Figure 7. Chromatograms of chlordane constituents from soil/water system. Incubation period ten months. A: residue in (clayey) Times Beach soil (TBS); B: residue in Eglin (sandy) soil (EGS).

ents were more stable, with a half-life of more than a year. Thus, the relative concentration of chlordane constituents in sediment samples can be used to establish the time of introduction of these contaminants into the aquatic system.

The detection of heptachlor and chlordenes in the sediment from certain streams in Missouri indicates that chlordane contamination in Missouri is of

Table 5. Relative Stability of Tech-Chlordane Constituents in Soil/Water Systems.

	Eglin Soil		Times Beach Soil		Visalia Soil	
Chlordane Constituents	Percent Residue	K_{oc}	Percent Residue	K_{oc}	Percent Residue	K_{oc}
Component C	17.6	5.0	25.0	5.0	33.0	5.3
Heptachlor	3.4	ND	<0.5	ND	<0.5	ND
α-Chlordene	8.0	ND	0.8	ND	1.0	ND
β-Chlordene	17.6	5.1	24.5	5.0	48.0	5.4
γ-Chlordene	13.4	5.0	23.0	5.0	45.0	5.4
Heptachlor isomer	29.0	4.9	35.0	5.0	47.0	5.2
trans-Chlordane	51.0	4.9	47.0	5.1	61.0	5.4
cis-Chlordane	63.0	4.9	53.0	5.1	67.0	5.3
trans-Nonachlor	50.0	5.0	53.0	5.1	49.0	5.4
cis-Nonachlor	49.5	5.0	47.0	5.2	60.0	5.3
Component K	52.0	5.0	62.0	5.2	56.0	5.3
Hydroxychlordene		2.9		3.1		3.5

Note: Expressed as percent residual concentration. ND—not determined.

recent origin (Figure 8). The concentration of total chlordane constituents in sediments varied from 1.5 to 310 ppb. A correlation between urban development and chlordane level was clearly revealed by the land use survey. Residue levels from predominantly agricultural areas were approximately two orders of magnitude lower than those from urban areas (Table 2).

The log of the partition coefficient (log k_d) of the chlordane constituents ranged from 2.9 to 3.7 in surfactant-free systems; addition of 0.01% lauryl sulfate resulted in an increase in the concentration of "dissolved" chlordane. This increase was most pronounced under basic pH conditions, and a tenfold increase in aqueous concentration chlordane was observed. The partitioning of chlordane constituents applied to sand was generally unaffected by the difference in pH. It is likely that the apparent increase in dissolved chlordane is related to an increase in dissolution of soil organic matter at the higher pH. The effect of surfactant on the dissolution of chlordane constituents decreased with an increased time interval. The effect was directly connected to the disappearance of lauryl sulfate through hydrolysis. The rate of decrease was found to be dependent on the soil system and occurred most rapidly under alkaline soil conditions.

ACKNOWLEDGMENTS

The study was supported in part by a grant from the U.S. Department of Interior, which was administered through the Water Resources Research Center, University of Missouri. The graphical assistance of Paul Nam and the secretarial assistance of Karen Bick is also acknowledged. Assistance provided by Mr. John Wingage in sampling soil samples from different sites of UMC campus is greatly appreciated.

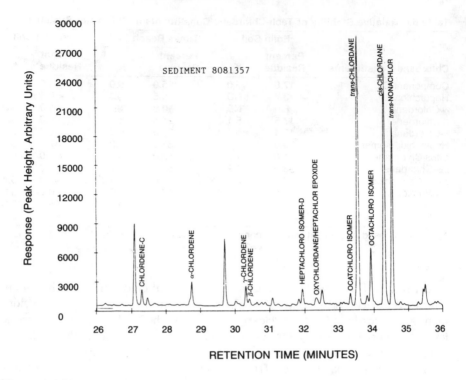

Figure 8.   Chromatographic profile of chlordanes and other contaminants in a sediment sample from the Meramec River.

REFERENCES

1. Velsicol Chemical Corporaton, Chicago, IL. "Standard for Technical Chlordane," Technical Bulletin (1971).
2. Saha, J., and Y. W. Lee. *Bull. Environ. Contam. Toxicol.* 4:285 (1969).
3. Cochrane, W. P., H. Parlar, S. Gäb, and F. Korte. *J. Agric. Food Chem.* 23(5):882–86 (1975).
4. Cochrane, W. P., and R. Greenhalgh. *J. Assoc. Off. Anal. Chem.* 59(3):696–702 (1976).
5. Gäb, S., H. Parlar, and F. Korte. *J. Agric. Food Chem.* 25(5):1224–25 (1977).
6. Gäb, S., L. Born, H. Parlar, F. Korte. *J. Agric. Food Chem.* 25(6):1365–70 (1977).
7. Sovocool, G. W., R. G. Lewis, R. L. Harless, N. K. Wilson, and R. D. Zehr. *Anal. Chem.* 49(6):734–40 (1977).
8. Parlar, H., K. Hustort, S. Gäb, and F. Korte. *J. Agric. Food Chem.* 27(2):278–82 (1979).
9. Miyazaki, T., T. Yamagishi, and M. Matsumoto. *Arch. Environ. Contam. Toxicol.* 14(6):475–83 (1985).
10. Hoff, R. M., and K. W. Chan. *Chemosphere* 15(4):449–52 (1986).
11. Kawano, M., S. Tanabe, T. Inoue, and R. Tatsukawa. *Trans. Tokyo Univ. Fish.* 6:59–66 (1985). *CA* 104(22):192066m.

12. Miyazaki, T., T. Yamagishi, and M. Matsumoto. *Shokuhin Eiseigaku Zasshi* 27(1):49–58 (1986). *CA* 105(4):29546e.
13. Arruda, J. A., M. S. Crigan, W. G. Layher, G. Kersh, and C. Bever. *Bull. Environ. Contam. Toxicol.* 41(4):617–24 (1988).
14. Johnson, M. G., J. R. Kelso, and S. E. George. *Can. J. Fish. Aquat. Sci.* 45(suppl. 1):170–78 (1988).
15. Saito, I., N. Kawamura, K. Uno, and N. Hiranaga. *Int. Arch. Occup. Environ. Health* 58(2):91–97 (1986).
16. Tojo, Y., M. Wariishi, Y. Suzuki, and K. Nishiyama. *Arch. Environ. Contam. Toxicol.* 15(4):327–32 (1986).
17. Mussalo-Rauhamaa, H., H. Pyysalo, and K. Antervo. *J. Toxicol. Environ. Health* 25(1):1–19 (1988).
18. U.S. Food and Drug Admin., *Federal Register* 45(10):2904 (1980).
19. Omija, T. *Okinawa-Kenkogai Eisei Kenkyushoho* 19:63–73 (1985). *CA* 105(5):37154m.
20. Beeman, R. W., and F. Matsumura. *J. Agri. Food Chem.* 29(1):84–89 (1981).
21. Bennett, G. W., D. L., Ballee, R. C. Hall, J. E. Fahey, W. L. Butts, and J. V. Osmen. *Bull. Environ. Contam. Toxicol.* 11(1):64–69 (1974).
22. Harris, C. R., and W. W. Sans. *Proc. Entomol. Soc. of Ontario* 106:34–38 (1975).
23. Glotfelty, D. E., A. W. Taylor, B. C. Turner, and W. H. Zoller. *J. Agri. Food Chem.* 32:638–43 (1984).
24. Paulausky, J., S. Kapila, S. E. Manahan, A. F. Yanders, R. K. Malhotra, and T. E. Clevenger. *Chemosphere* 15(9–12):1389–96 (1986).
25. Kapila, S., D. D. Duebelbeis, R. Malhotra, A. F. Yanders, and S. E. Manahan. In *Pesticide Science and Biotechnology: Proceedings of the Sixth International Congress of Pesticide Chemistry,* R. Greenhalgh and T. R. Roberts, Eds. (Oxford, England: Blackwell Scientific Publications, 1987).
26. Duebelbeis, D. D., S. Kapila, T. E. Clevenger, A. F. Yanders and S. E. Manahan. *Chemosphere 18,* (1–6):101–108 (1989).
27. Kapila, S., A. F. Yanders, C. E. Orazio, J. E. Meadows, S. Cerlesi, and T. E. Clevenger. *Chemosphere* 18(1–6):1297–1304 (1989).
28. Palausky, J., S. Kapila, S. E. Manahan, A. F. Yanders, and R. K. Malhotra. *Chemosphere* 15(9–12):1309–96 (1986).
29. Simon, J., and F. L. Parker. In *The Biosphere: Problems and Solutions,* T. N. Veziroglu, Ed. (Amsterdam: Elsevier Science Publishers, 1984), pp. 453–60.
30. Eichelblerger, J. W., and J. J. Lichtenberg. *Environ. Sci. Tech.* 5(6):541–44 (1971).
31. Rickabaugh, J., S. Clements, and R. F. Lewis., *Proceedings of the 41st Industrial Waste Conference,* Purdue University (1986), pp. 377–82.

## CHAPTER 19

Mass Balance Accounting of Chemicals in the Great Lakes

William M. J. Strachan and Steven J. Eisenreich

INTRODUCTION

In the Great Lakes, a number of chemicals have been identified as "Priority Pollutants" (Table 1) by the International Joint Commission (IJC).[1] This binational agency is charged with overseeing the Great Lakes Water Quality Agreement and ensuring that its provisions are carried out. The chemicals that they have identified in their Priority List are those which, in the main, are already substantially restricted as to use or release. Some are totally banned in both countries, and many of them have been monitored in one or more parts of the aquatic system for a number of years. The particular concern in the IJC is that the levels of the substances are not decreasing as rapidly as expected, and the IJC would therefore like to establish the significance of the several possible routes that these compounds may use to enter the system. This would seem to

Table 1. International Joint Commission Priority Pollutants

Metals:	alkylated lead[a]
	mercury (Hg)
	cadmium (Cd)[b]
	arsenic (As)[b]
Industrial Organics:	benz(a)pyrene (B(a)P)
	polychlorinated biphenyl (PCBs)
	2,3,7,8-tetrachlorodibenzo-p-dioxin (TCDD)[c]
	2,3,7,8-tetrachlorodibenzofuran[c]
	hexachlorobenzene (HCB)
	mirex
Pesticides:	dieldrin
	lindane[b]
	alpha-hexachlorocyclohexane (HCH)[b]
	DDT and metabolites
	toxaphene

[a]Changed to total lead for the workshop.
[b]Added to the list for the workshop.
[c]Deleted from the list for the workshop.

be a prerequisite to recommending additional, possibly more effective, ways to control them and to speed up the trend to lower levels.

The IJC, in October 1986, sponsored a workshop to deal with the particular question of whether the atmosphere is a significant source of any of the compounds on the Priority List for any of the five lakes (four international and one U.S.).[2] They invited approximately 40 technical and scientific experts active in fields relevant to the transfer of substances between the atmosphere and water. A working paper was prepared and served as the basis for discussion of the various issues; a revised version of this working paper was published by the commission as part of the report of the meeting.[3]

One of the conclusions of the workshop was that, despite a decade and a half of intensive monitoring for the compounds of concern, there was an insufficiency of data to undertake a reliable mass balance account of these compounds in any of the lakes. This was considered to be the consequence of a focus upon the biota of the system—with some justification—and an inability to analyze for many of the chemicals in several of the media subcompartments. Several compounds, however, did have databases that permitted semi-quantitative estimates, and among these, PCBs and Pb (total) were better than the others. This chapter provides a summary of some of the aspects of the mass budgeting that were discussed at the meeting.

GENERAL CONSIDERATIONS

Modern approaches to describing the behavior and fate of chemicals in any part of the environment take into account all processes and relationships within and among interacting parts of the ecosystem. This sort of "ecosystem approach" is one firmly entrenched in the Great Lakes Water Quality Agreement activities which the IJC administers.[4] For mass balancing of the Priority Pollutants (Table 1), however, only the abiotic components of the lakes were considered since it was the consensus of the workshop that the amounts in the biota would represent only a very small fraction of the chemical in the aquatic system. At least for the persistent compounds, it was felt that this assumption would not introduce any major error. The same statement cannot be made for nonpersistent compounds where degradation may be appreciable and where levels in the biota may therefore be significant.

The mass balancing of the chemicals can be represented by the fluxes in Figure 1. For each process indicated by the arrow movements of material, a concentration term and a flux rate constant is required. For some of these, there are methods of estimating the values and constants, but many of the methods have never been tested rigorously and the results are therefore speculative. The model is admitted to be simple and additional refinements could be inserted; the data available, however, do not justify this, and it is felt that no improvement in reliability could be achieved in this manner at this time. The class "compound," PCBs, and the metal lead (total) had more extensive, and

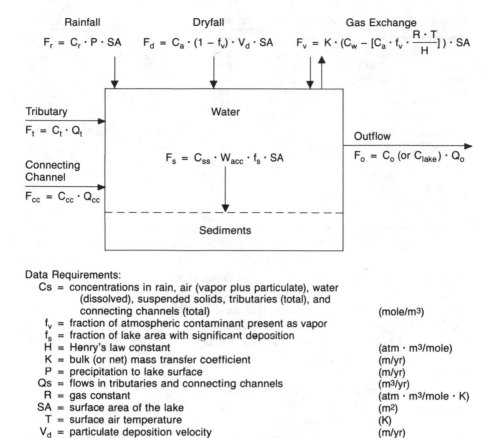

Figure 1.   Flux estimates (F) for chemicals in the Great Lakes.

hence reliable, databases for the concentration terms than did the other compounds. The databases for DDT residues, benzo(a)pyrene, and mirex were sufficient to run the model and to assess qualitatively the significance of the atmospheric route of input.

CHEMICAL AND SYSTEM PARAMETERS

The Great Lakes system is well known, and the relevant physical parameters (e.g., volumes, flows, areas, precipitation, etc.) are available. Table 2 summarizes the data used during this evaluation of the budgets of the various chemicals in the lakes. Also in this table are several of the important property values for the organic chemicals; the corresponding lead values were assumed to be zero, i.e., none appears in the vapor state.

The participants at the workshop attempted to describe the error associated

Table 2.   Parameters for Mass Balance Calculations

Physical Parameters	Superior	Huron	Michigan	Erie	Ontario
Precipitation (m/yr)	0.76	0.76	0.79	0.84	0.89
Surface area ($10^{10}/m^2$)	8.21	5.97	5.78	2.57	1.95
Sediment Acc. ($g/m^2/yr$)	200	220	400	1000	400
Q_{trib} ($10^{10}/m^3/yr$)	5.4	5.1	2.9	2.2	3.0
Q_{cc} ($10^{10}/m^3/yr$)	—	12.0	—	19.0	21.0
Q_{out} ($10^{10}/m^3/yr$)	7.1	18.0	4.9	21.0	25.0

Chemical Parameters	PCBs	DDT	BaP	Mirex
Henry's law constant (H, atm/m^3 • yr)	200	120	7.6	790
Vapour fraction (f_v)	0.8	0.7	0.2	0.2
Mass transfer coeff. (K, m/day)	0.24	0.21	0.032	0.30
Dissolved fraction	0.8	0.7	0.2	0.2

with the chemical parameters and stated that it seemed possible that they could individually be in error by as much as 400%, except for the mass transfer coefficient which they believed might err by up to 1000%! The values employed in the model calculations were taken from recent estimates and were generally considered to be appropriate at the workshop. It was also the opinion of the workshop that temperature-dependent parameters should be employed when estimating annual fluxes but that these were generally not available; the ones used were those experimentally determined, usually for the 20–25°C range. Despite these caveats, the workshop felt that there was value in proceeding with the estimations since it could provide information on the relative significance of the different inputs and outputs.

There are, in addition to the parameters in Table 2, several others that influence the fluxes and, therefore, the apparent significance of specific inputs to the mass balance budget of the lakes. The particulate deposition velocity employed, 0.1 cm/sec (about 3200 m/year), is representative of the chemicals investigated and corresponds to a particle size range of 0.1–2.0 μm; a recent review of the topic[5] indicates that such a value is on the low end of the reported range. Other parameters that enter into the flux equations include the fraction of the year, and the extent, when there is ice cover (this was taken as a combined 0.1 for all lakes), the fraction of the lake area that has active sedimentation (0.5 for all lakes except Erie, which was assigned 0.7), and the fraction of the year without rain/snowfall (0.9 for all lakes). Selection of these values by the authors was based upon their experience with the system, the literature, and the advice of colleagues.

ENVIRONMENTAL CONCENTRATIONS

In the workshop paper,[3] 67 references to published and unpublished material were examined for concentrations of the different target chemicals in one or more of the compartments of the system. Some of these were compilations of other works. All values reported for more-or-less open waters were used to guide the selection of values that were considered representative of ambient

Table 3.   Polychlorinated Biphenyls Concentrations in the Great Lakes

	Water (ng/L)	Susp. Solid (μg/g)	Sediment (μg/g)	Air (ng/m^3)	Rain (ng/L)
Superior					
Lake	0.4	0.2	0.03	0.5	5.0
Tributaries	1[a]	—	—	—	—
Huron					
Lake	0.7	0.3	0.1	0.5[b]	5.0
St. Mary's R.	0.6[a]	—	—	—	—
Mackinaw Str.	2.0[a]	—	—	—	—
Tributaries	1[a]	—	—	—	—
Michigan					
Lake	0.7	0.3	0.2	0.5	5.0
Tributaries	10[a]	—	—	—	—
Erie					
Lake	0.7[b]	0.3[b]	0.06	0.5	5.0
Detroit R.	10[a]	—	—	—	—
Tributaries	20[a]	—	—	—	—
Ontario					
Lake	0.6	0.3	0.1	0.5	5.0
Niagara R.	10[a]	—	—	—	—
Tributaries	10[a]	—	—	—	—

[a]Includes suspended solids.
[b]An estimate based on data from other lakes and pollution levels in other compartments of the five lakes.

conditions rather than those of point sources or particularly polluted areas. Those for PCBs and Pb (total) are presented in Tables 3 and 4, where, in general, the values were derived from samplings during the period 1980–85. Older data were used only when more current information was not available.

One of the compartments where data availability was very poor was the tributaries, especially for compounds other than PCBs and Pb. There has been some determination of the levels of PCBs in dissolved and suspended matter in the tributaries,[6] but on the whole, even for these compounds, such data are pre-1980 and usually correspond to a single sample. There was also little information to estimate the effect of the large flow differences that occur seasonally or the variability that may exist in upstream discharge patterns. Air concentrations, for the most part, were assumed to be constant throughout the basin—which, given the rate of tropospheric mixing at the same latitude, is probably reasonable.[7] Open lake water concentrations, both "dissolved" and adsorbed to suspended matter, have only recently become available.[8] These are operationally defined by filtration (0.5 μm usually) or centrifugation (continuous flow at a rate and speed equivalent to the filtration). It is uncertain how representative these values are since they correspond to only a few samples in each of the lakes investigated; they are accepted and used as the best available. Little recent information exists on the levels of the compounds under study for the sediments of the depositional parts of the basins; older data from the early 1970s have to be used.[6]

Table 4.   Total Lead Concentrations in the Great Lakes

	Water (ng/L)	Susp./Solid (μg/g)	Sediment (μg/g)	Air (ng/m^3)	Rain (ng/L)
Superior					
Lake	75	25	100	20	3000
Tributaries	50[a,b]	—	—	—	—
Huron					
Lake	150	50	70	50[b]	7000
St. Mary's R.	100[a,b]	—	—	—	—
Mackinaw Str.	200[a,b]	—	—	—	—
Tributaries	24	—	—	—	—
Michigan					
Lake	150	50	40	50	10000[b]
Tributaries	100[a,b]	—	—	—	—
Erie					
Lake	750	250	100[b]	70[b]	8000
Detroit R.	1800[a]	—	—	—	—
Tributaries	10[a,b]	—	—	—	—
Ontario					
Lake	300	100	100	75[b]	10000
Niagara R.	1000[a]	—	—	—	—
Tributaries	11[a]	—	—	—	—

[a]Includes suspended solids.
[b]An estimate based on data from other lakes and pollution levels in other compartments of the five lakes.

PROCESS RATES

Strictly speaking, from an ecosystem perspective, the model should allow for interaction with the terrestrial part of the biosphere. In a practical sense, however, there is very little relevant data available for this compartment — either for concentrations or process rates — and the interchange of chemical is provided, to a degree, through the input of the tributaries, connecting channels, and the several atmospheric processes. Direct discharges to the lakes do not appear as a process in Figure 1, but where data were available, this input was treated in the same fashion as the tributaries; generally, there were few data for this. When it was possible to include this term, the results indicated that this input represented only a few percent, at most, for the compounds under study. Indeed, most direct waste discharges occur via the tributaries and connecting channels, and consequently, such inputs to the system are accounted for.

The atmospheric deposition processes from Figure 1 include wetfall ("washout" of the chemical from both the vapor and particle-adsorbed states by both rain and snow), dryfall (the fallout of the chemical adsorbed to particles of various sizes) and vapor exchange (a two-way process involving absorption at the water surface and volatilization from the water). For the most part, the rate constants for these processes are experimentally unknown, although there are various estimates of the values. For wetfall, there is some information on the washout coefficients for rain;[9] those for snow are largely unknown. The deposition velocity for various particle size fractions has been estimated, and a representative value of 0.1 cm/sec assigned.[5] The distribution of the chemicals among the different size fractions is unknown, and therefore it is uncertain whether this value is representative of the chemical being deposited in this mode. For vapor-phase exchange, a much needed parameter is the mass transfer coefficient, which is experimentally unknown although estimates have been made.[10,11] Also largely unknown is the fraction of the chemical in the atmosphere that is adsorbed to the particle fraction. This is required to provide for the calculation of the concentrations of chemical in both the vapor and adsorbed states since data are available only for "whole" air samples. There are, at present, efforts underway in the basin to determine each of these concentrations independently.

Within the aquatic part of the system, there are some data for each lake on the sediment accumulation rates,[12,13] which, together with surficial concentrations of the chemicals and the fraction of the lake area represented by active deposition of the suspended material, can permit a coarse estimate of the net flux to the sediment. In a more refined model, the flux to the sediment would be dealt with as two separate terms — settling and resuspension. Unfortunately, adequate data for the lakes on these two processes are not available and the accumulation term has been used. The fraction of the lake area in which active accumulation is occurring is somewhat arbitrary but is based on the scientific judgment of the authors.

MASS BALANCES

When the mass balance model indicated by Figure 1 is applied to the data for concentrations and processes described, the results in Tables 5 (sources) and 6 (fate) are obtained. Any conclusions or conjectures based on these tables must be made with the knowledge that these are the best conclusions that can be drawn at the present, given the uncertainties described. The results should only be considered indicative of probable sources and sinks of the chemicals, and the direction of any actions based on these assignments should be confirmed by further investigation of some of the contentious parameters and model assumptions.

The percentages of the inputs arising from the atmosphere, as presented in Table 5, clearly show that the major sources for PCBs entering the upper lakes (Superior, Michigan, and Huron) are atmospheric. Atmospheric inputs would probably also be equally important in the downstream lakes (Erie and Ontario) were it not for the substantial inputs along the connecting channels of the St. Clair/Detroit and Niagara River systems, respectively. In the case of lead, the atmosphere is clearly indicated as the most important single source for all of the lakes except Lake Erie, where sources along its input connecting channel reduce the atmospheric contribution to slightly less than 50%.

From Table 6, it is apparent that volatilization is the primary loss mechanism for PCBs from the lakes. For the upper lakes, where the connecting

Table 5.   Sources of PCBs and Lead in the Great Lakes

		Atmospheric Input				Connecting Channel Other (%)	Tributaries Input (%)	Direct Waste (%)
	Total Input	Wetfall (%)	Dryfall (%)	Connecting Channel (%)	Total (%)			
			Polychlorinated Biphenyls					
	(kg/yr)							
Superior	606	52	39	—	91	—	9	<1
Huron	636	36	27	15	78	7	8	7
Michigan	685	33	25	—	58	—	42	?
Erie	2520	4	3	6	13	69	18	?
Ontario	2540	4	2	1	7	81	12	?
			Lead					
	(metric tons/yr)							
Superior	241	77	20	—	97	—	1	2
Huron	430	74	20	4	98	<1	<1	2
Michigan	543	84	15	—	99	—	<1	?
Erie	567	31	9	6	46	54	<1	?
Ontario	426	41	10	23	74	26	<1	?

channel inputs are of lesser or no significance, the quantity volatilized greatly exceeds the total quantity entering the lakes; they are said to be degassing their existing PCB burdens. This process has been suggested by others[14] to be the present-day situation; if true, it would result in transferring the burden from the lakes to the global atmosphere. Lead is not assigned any volatilization component (alkyl leads might contribute but are considered insignificant in quantity in the "total" lead picture), and there is no comment to be made about

Table 6.   Fate of PCBs and Lead in the Great Lakes

	Total Output	Volatilization (%)	Fate Sedimentation (%)	Export Connecting Channel (%)
		Polychlorinated Biphenyls		
	(kg/yr)			
Superior	2190	87	11	2
Michigan	7550	68	31	1
Huron	3400	75	19	6
Erie	2390	46	45	9
Ontario	1320	53	30	17
		Lead		
	(metric tons/yr)			
Superior	828	—	99	1
Michigan	472	—	98	2
Huron	496	—	93	7
Erie	2010	—	90	10
Ontario	490	—	80	20

the atmosphere in any loss mechanism — sedimentation and consolidation would seem to be the fate of this material.

Little comment has been made about the other substances on the IJC's Priority Pollutant list. The data for these materials are much poorer in quantity and geographic coverage than for PCBs and Pb. Mirex was examined and found to have only a marginal atmospheric involvement in Lake Ontario. Indeed, the small flux indicated for the atmosphere was more the consequence of uncertainties in the concentrations, chemical properties, and rate determinations than of any real appearance in the atmospheric part of the system. Mirex was not significant in any of the other lakes.

DDT and benzo(a)pyrene were the only other compounds for which sufficient data on environmental concentrations in enough compartments and rate/property constants existed to permit an attempt at mass balancing. Even for these, the environmental database was much poorer than for PCBs and Pb, and the conclusions from the model are even more tentative than for PCBs and lead. In the case of DDT, the contribution of the atmosphere to the lake loadings, directly plus indirectly, was 97–98% in the upper lakes and 22–31% in the lower ones; the volatilization losses ranged from 27–89%, but with much less of a upper/lower lake pattern. B(a)P had a similar input pattern for all of the lakes, 72–96% atmospheric contribution; its volatilization losses were low, ranging from 2 to 19%.

SUMMARY AND CONCLUSIONS

The atmosphere plays an important, if not dominant, role in the loading of many toxic chemicals to surface waters of large lakes with long retention times. PCBs arrive in the Great Lakes from atmospheric sources at the rate of 180–500 kg/year and are volatilized from there at 3–12 times this rate. Removal via sedimentation is much less than via the atmosphere, and export via the connecting channels is significant only for Lake Ontario (17%).

Lead enters the lakes at the rate of 240–570 metric tons/year, and its main fate is consolidation in the sediments at similar loading rates except for Lake Erie, where the sediments accumulate nearly seven times that added from the atmosphere. Most of this latter burden is derived from sources along the input connecting channel.

The conclusions drawn from the mass balance accountings of the chemicals investigated in the Great Lakes must be viewed as tentative since several of the key parameters are estimates and experimental evidence is needed to confirm them. Included among these are

- snow concentrations of the chemicals
- deposition velocities of the different airborne particle size fractions
- distribution of the chemicals between the vapor state and the different size fractions
- representative air concentrations

- mass transfer coefficients for absorption-volatilization of the chemicals at the air-water interface
- representative concentrations for the chemicals in tributaries and direct discharges to the lakes
- current data on open lake sediment and water concentrations

For the chemicals themselves, it is important that temperature-dependent values for the physicochemical properties be used or that an appropriately representative value for all of the parameters be used. For compounds that may not be persistent, the degradation rates will be required for the relevant compartments and the processes included in the mass balance model.

It is certain that releases of PCBs to different parts of the environment have all contributed to the present-day load in the atmosphere and that cycling of this compound among the several compartments will continue the transport and deposition of such material for many years. In the case of PCBs and several other persistent organic compounds, this atmospheric burden is probably derived from global sources; any controls will therefore need to be made on the same scale. Lead is present in the atmosphere in particulate form and is derived, in large part, from local sources; controls can effectively be applied on a national or regional basis.

REFERENCES

1. "Report on Great Lakes Water Quality," International Joint Commission, Water Quality Board, Windsor, Ont. (1985).
2. "Summary Report of the Workshop on Great Lakes Atmospheric Deposition," International Joint Commission, Science Advisory Board, Windsor, Ont. (1987).
3. Strachan, W. M. J., and S. J. Eisenreich. "Mass Balancing of Toxic Chemicals in the Great Lakes: The Role of Atmospheric Deposition," International Joint Commission, Windsor, Ont. (1988).
4. "The Ecosystem Approach," International Joint Commission, Science Advisory Board, Windsor, Ont. (1978).
5. Bidleman, T. F. "Atmospheric Processes: Wet and Dry Deposition of Organic Compounds are Controlled by Their Vapour-Particle Partitioning," *Environ. Sci. and Technol.* 22:361–67 (1988).
6. Thomas, R. L., and R. Frank. "PCBs in Sediment and Fluvial Suspended Solids in the Great Lakes," in *Physical Behaviour of PCBs in the Great Lakes*, D. Mackay, S. Paterson, S. J. Eisenreich, and M. Simmons, Eds. (Ann Arbor, MI: Ann Arbor Science, 1983), pp. 245–68.
7. Dilling, W. L. "Atmospheric Environment," in *Environmental Risk Analysis for Chemicals*, R. A. Conway, Ed. (New York, NY: Van Nostrand Reinhold, 1982), pp. 154–97.
8. Biberhofer, J., and R. J. Stevens. "Organochlorine Contaminants in Ambient Waters of Lake Ontario," Inland Waters Scientific Series No. 159, Environment Canada, Ottawa, Ont. (1987).
9. Ligocki, M. P., C. Leuenberger, and J. F. Pankow. "Trace Organic Compounds in

Rain. III: Particle Scavenging of Neutral Organic Compounds," *Atmos. Environ.* 19:1619–26 (1985).

10. Mackay, D., and A. T. K. Yeun. "Mass Transfer Coefficient Correlations for Volatilization of Organic Solutes from Water," *Environ. Sci. and Technol.* 17:211–17 (1983).

11. McVeety, B. D. "Atmospheric Deposition of PAHs to Water Surfaces: A Mass Balance Approach," PhD Thesis, University of Minnesota (1986).

12. Kemp, A. L. W., C. I. Dell, and N. S. Harper. "Sedimentation Rates and a Sediment Budget for Lake Superior," *J. Great Lakes Res.* 4:276–87 (1978), and references therein.

13. Edgington, D. N., and J. A. Robbins. "Records of Lead Depositions in Lake Michigan Sediments Since 1800," *Environ. Sci. and Technol.* 10:266–73 (1976).

14. Murphy, T. J., J. C. Pokojowczyk, and M. D. Mullin. "Vapour Exchange of PCBs with Lake Michigan: The Atmosphere as a Sink for PCBs," in *Physical Behaviour of PCBs in the Great Lakes* (Ann Arbor, MI: Ann Arbor Science, 1983), pp. 49–58.

## CHAPTER 20

Limitations of the Compartmental Approach to Modeling Air-Soil Exchange of Pesticides

Warren Stiver and Donald Mackay

INTRODUCTION

Multimedia environmental fate models commonly assume that each compartment or medium is uniform in concentration. The compartmental approach was first introduced by Baughman and Lassiter[1] and has been extensively used by Mackay and coworkers in the fugacity series of models.[2-4] Models that use this assumption retain mathematical simplicity and yield equations that may be readily solved. However, it is uncertain what effect this uniform concentration assumption has on the prediction of the ultimate fate of a chemical. This work addresses the effect that the uniform concentration assumption in soil has on the predicted atmospheric concentration of a chemical. A model-to-model comparison is performed to isolate the significance of the assumption and therefore provide model users and developers with a fuller understanding of the significance of this modeling simplification.

It is recognized that the physical environment does not maintain uniform concentrations within compartments; therefore, models that do not constrain their compartments in this manner are closer physical representations of the environment. However, the quality of the final prediction by these models may not be superior due to other factors, assumptions, or errors within their structure.

A comparison is made between two steady-state models differing only in the uniform concentration assumption, and two single-dose unsteady-state models also differing only in the uniform concentration assumption.

STEADY-STATE MODELS

Steady-state models represent situations in which there is no accumulation or depletion of a chemical in a system; this does *not* imply that the system is at equilibrium. Essentially, a steady-state situation is one that is independent of time. Such situations are eventually attained if the emissions remain constant.

Table 1.   Compartment Volumes and Contact Areas for Level III Model.

Compartment	Number	Volume (m^3)
Air	1	6.0×10^9
Water	2	7.0×10^6
Soil	3	4.5×10^4
Sediment	4	2.1×10^4
Suspended sediment	5	3.5×10^1
Biota	6	7.0
Air particulates	7	1.0

Interface	Area (m^2)
Air-water	7.0×10^5
Air-soil	3.0×10^5
Air-particulates	—
Water-sediment	7.0×10^5
Water-suspended sediment	2.0×10^6
Water-biota	4.0×10^3

Source: Mackay and Paterson.[4]

The steady-state compartmental model presented by Mackay and Paterson[4] is used as the base model in this work. It is referred to as the fugacity level III model. The compartments in the fugacity level III model are air, water, soil, sediment, suspended sediment, and biota. Other compartments can be added; in this analysis air particulates are included. Each compartment is of a defined volume, is homogeneous in concentration, and is in contact with some of the other compartments. The volumes and contact areas used in this analysis are given in Table 1.

The models are referred to as "fugacity" models because the equations describing the fate of chemicals in the system are written in terms of fugacity as opposed to concentration. Fugacity merely facilitates calculation, it does not alter the results of the model. Fugacity, f (Pa), is a thermodynamic property related to the chemical potential of a chemical in a compartment. At chemical equilibrium between two compartments, the fugacities in each are equal. Fugacity is linearly related to concentration, C (mol/m^3), by the proportionality constant termed the fugacity capacity, z (mol/m^3/Pa).

$$C = zf \tag{1}$$

The fugacity capacities are a property of the chemical in a given compartment. For all chemicals the fugacity capacity in air is equal to 1/RT, as can be derived from Equation 1 with the fugacity in air equaling the partial pressure. The other capacities are determined from the partition coefficient between two compartments, K_{12}, and the knowledge that the partition coefficient is the ratio of the two fugacity capacities, z_1/z_2.

In the steady-state model, it is necessary to construct a mass balance around each compartment. This requires a prediction of the transport rates between compartments and a prediction of losses due to reaction. Transport of chemical between compartments is by one of two physical processes:

1. by diffusion due to differences in the chemical potential between compartments
2. by the bulk motion of one phase from one compartment to another

The diffusional transport rate is calculated based on the Whitman two-film theory[5] for intermedia mass transport. Liss and Slater[6] have described the use of the two-film theory for the mass transfer of chemicals between air and water in the environment. The transport rate, N (mol/hr), is dependent on the mass transfer coefficients, k (m/hr), on the concentration difference, C (mol/ m^3), and on the contact area, A (m^2). Expressions can be written for both sides of the interface and then combined based on an equilibrium relationship at the interface. To illustrate, the transport rate equations are developed in the fugacity format starting from initial concentration expressions. The subscripts 1 and 2 refer to the two phases in contact, and the subscript i refers to the interface location.

$$N = k_1 A (C_1 - C_{i1})$$
$$N = k_2 A (C_{i2} - C_2) \tag{2}$$

rewritten in fugacity format,

$$N = k_1 A z_1 (f_1 - f_{i1})$$
$$N = k_2 A z_2 (f_{i2} - f_2) \tag{3}$$

Based on the equilibrium relationship of equal fugacities at the interface,

$$f_{i1} = f_{i2} \tag{4}$$

Equation 3 becomes

$$\frac{N \left[\dfrac{1}{k_1 z_1} + \dfrac{1}{k_2 z_2} \right]}{A} = (f_1 - f_2) \tag{5}$$

which upon rearrangement becomes

$$N = \frac{A}{\left[\dfrac{1}{k_1 z_1} + \dfrac{1}{k_2 z_2} \right]} (f_1 - f_2) \tag{6}$$

or

$$N = D_{12} (f_1 - f_2) \tag{7}$$

Equation 7 introduces the "D value" (mol/hr/Pa), used throughout the fugacity level III model to calculate transport rates. The D value is a combination of the mass transfer characteristics (k's), the mass transfer area (A), and the phase affinities (z's). It linearly relates the diffusion transport rate between compartments to the fugacity difference. For each pair of compartments, an expression equivalent to Equation 7 may be written.

In the fugacity level III model, D values are also used to describe the transport of chemical associated with the bulk motion of one phase between compartments. However, in this situation, the D value is defined by the volume flow rate of the bulk motion (m³/hr) multiplied by the fugacity capacity corresponding to the phase that is contributing to the bulk motion. The transport rate between compartments is thus defined as

$$N = D_{Uj}f_j \tag{8}$$

where U signifies the unidirectional nature of the transport mechanism and j specifies the phase involved in the bulk motion.

The chemical in a given compartment may also leave the local environment as a result of the advective flow of that compartment. In this case, a D value is also used to describe the process, and it is defined by the advective flow rate (m³/hr) multiplied by the corresponding fugacity capacity for the compartment of interest. The loss of chemical is defined by

$$N = D_{Ai}f_i \tag{9}$$

where A signifies the advective flow.

Finally, a chemical may also react within a compartment. The loss rate is given by

$$N = V_i\mu_iz_if_i \tag{10}$$

where μ is the first-order reaction rate constant and V is the volume of the compartment.

By performing a mass balance around each compartment and incorporating reaction rate, emission rate, and transport rate expressions, an algebraic equation is developed for each compartment in which the rate of chemical flow into a compartment is equal to the rate of chemical flow out. The result is seven equations and seven unknowns. Table 2 gives the resulting seven equations for this analysis. These equations can be solved analytically or by a matrix technique.

For example, the fate of lindane in the environment may be predicted from its physicochemical properties and the fugacity level III model. Lindane's properties are given in Table 3.[7,8] Based on these properties, the fugacity capacities and D values are determined; they are given in Table 4.[8,10] The seven equations in Table 2 are then solved for the seven unknown fugacities, and the corresponding concentrations are determined based on Equation 1. For a steady atmospheric emission of 0.001 mol/hr (E_1), the model predicts the following concentrations: air 3.3×10^{-11}, water 4.2×10^{-8}, soil 3.5×10^{-6}, sediment 5.2×10^{-6}, biota 1.0×10^{-5}, suspended sediment 8.0×10^{-6}, air particulates 7.4×10^{-1} mol/m³. The input of 0.001 mol/hr leaves the system at a rate of 4.4×10^{-5} mol/hr through reaction and 9.5×10^{-4} mol/hr through advective

Table 2. Fugacity Level III Model Equations.

Basic Equations:

$$E_1 = v_1 f_1 - D_{12} f_2 - D_{13} f_3 - D_{17} f_7$$
$$E_2 = -D_{12} f_1 - D_{U1} f_1 + v_2 f_2 - D_{24} f_4 - D_{25} f_5 - D_{26} f_6$$
$$E_3 = -D_{13} f_1 - D_{U2} f_1 + v_3 f_3 - D_{U4} f_7$$
$$0 = -D_{24} f_2 + v_4 f_4$$
$$0 = -D_{25} f_2 + v_5 f_5 - D_{U3} f_7$$
$$0 = -D_{26} f_2 + v_6 f_6$$
$$0 = -D_{17} f_1 + v_7 f_7$$

where
v_i = the loss mechanism from a given compartment including reaction, advection, and transport

$$v_1 = V_1 \mu_1 Z_1 + D_{12} + D_{13} + D_{17} + D_{A1} + D_{U1} + D_{U2}$$
$$v_2 = V_2 \mu_2 Z_2 + D_{12} + D_{24} + D_{25} + D_{26} + D_{A2}$$
$$v_3 = V_3 \mu_3 Z_3 + D_{13}$$
$$v_4 = V_4 \mu_4 Z_4 + D_{24}$$
$$v_5 = V_5 \mu_5 Z_5 + D_{25}$$
$$v_6 = V_6 \mu_6 Z_6 + D_{26}$$
$$v_7 = V_7 \mu_7 Z_7 + D_{17} + D_{U3} + D_{U4}$$

E_i = the emission into the environment into a particular compartment
D_{ij} = transport between compartments
D_{Ai} = advective flow D values
D_{Ui} = depositional flow D values
μ_i = first-order reaction rate constant

flow of water and air. The total amount of lindane that accumulates in this local environment is 0.77 mol, and therefore the residence time of lindane is 770 hr.

This predicted fate of lindane is sensitive, to varying degrees, to the assumptions made within the model and to the accuracy of the input data.

To test the sensitivity of the predicted atmospheric concentration to the uniform soil concentration assumption, it is necessary to hold all other parameters constant while allowing the concentration in the soil to be nonuniform. The simplest manner in which to do this is to modify the one soil compartment base model to have four soil compartments as four layers. Concentrations may differ between layers as a result.

In the base level III model, the soil compartment is 15 cm deep, and therefore the chemical is restricted from penetrating beyond this 15 cm depth. In

Table 3. Selected Chemicals and Their Properties.

Compound	Vapor Pressure (Pa)		Water Solubility (g/m³)		Log K_{ow}		Soil Reaction Half-Life	
Lindane	0.027	(8)	7.5	(8)	3.7	(8)	230 days	(7)
DDT	2.0×10^{-5}	(8)	0.005	(8)	6.0	(8)	10 years	(9)
Aldrin	0.003	(8)	20	(8)	5.3	(8)	2 years	(9)
Heptachlor	0.04	(8)	0.2	(8)	4.4	(8)	2 years	(9)
2,4-D	5.5×10^{-5}	(7)	900	(7)	2.8	(7)	14 days	(7)

Note: References are in (#).

Table 4.   D Value Definitions Used in the Basic Level III Model.

z Values (mol/m³/Pa)	Value for Lindane	Basis
Air	4.2×10^{-4}	$1/RT$
Water	0.95	C_s/P_s
Soil	59	$0.012\ K_{ow}\ z_2$
Sediment	120	$0.024\ K_{ow}\ z_2$
Suspended Sediment	120	$0.024\ K_{ow}\ z_2$
Biota	230	$0.05\ K_{ow}\ z_2$
Air particulates	91,000	$6 \times 10^6/(P^s_L RT)$

D Values (mol/hr/Pa)		Lindane	Value for Basis
Air-water	D_{12}	2650	mtc[a]
Air-soil	D_{13}	1.0	mtc
Air-particulates	D_{17}	417000	estimate
Water-sediment	D_{24}	617	mtc
Water-suspended sediment	D_{25}	1770	mtc
Water-biota	D_{26}	27.0	mtc
Advection in air	D_{A1}	10300	10%/day
Advection in water	D_{A2}	2780	1%/day
Rain deposition to water	D_{U1}	39.0	0.5 m/yr
Rain deposition to soil	D_{U2}	16.7	0.5 m/yr
Particle deposition to suspended sediment	D_{U3}	533	5 days[b]
Particle deposition to soil	D_{U4}	228	5 days

[a]Using mass transfer coefficients listed below and Equations 6 and 7.

k_{12} 10
k_{21} 0.05
k_{13} 2.0
k_{31} from Millington-Quirk formula (Jury et al.,[8] Millington and Quirk,[10] and Table 7)
k_{24} 0.001
k_{42} 0.0001
k_{25} 0.001
k_{52} 0.0001
k_{26} 0.01
k_{62} 0.0001

[b]Average residence time in the atmosphere in days.

reality, a chemical may penetrate beyond this depth to some extent dependent on the chemical's mobility and reactivity in the soil media. In the modified model each of the four soil layers is 7.5 cm. This addresses the two shortcomings of the uniform soil concentration assumption:

1. The soil concentration is able to vary with depth.
2. The chemical is able to penetrate to depths dependent on its characteristics, not dependent on an arbitrary boundary.

The modification of the base model to include four soil layers leads to a system of ten equations and ten unknowns. A corresponding modification of the D values relating to the transfer of chemical within the soil is required. The additional equations and D values are given in Table 5.

For lindane the predicted atmospheric concentration by the modified model is 3.3×10^{-11} mol/m³, the same as predicted by the base model. This sensitivity test has been repeated for four other chemicals under the same conditions. The

Table 5.   Modified Level III Equations and D Values.

	Compartment Number	Volume (m³)
Soil I	3	22500*
Soil II	8	22500
Soil III	9	22500
Soil IV	10	22500

Equations:

$$E_3 = -D_{13}f_1 - D_{U2}f_1 + v_3f_3 - D_{U4}f_7 - D_{38}f_8{}^*$$
$$E_8 = -D_{38}f_3 + v_8f_8 - D_{89}f_9$$
$$O_3 = -D_{89}f_8 + v_9f_9 - D_{910}f_{10}$$
$$O_3 = -D_{910}f_9 + v_{10}f_{10}$$

Parameters:

$$D_{13} = 2.07{}^*$$
$$D_{38} = D_{89} = D_{910} = 1.05$$
$$v_8 = V_8\mu_8z_8 + D_{38} + D_{89}$$
$$v_9 = V_9\mu_9z_9 + D_{89} + D_{910}$$
$$v_{10} = V_{10}\mu_{10}z_{10} + D_{910}$$

*Changed value in equation from that presented in Table 1.

other chemicals are DDT, Aldrin, Heptachlor, and 2,4-D. Their properties are given in Table 3.

Figure 1 is a bar chart with the logarithm of the predicted atmospheric concentrations for each of the five chemicals for the base and modified

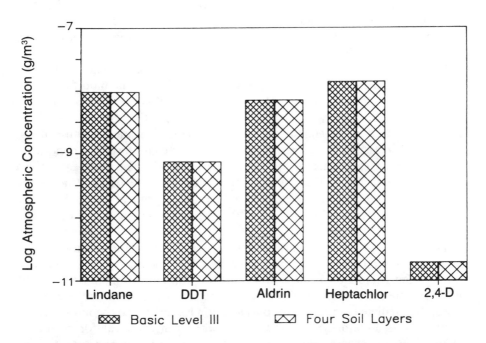

Figure 1.   Predicted atmospheric concentrations for steady-state air emission of the chemicals.

models. The predicted atmospheric concentrations range from 0.02 ng/m³ for 2,4-D to 14 ng/m³ for heptachlor. Clearly, the atmospheric concentration is strongly dependent on the physicochemical properties of the chemical and, therefore, the nature of the chemical itself. It is also evident that the predicted air concentrations are insensitive to the number of soil layers used in the model. The uniform soil concentration assumption does not have an effect on the predicted steady-state atmospheric concentration for a chemical emitted directly into the atmosphere.

This insensitivity to the assumption is limited to situations of atmospheric emissions of the chemical. In the situation where a chemical is introduced into soil, the sensitivity is quite different. To illustrate this, it is necessary in the calculations to set E_1 to zero and the soil emission, E_3, to 0.001 mol/hr. For the soil emission of lindane, the predicted concentrations by the base model in each compartment are air 1.0×10^{-13}, water 1.3×10^{-10}, soil 1.8×10^{-4}, sediment 1.6×10^{-8}, suspended sediment 2.5×10^{-8}, biota 3.1×10^{-8}, and air particulates 2.3×10^{-5} mol/m³. The modified model predicts an atmospheric concentration for this situation of 4.1×10^{-13} mol/m³, or approximately four times the prediction by the base model. However, the two situations modeled here are not identical. In the base model, the emission rate into soil is 0.001 mol/hr to a soil layer 15 cm deep, but in the modified model, the same emission rate is into a soil layer 7.5 cm deep. The surface soil in the modified model is experiencing an effective emission rate double that for the base model. A true comparison would have an emission of 0.0005 mol/hr into compartments 3 and 8 ($E_3 = E_8$ = 0.0005 mol/hr). For lindane the predicted atmospheric concentration for this scenario is 2.1×10^{-13} mol/m³, a factor of two higher than the base model and therefore still dependent on the model assumption.

Figure 2 gives the predicted atmospheric concentrations for each of the five chemicals for each of three cases: (1) the base model, (2) the modified model with emission into the surface layer only, and (3) the modified model with emission into the top two soil layers. It is apparent that the results of the prediction based on an emission into soil are dependent on the uniform soil concentration assumption. The range of the predicted atmospheric concentration for the five chemicals is seven orders of magnitude. The range for a given chemical is a factor of four for the three cases. The predicted atmospheric concentration remains more sensitive to the physicochemical properties of the chemical and, therefore, the nature of the chemical itself than to the uniform concentration assumption.

The ratio of the predicted atmospheric concentration by the modified model with emission into the top two soil layers to the prediction by the base model ranges from 1.0 to 2.0 for the five chemicals (a ratio of 1.0 signifying that the uniform concentration assumption has no effect). Clearly under these conditions, the predicted atmospheric concentration is sensitive to the uniform concentration assumption. Sensitivity varies with each chemical for reasons — although not yet exactly determined — that are related to the proportion of the chemical that is eliminated by the soil compartment.

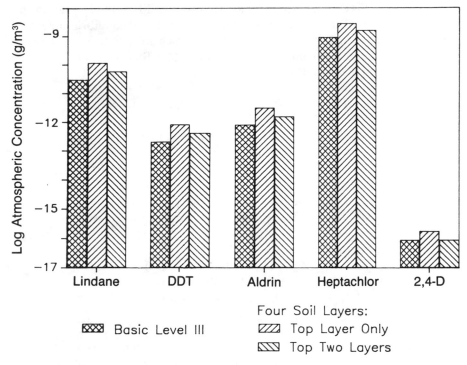

Figure 2. Predicted atmospheric concentrations for steady-state soil emission of the chemicals.

The ratio of the predicted atmospheric concentration by the modified model with emission into the surface layer only to the prediction by the base model ranges between 2.0 and 4.0 for the five chemicals. As in the lindane case, the surface-layer-only emission model results in a doubled effective emission rate of the chemical to the surface soil. Taking this factor into consideration, the comparison of the two modified models with the base model is similar.

Predicting the atmospheric concentrations of chemicals emitted or introduced into the soil environment is improved by models that allow for nonuniform soil concentrations. The difference between four soil layers and one soil layer is up to a factor of two. A larger difference may be expected with a larger number of soil layers.

SINGLE-DOSE MODELS

In single-dose models, a single dose of chemical is added to the soil, and the subsequent dynamic fate of this chemical is predicted. A rigorous model presented by Jury et al.[7] predicts the fate of soil-applied organic chemicals without making the uniform soil concentration assumption. The soil compartment concentration may vary continuously in time and depth within Jury's model.

The Jury model is an analytical solution to a partial differential equation that incorporates volatilization, reaction, and leaching. Local equilibrium between soil-solid, soil-air, and soil-water is assumed. The chemical moves by diffusion in both the air and water pores, and by a net water flux.

The differential equation is

$$\partial C_T/\partial T = D_E (\partial^2 C_T/\partial Z^2) - V_E (\partial C_T/\partial Z) - \mu C_T \tag{11}$$

where
C_T = total bulk soil concentration (g/m³)
D_E = effective diffusivity (m²/hr)
T = time (hr)
Z = depth (m)
V_E = effective bulk water velocity (m/hr)
μ = first-order reaction rate constant (hr⁻¹)

The boundary conditions are

$$C_T(Z,0) = C_o \qquad \text{if } 0 < Z < L$$
$$C_T(Z,0) = 0 \qquad \text{if } Z > L$$
$$-D_E \, \partial C_T/\partial Z + V_E C_T = -H_E C_T \qquad \text{at } Z = 0$$
$$C_T(\infty,t) = 0$$

where
L = depth of initial application of chemical (m)
H_E = effective transport coefficient of the stagnant air boundary layer (m/hr)

The analytical solution gives the soil concentration as a function of depth and time. It is a complex combination of exponentials and error functions and is not reproduced here (see Equations 24 and 25 in Jury et al.[8,11]). The fraction of chemical that evaporates is calculated by a numerical integration of the evaporation flux solution from time zero to time infinity.

For an initial application of lindane of 10 g/m² to the top 2 cm of soil, the Jury model predicts 29% (2.9 g/m²) loss of the lindane from soil to the atmosphere; the remaining lindane either reacts or leaches from the surface soil. Table 6 gives the specific parameters and conditions used.

To test the uniform concentration assumption, it is necessary to model the same lindane application using a model that has the same features as the Jury model except that the soil concentration remains uniform. Mackay and Stiver[12] present a single-dose model that satisfies these requirements.

The simple model uses a single, uniform concentration soil compartment. The chemical may volatilize, react, and leach from the soil. Volatilization is dependent on the mobility of the chemical in the soil-air and soil-water pores, on the resistance at the air-soil interface, and on the average depth of the chemical in the soil. Reaction rates are dependent on the reaction rate constant for the chemical in soil. The leaching process depends on the water percolation rate through the soil column. Each process is represented by a transport rate as

Table 6. Jury Model Input Parameters.

f_{oc}	fraction organic carbon in the soil	0.0125 kg/kg
ρ_b	soil bulk density	1.35×10^3 kg/m^3
μ	first-order reaction rate constant	4.2×10^{-5} hr^{-1}
D_E	effective diffusivity Jury definition (Jury et al.,[7] Equation 18) based on	4.8×10^{-9} m^2/hr
	a air-filled porosity	0.2
	w water-filled porosity	0.3
	D_A molecular diffusivity in air	1.8×10^{-2} m^2/hr
	D_W molecular diffusivity in water	1.8×10^{-6} m^2/hr
V_E	net effective water flux	0.0 m/hr
L	chemical's incorporation depth	0.02 m
H_E	effective transport coefficient for the air boundary layer	2.8×10^{-5} m/hr

a linear function of the fugacity in the soil. The proportionality constant is a D value similar to that used in the level III steady-state model.

$$N_V = D_V f_S$$
$$N_R = D_R f_S$$
$$N_L = D_L f_S \tag{12}$$

Table 7 gives the D values used in the analysis in this work. The fraction of chemical that will volatilize is given by the ratio of D_V to the sum of the D values ($D_V + D_R + D_L$).

For lindane, the D values for volatilization, reaction, and leaching are 2.5×10^{-5}, 1.2×10^{-4}, and 8.8×10^{-5} mol/hr/Pa, respectively. Therefore, this simple model predicts that 11% of the lindane will evaporate from the soil. Table 8 gives the predictions for the five chemicals for both the Jury model and the simple model. The discrepancy between the results from the two models indicates some sensitivity to the uniform soil concentration assumption. The rigorous Jury model has been validated;[13] therefore, the discrepancy may be attributed to the simple model and its uniform soil concentration assumption.

The ranking of the five chemicals is the same for the two models, and the percentages match well at the extremes. However, for chemicals for which a moderate fraction evaporates, the simple model significantly underpredicts the fraction evaporated as calculated by the Jury model. This underprediction by the simple model is a result of using a resistance to evaporation that is representative of chemical at the 1 cm soil depth location. Any chemical above this location experiences less resistance and, therefore, is more likely to evaporate than would be predicted.

Given time and capability on the part of the user, the Jury model is the better model, but its implementation requires more effort. If only an indication of evaporation potential is needed, the simple model has considerable merit.

Table 7. Simple Model Input Parameters.

Volatilization D

$$1/D_v = 1/(k_{AS}Az_A) + Y/(A(D_{EA}z_A + D_{EW}z_W))$$

k_{AS}	air side air-soil mass transfer coefficient	3.8 m/hr
A	contact area between air and soil	1.0 m^2
Y	half the depth of incorporation of chemical	0.01 m
D_{EA}, D_{EW}	effective diffusivity in air and water pores, respectively, calculated from Millington-Quirk[10] formula	

$$D_{EA} = (a^{10/3}/\phi^2)D_A$$

a	air-filled porosity	0.2
ϕ	soil porosity	0.5
D_A	molecular diffusivity in air	1.8×10^{-2} m^2/hr

Similarly for D_{EW} except the air-filled porosity is replaced by the water-filled porosity and the molecular diffusivity in water is used.

w	water-filled porosity	0.3
D_w	molecular diffusivity in water	1.8×10^{-6} m^2/hr

D_v (Lindane) $= 2.5 \times 10^{-5}$ mol/hr/Pa

Reaction D

$$D_R = \mu_s V_s z_s$$

Vs	volume of soil	0.02m^3
μs	first order reaction rate constant	

D_R (Lindane) $= 1.2 \times 10^{-4}$ mol/hr/Pa

Leaching D

$$D_L = G_L z_w$$

G_L	precipitation rate	9.1×10^{-5} m^3/hr

D_L (Lindane) $= 8.8 \times 10^{-5}$ mol/hr/Pa

Table 8. Predicted Percentage Evaporated After Batch Application.

Compound	Simple	Jury
Lindane	10.8	29
DDT	1.8	8.8
Aldrin	0.7	2.7
Heptachlor	77	71
2,4-D	0.006	0.007

CONCLUSIONS

The implications of predicting atmospheric concentrations using models with a uniform soil concentration have been tested. These tests included both steady-state models and single-dose application models. The steady-state analysis included air emission and soil emission scenarios.

Atmospheric concentrations resulting from the emission of chemical into the air can be predicted equally well by a model containing a single soil compartment and a model containing four soil compartments as four distinct layers.

The prediction of atmospheric concentrations is dependent on the uniform soil concentration assumption if the emission is into soil. The use of a larger number of layers gives a more physically realistic model, and therefore more reliable predictions of atmospheric concentrations. For the five chemicals tested, the uniform soil concentration assumption can lead to the underestimation of atmospheric concentration by up to a factor of two. The depth of the incorporation of the chemical in the soil has a direct effect on the predicted atmospheric concentrations.

For single-dose application models characterizing the fate of a soil-applied chemical, a simple model with one soil compartment is less successful than the rigorous Jury model. However, the simple model has merit for quickly characterizing the likelihood of evaporation.

REFERENCES

1. Baughman, G., and R. Lassiter. "Prediction of Environmental Pollutant Concentration," ASTM STP 657 (1978), p. 35.
2. Mackay, D. "Finding Fugacity Feasible," *Environ. Sci. Technol.* 13:1218–23 (1979).
3. Mackay, D., and S. Paterson. "Calculating Fugacity," *Environ. Sci. Technol.* 15:1006–14 (1981).
4. Mackay, D., and S. Paterson. "Fugacity Revisited," *Environ. Sci. Technol.* 16:654A-60A (1982).
5. Whitman, W. G. "The Two-Film Theory of Gas Absorption," *Chem. Metallurg. Eng.* 29:146–48 (1923).
6. Liss, P. S., and P. G. Slater. "Flux of Gases Across the Air-Sea Interface," *Nature* 247:181–84 (1974).
7. Jury, W. A., W. F. Spencer, and W. J. Farmer. "Behavior Assessment Model for Trace Organics in Soil: I. Model Description," *J. Environ. Qual.* 12:558–64 (1983).
8. Callahan, M. A., M. W. Slimak, N. W. Gabel, I. P. May, C. F. Fowler, J. R. Freed, P. Jennings, R. L. Durfee, F. C. Whitmore, B. Maestri, W. R. Mabey, B. R. Holt, and C. Gould. "Water-Related Environmental Fate of 129 Priority Pollutants," U.S. EPA Report-440/4–79–029a (1979).
9. Verschueren, K. *Handbook of Environmental Data on Organic Chemicals* (New York, NY: Van Nostrand Reinhold Co., 1977).

10. Millington, R. J., and J. P. Quirk. "Permeability of Porous Media," *Nature* 183:387–88 (1959).
11. Jury, W. A., W. F. Spencer, and W. J. Farmer. "Errata to J. Environ. Qual. 12:558–564," *J. Environ. Qual.* 16:448 (1987).
12. Mackay, D., and W. Stiver. "Predictability of Herbicide Behavior," in *Environmental Chemistry of Herbicides, Vol. II*, R. Grover, Ed. (Boca Raton, FL: CRC Press, 1989), Chapter 15.
13. Jury, W. A., W. F. Spencer, and W. J. Farmer. "Behavior Assessment Model for Trace Organics in Soil: IV. Review of Experimental Evidence," *J. Environ. Qual.* 13:580–86 (1984).

CHAPTER 21

Atmospheric Contributions to Contamination of Lake Ontario

Donald Mackay

INTRODUCTION

It is well established that certain organochlorine chemicals, such as DDT, PCBs, and toxaphene, can contaminate lakes some distance from the source of application.[1,2] Apparently these chemicals evaporate, are transported in the atmosphere, then enter lakes directly by absorption and by wet and dry deposition, and possibly also indirectly by washoff from the surrounding catchment area. The purpose of this chapter is to present a quantitative model of the processes by which chemicals migrate between the atmosphere, water column, and bottom sediments. It is expected that this will shed some light on the relative extents that contamination is contributed from the atmosphere and nearby emission sources such as industries and local sewage treatment plants. Finally, a graphical method is suggested by which the detailed mass balance can be presented, facilitating interpretation of what proves to be a rather complex system of interacting transport processes.

In 1986 the Task Force on Chemical Loadings of the Toxic Substances Committee of the International Joint Commission's Great Lakes Water Quality Board sponsored a modeling effort to investigate the feasibility of using mathematical models to establish the relationships between loadings of toxic chemicals and their concentrations in various compartments of the Great Lakes ecosystem. A workshop was held at the University of Toronto in February 1987 in which three models were presented describing the behavior of PCBs in Lake Ontario over the period 1940 to the present. This chemical and lake were selected because they probably represent the best-known toxic contamination situation in any of the Great Lakes. Further, PCBs are convenient chemicals because they are subject to appreciable transport to and from the atmosphere. A report describing the findings of this workshop has been compiled by the International Joint Commission.[3]

The first model was prepared by Connolly and DiToro of Manhattan College, New York; the second by Rodgers and DePinto of Limno-tech, Inc., Ann

Arbor, Michigan; the third by the present author. As a result of comparisons of the three models and constructive criticisms made by viewers at that workshop, a modified version of the University of Toronto model was prepared and published in 1989.[4] The model described in this chapter is a sequel to that model, modified by including an atmospheric compartment rather than assuming the atmosphere to be a semi-infinite volume. The present study has benefited greatly by these studies, and also by the report by Strachan and Eisenreich[5] in which an attempt was made to assemble a model for a number of chemicals in all the Great Lakes and thus elucidate the contribution of the atmosphere toward Great Lakes contamination.

The most accurate and useful models describing the dynamics of toxic chemicals in the Great Lakes are unsteady-state or time-dependent models. These usually involve numerical solution of differential equations that describe the pathways of chemicals as they (1) enter the system, (2) partition between water, suspended matter, biota, and sediments, and (3) are subject to removal processes by outflow, reaction, sediment burial, and evaporation to the atmosphere. In the original unsteady-state model,[4] the atmosphere was treated as a source and sink of chemicals, but the concentration in the atmosphere was assumed to be unaffected by the presence of the lake. The primary input variables were the inflow concentration in water, the atmospheric concentration, and the discharge rate of chemical to the lake. The solution was obtained by numerical integration.

The principal difficulty with numerical solutions is that the computer model generates a large volume of data, which is difficult to assess and interpret. A much simpler, more readily comprehensible approach is to assemble a steady-state model that reflects the condition that the lake would adopt after prolonged constant exposure to a steady input of chemical. The mathematical advantage of this approach is that the differential equations become algebraic and solution is easy. It is the latter approach that we adopt in this study. The model should thus not be viewed as depicting precisely what occurs in Lake Ontario, but rather as depicting how the Lake Ontario system would respond if subjected to these input conditions for a prolonged period. It should further be emphasized that the rates of many Lake Ontario processes such as sediment deposition and resuspension are not well quantified. Thus, there is considerable doubt about certain parameter values used in the model. In total, therefore, the model should be viewed as illustrative of the processes that occur, giving only order of magnitude estimations of rate. We believe, however, that the model elucidates the dominant processes and their approximate magnitudes.

MODEL STRUCTURE

Figure 1 shows the three-compartment model structure used in this study. The sediment depth is 5 mm, the water column mean depth 86 m, and the

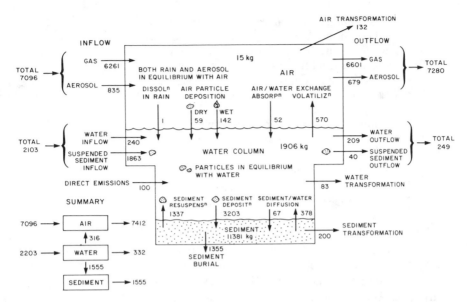

Figure 1. Steady-state mass balance for PCBs in Lake Ontario for compartments (kg) and for transfer processes between compartments (kg/year).

atmosphere has a height of 1500 m. The horizontal area is the area of the lake, 1.95×10^{10} m², from which the volumes are deduced. The atmospheric height of 1500 m is somewhat arbitrary but was selected on the basis that it is typical of the volume of atmosphere "accessible" to the lake during the time that the air resides over the lake. If desired, the effect of changing this volume can be readily explored. All three compartments are treated as being well-mixed, and each is at a state of internal equilibrium (e.g., gaseous pollutant is in equilibrium with aerosol particles, chemical dissolved in water is in equilibrium with suspended matter, and in the sediments chemicals sorbed to solid particles and present in pore water are at equilibrium). The three primary compartments are not, however, in equilibrium with each other because of "uneven" inputs and "resistances" to transfer between compartments.

The model includes expressions for 22 chemical transport processes as follows:

- discharges into water, e.g., from sewage treatment plants, spills, and industrial facilities
- inflow in dissolved form from the Niagara River and other tributaries
- inflow in particulate form from the Niagara River and other tributaries
- resuspension from sediments to the water column
- deposition from water to sediments
- diffusion from water to sediments
- diffusion from sediment to water
- sediment burial
- transformation of chemical in the sediments

- transformation of chemical in the water column
- outflow to the St. Lawrence River in dissolved form
- outflow to the St. Lawrence River in particulate form
- deposition from the atmosphere by dissolution in rain
- dry deposition from the atmosphere
- wet deposition from the atmosphere, i.e., aerosol scavenging by rain
- absorption from the atmosphere
- volatilization from water to the atmosphere
- inflow in gaseous form to the atmosphere
- inflow in aerosol form to the atmosphere
- transformation of chemical in the atmosphere
- outflow in gaseous form from the atmosphere
- outflow in aerosol form from the atmosphere

The approach taken in this study was to quantify these process rates using the fugacity approach. The rates of all transport and transformation processes are described by the product of the fugacity in the source phase and a D value. These D values are deduced from flow rates, mass transfer coefficients, diffusivities, reaction rate constants, etc., as described by Mackay.[4] D values are also affected by the physical/chemical properties of the substance, notably the solubility in water, vapor pressure, octanol water partition coefficient, and other sorption partition coefficients in the various media.

Steady mass balance equations can then be assembled for each compartment and solved to yield the three fugacities—those in air, water, and sediment. From these fugacities the process rates can be calculated, concentrations can be estimated, and amounts of chemicals residing in each phase deduced. Table 1 gives the equations and the D values used in the study. These are identical to the values published previously, except for the additional values for gas and aerosol inflow and outflow and air transformation. The inflow rate to the atmospheric compartment was taken as 1.62×10^{12} m³/hr, which corresponds to a velocity of 3 m/sec over a width of 100 km to a height of 1500 m. Since the volume of the air compartment over Lake Ontario is 29.2×10^{12} m³, this corresponds to an atmospheric residence time of air of 18 hr. It is recognized that this number is highly variable. However, any conclusion's sensitivity to the magnitude of this quantity can be easily tested. It is regarded as a reasonable first estimate. The concentration of aerosol flowing into the region is 30 µg/m³, or a volume fraction of 20×10^{-12}. The assumed rain rate was 761 mm/year. A mass balance was assembled for aerosol, i.e., the quantity of aerosol entering equaled the quantities removed by wet and dry deposition and by outflow. A transformation rate was included, largely for illustrative purposes, by assuming an arbitrary reaction half-life of PCB in the atmosphere of 29 days. Since this half-life considerably exceeds the atmospheric residence time, this process of transformation is relatively unimportant to the overall mass balance.

Selected output of the model is given in Table 2. Figure 1 is a presentation of these data in graphical form. All rates are expressed in kilograms per year, and

Table 1.   D Values and Equations Used in the Model.

D Values (mol/Pa/hr)	
Burial (DB)	2.29E + 06
Sediment transformation (DS)	3.36E + 05
Sediment resuspension (DR)	2.26E + 06
Water-to-sediment diffusion (DT)	6.38E + 05
Sediment deposition (DD)	3.05E + 07
Water transformation (DW)	7.94E + 05
Volatilization (DV)	5.43E + 06
Water inflow (DI)	1.96E + 06
Water particle inflow (DX)	1.52E + 07
Water outflow (DJ)	1.98E + 06
Water particle outflow (DY)	3.81E + 05
Rain dissolution (DM)	1.39E + 05
Wet particle deposition (DC)	1.47E + 07
Dry particle deposition (DQ)	6.10E + 06
Air inflow and outflow (DG)	6.84E + 08
Aerosol inflow (DA)	9.12E + 07
Aerosol outflow (DE)	7.03E + 07
Air transformation (DZ)	1.36E + 07

Mass Balance Equations

Air $\quad f_U(DG + DA) + f_W DV = f_A(DG + DE + DV + DQ + DC + DM + DZ)$

Water $\quad f_I(DI + DX) + E + f_A(DV + DQ + DC + DM) + f_S(DT + DR)$
$\qquad = f_W(DV + DJ + DY + DW + DD + DT)$

Sediment $\quad f_W(DD + DT) = f_S(DR + DT + DS + DB)$

Fugacities (f) are subscripted A air, W water, S sediment, U inflow air, I inflow water. E is direct emissions to water.

The mass balance equations can be readily solved for the unknowns or output variables f_A, f_W, and f_S.

masses in the system in kilograms. Although it is difficult to assign an accuracy or confidence limit to these numbers, they should be within a factor of 2 to 3 of the "correct" values. The number of significant figures in Figure 1 and Table 2 implies an accuracy far in excess of that claimed. The primary purpose is to permit order-of-magnitude assessments.

DISCUSSION

The balance of this chapter is essentially a point-by-point discussion of the magnitude of the various terms in Figure 1. A number of conclusions can be drawn.

Cycling

The first notable feature of Figure 1 is the extent of cycling between the three compartments. It is highly misleading to view atmosphere-to-water or

Table 2.   Selected Output from the Model.

	Fugacity (Pa)	Concentrations (mol/m³)	Natural Units
Atmosphere		1.43E−12	
Aerosol particles	3.38E−09	1.47E−13	
Total air		1.57E−12	0.51 ng/m³
Water		3.01E−09	
Water particles	3.68E−08	4.91E−10	
Total water		3.50E−09	1.14 ng/L
Particulates		2.36E−03	320 ng/g
Inflow water	4.28E−08	3.07E−08	10.0 ng/L
Sediment	2.08E−07	2.39E−03	324 ng/g
Inflow air		1.35E−12	
Inflow aerosol	3.21E−09	1.80E−13	
Total air inflow		1.53E−12	0.50 ng/m³
Fish (TL1)	7.36E−08	1.20E−03	0.39 μg/g
Fish (TL2)	1.84E−07	2.99E−03	0.97 μg/g
Fish (TL3)	4.42E−07	7.19E−03	2.34 μg/g
Gull eggs	3.68E−06	9.59E−02	31.2 μg/g
Rain	3.38E−09	2.96E−08	9.7 ng/L

water-to-sediment processes as merely a one-way transfer process. In fact, there is considerable cycling in both directions at both interfaces. There is deposition from the atmosphere of 254 kg/year, but also volatilization of 570 kg/year, giving a net transfer from water to atmosphere of 316 kg/year. If the volatilization rate were in error by a factor of 2.3 and is only 248 kg/year, this could dramatically alter the picture and result in net deposition. Similarly, of the 3270 kg/year moving from water to sediment, 1715 return to the water, giving a net movement from water to sediment of 1555 kg/year. Again, any error in deposition or resuspension rates would dramatically affect the mass balance.

The atmosphere, water, and sediment are closely linked, since there is active cycling of chemical between these phases. Often there will be difficulty deciding on the magnitude and the direction of net transfer because it is experimentally impossible to measure all the contributing fluxes. For example, measurement of wet deposition gives only part of the air-to-water deposition picture. The cycling between water sediments is primarily controlled by the fate of organic carbon—introduced into water and formed in it—which settles to the bottom, degrades there, and is partially recycled. Hydrophobic chemicals "piggyback" on this organic carbon. We thus need a far better quantification of the rates of flow of organic carbon in these lake systems if we are to understand the dynamics of hydrophobic chemicals.

Residence and Response Times

The second point concerns the residence times or response times and amounts in each compartment. It is striking that the amounts in air, water, and sediment differ so dramatically — 15 kg in the atmosphere, 1900 kg in the water, and 11,400 kg in the sediments. Further, the net input and removal rates from each compartment also differ greatly — from the atmosphere, the total removal rate is 7400 kg/year; from water, it is 2200 kg/year; for sediment, it is 1500 kg/year. Dividing the mass present in each compartment by the net removal rate gives an indication of the residence time of the chemical in the compartment. In the case of air, this is 0.002 years (or 18 hr); in water, 0.87 year; and in sediments, 7.3 years.

The atmospheric residence time is, of course, dominated by the advective flow rate of air from the region. Water residence time is dominated by sediment deposition rate, and the sediment residence time by sediment burial. This leads to an important conclusion. If it were possible to alter the inflow rates of chemical to the various compartments, the concentration in each compartment would respond in a time similar in magnitude to the residence times or response times. For example, the atmosphere would substantially change concentration within a day. The water concentration would respond within a year or two, and the sediments within seven years.

The water column response time is, however, complicated by its interaction with sediment. In reality, the water column would respond more slowly because it is hindered or "buffered" by the slow response of the sediment. Treating the water and sediment as one compartment with a mass of 13,300 kg and a removal rate of 2200 kg/year, the response time would be about 6 years. Thus we can conclude that emission changes to the water can produce significant changes within about a year, followed by a more gradual change over 6 years as the sediment becomes decontaminated.

This point has also been made by Eisenreich[6] in his observations on the response of Lake Superior, which has a hydraulic residence time of about 180 years. It is important to appreciate that the water column response time is controlled not only by hydraulic residence time, but also by other processes such as deposition and degradation. These response times are, of course, chemical-specific, since they depend on the magnitude of the rates of transport by various processes. It is encouraging that measures to reduce chemical discharges can produce beneficial effects in times shorter than the hydraulic residence time.

Can the Lake Affect Its Atmosphere?

An important question is: Can the lake affect the concentration of chemical in the atmosphere above it? Figure 1 suggests that the answer is "Only to the extent of 5–10%." The total inflow of chemical to the atmosphere is 7100 kg/year. The rate of volatilization is 570 kg/year, or 8% of that quantity. The total rate of deposition is 254 kg/year, or 3.6%. It appears that of the chemical flowing into the atmosphere above Lake Ontario, there is potential for less than 10% to be removed or added. The atmosphere has such rapid advective exchange characteristics, and air-water transport resistances are so large, that the lake is unable to affect the composition of the atmosphere above it significantly. There is, therefore, little merit in looking in the atmosphere for measurements of how the lake is behaving because the air system is so heavily dominated by advective flows.

There is, however, one situation that merits closer inspection. The amount of wet deposition is 142 kg/year, 17% of the aerosol-associated material entering the atmosphere. This is an appreciable fraction. The model assumes that it is gently raining all the time — far from the truth, of course. It rains intensively for short periods of time. It is thus interesting to examine in detail the nature of rainfall characteristics over Lake Ontario as estimated from measurements at Toronto International Airport.[7]

In 1987 there was 631 mm of rain and 130 mm of snow (expressed as rain), totaling 761 mm. There were 161 dry days, 73 days with a trace of precipitation, and 131 with measurable precipitation. There were about 15 spells of about 2 days each when it rained quite intensely, and about half the total rain fell during these periods. The other half of the rain fell in about 40 spells of light rain and a number of isolated showers. During the periods of fairly intense rainfall, the rain rate averaged about 0.5 mm/hr, or 6 times the yearly average of .086 mm/hr. (In fact, the rain rate exceeded 1 mm/hr frequently during these periods.) During these periods, the wet deposition rate may rise by a factor of six to approximately 850 kg/year, which exceeds the inflow rate of aerosols. There is a possibility that a high proportion of the aerosols and the associated PCBs introduced into the region will be deposited. This raises a question about the suitability of using washout ratios and annual rainfall rates to calculate total annual wet deposition rates. Much of the rain that falls may not have the opportunity to transport PCBs because it falls towards the end of periods of intense rainfall when there is little or no aerosol left in the atmosphere. Such days are directly observable as times of remarkably high visibility following intense rainfall.

Can the Atmosphere Affect Lake Concentrations in the Short Term?

Another question is: Can changes in atmospheric deposition rate affect the lake concentrations in the short term, i.e., after a period of intense deposition? Could we observe concentration changes in the water? The amount in the atmosphere is about 20 kg, of which perhaps 2 kg is associated with particles. If, for example, for 2.6 atmospheric residence times (two days) all of this aerosol-associated PCB were conveyed to the water column, then this would represent an increase in the mass PCB in the water column from approximately 1900 to 1905 kg, or 0.2%. It is thus tempting to conclude that the amount of material in the atmosphere is so small compared to the amount of material in the water that even the addition of all the chemical in the atmosphere to the water would produce a relatively small change in water concentration that would probably be undetectable analytically given the variation in concentration and analytical error. In practice, however, the water column is not well mixed, and intense deposition from the atmosphere would probably affect only the water in the top few meters. Since the top 2.6 m of water contain about 3% of the PCB, or 58 kg, deposition of 5 kg would have about a 10% effect. Hence, it is possible that during periods of intense wet deposition, concentration changes could be observed in the surface waters of the lake. However, this would require very precise analytical techniques.

Equilibrium Status and Physical/Chemical Properties

The next point concerns the interpretation of the behavior of chemicals in the lake in terms of equilibrium status, and the question of which chemical properties tend to favor atmospheric transport and accumulation in lakes. The fugacities in the air, water, and sediment are respectively 3.3, 37, and 207 nano-Pascals. The gaseous concentration in the air is 0.47 ng/m^3. Air brought into contact with the water to achieve equilibrium will have a concentration of 5.2 ng/m^3; with the sediment, 29 ng/m^3. Examining the air, water, and sediment concentrations in terms of the equilibrium status is very useful, not only because it gives an impression of the direction of diffusive transfer but also because it conveys an impression of the extent to which the system departs from equilibrium. This nonequilibrium state is caused by the presence of nondiffusive processes such as deposition. Apparently the chemical is being transported from air to water and from water to sediment by nondiffusive processes. It is then striving to return by diffusion from sediment to water, and water to air. Concentrations are established that reflect a balance between these rates. Recently Baker and Eisenreich[8] have presented data for PAHs in Lake Superior that support the contention that steady-state, nonequilibrium conditions can be established.

This nonequilibrium situation will be most pronounced for chemicals that have substantial nondiffusive deposition rates, i.e., they sorb appreciably to depositing material because of a high octanol water partition coefficient or, in the case of aerosols, a very low vapor pressure. This may explain why compounds such as PCBs, DDT, and toxaphene are observed as the primary contaminants in remote lake systems where the atmosphere is the only source. The physical/chemical properties that drive these chemicals into the lakes are low vapor pressure, high octanol water partition coefficient, and, of course, persistence. It may therefore be possible to examine other chemicals and agrochemicals for their susceptibility to these transport characteristics simply by examining these properties. A strong case can be made against introducing chemicals with this combination of properties into the environment for agricultural and other purposes because they have the potential to migrate long distances through the atmosphere, associate with particulate matter, and enter

Table 3.   Response of the Lake System to Changes in Air and Water Inputs.

PCB Inputs to Lake		Computed Response Concentrations			
Air Conc. (ng/m³)	Loadings to Water (kg/year)	Water (ng/L)	%	Salmonid (μg/g)	Gull Egg (μg/g)
0.500	2202	1.14	100	2.34	31.3
0.500	0	0.11	9.6	0.22	3.1
0	2202	1.03	90.4	2.12	28.2

lakes, where they will tend to bioconcentrate in fish and other organisms and concentrate in sediments. The model suggests that the key properties are vapor pressure, octanol water partition coefficient, and persistence, both in the atmosphere and the water column.

Roles of Atmospheric and Aquatic Sources

After establishing the model, it is easy to manipulate the input loading or atmospheric concentration data to explore the sensitivity of an output quantity such as water or fish concentration to aquatic or atmospheric changes. Table 3 gives the results of the base case in which the inflow air concentration is 0.5 ng/m³ and the total loadings to the lake are 2202 kg/year. Also given are the water concentrations that are predicted to result when each input is reduced to zero. It is clear that the atmosphere contributes about 10% to the total.

Thus, the model can be used to determine the relative importance of air and water sources. It is believed that air sources are much more important in Lake Superior.[5] This capability should help to resolve the concern that there is little merit in requiring direct discharge reductions in the lower Great Lakes because the atmosphere is a significant source. It is suspected that these concerns are motivated more by a desire for inertia than for scientific knowledge.

Graphical Presentation of the Data

Figure 2 is an attempt to depict the mass balance fluxes between air, water, and sediment in pictorial form. The length of the six bars represents flows of chemical in kg/year. For each compartment there is a pair of bars representing net movement to and from that compartment. In a steady-state model such as this, the lengths of these pairs of bars must be equal. Amounts leaving the atmosphere, e.g., by wet deposition, appear on the exit bar for air, and identically on the entry bar for water. Transformation processes appear only once. Intermedia transfer processes appear twice. The top two bars are shortened artificially because their length is much greater than could be accommodated on the diagram. Ideally, the length of each segment should be proportional to the quantity, but some license has been used to improve legibility. By examining the length of the various segments, it is possible to compare the magnitude of the various sources to, and the transport routes from, each medium. For example, the contribution from the atmosphere to contamination of the lake can be readily assessed by viewing the length of the atmospheric contribution

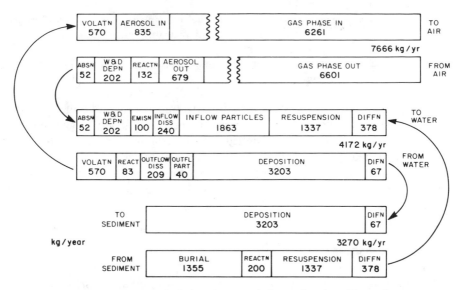

Figure 2. Illustrative bar graph depicting the mass balance data from Figure 1.

bar to the water in comparison to the total length, or to the length excluding the sources from the sediment.

Diagrams of this type could help to convey information about the relative importance of the processes contributing to lake contamination from the atmosphere, from direct discharges, and from sediments. The technique could be extended to include unsteady-state conditions by inserting a block for the inventory change during the period in question.

CONCLUSIONS

A model has been assembled involving three compartments — the atmosphere, water column, and sediment — which helps to elucidate the role of atmospheric sources of contamination of lake water and sediments. The model can be used to assess response times, compare throughputs and masses resident in the system, explore the extent of cycling, elucidate the equilibrium status, and determine if conditions prevailing in the atmosphere can affect the water column in the short term, and conversely if water conditions can affect atmospheric concentrations. When applied to PCBs in Lake Ontario, the model indicates the following:

1. There is appreciable air-water and sediment-water cycling of PCBs.
2. The PCB residence and response times vary greatly, from 18 hr in the air to 0.9 year in the water and 7 years in the bottom sediments.
3. Lake processes are unlikely to affect atmospheric concentrations.
4. At times of intense rainfall there may be nearly complete scavenging of aerosol-associated PCB from the air mass over the lake.

5. Changes in deposition rates from the atmosphere, for example during rainfall, will have only a slight, and probably undetectable, effect on surface water concentrations.
6. It is useful to examine the prevailing media concentrations in terms of their equilibrium status.

The model can help to elucidate the roles of atmospheric and aquatic sources with the novel method of presenting such data graphically suggested here. Approaches of this type can be applied to other contaminants and other lakes, and to both steady- and unsteady-state conditions. Such mass balance assessments may contribute to a more quantitative understanding of the role that atmospherically transported chemicals play in contaminating aquatic systems and how these systems may be best remedied.

ACKNOWLEDGMENTS

The financial support of the Wildlife Toxicology Fund and NSERC are gratefully acknowledged.

REFERENCES

1. Eisenreich, S. J., Ed. *Atmospheric Pollutants in Natural Wastes* (Ann Arbor, MI: Ann Arbor Science, 1981).
2. Hites, R. A., and S. J. Eisenreich, Eds. *Sources and Fates of Aquatic Pollutants*, ACS Advances in Chemistry Series, No. 216 (Washington, DC: American Chemical Society, 1987).
3. "Report on Modeling the Loading-Concentration Relationship for IJC Critical Pollutants in the Great Lakes," International Joint Commission, Windsor, Ont. (1988).
4. Mackay, D. "An Approach to Modelling the Long Term Behavior of an Organic Contaminant in a Large Lake: Application to PCBs in Lake Ontario," *J. Great Lakes Research 15* 283–297 (1989).
5. Strachan, W. M. J., and S. J. Eisenreich. "Mass Balancing of Toxic Chemicals in the Great Lakes: The Role of Atmospheric Deposition," International Joint Commission, Windsor, Ont. (1988).
6. Eisenreich, S. J. "The Chemical Limnology of Nonpolar Organic Contaminants: PCBs in Lake Superior," in *Sources and Fates of Aquatic Pollutants,* R. A. Hites and S. J. Eisenreich, Eds., ACS Advances in Chemistry Series, No. 216 (Washington, DC: American Chemical Society, 1987), pp. 393–469.
7. "Chemical Meteorological Summary. 1987. CYYZ. Toronto-Pearson International Airport, Ontario," Environment Canada Atmospheric Environment Service, Downsview, Ont. (1987).
8. Baker J. E., and S. J. Eisenreich. "Exchange of PAHs and PCBs across the Air-Water Interface: Evidence of Volatilization from the Great Lakes," paper presented at 32nd Conference on Great Lakes Research, University of Wisconsin, Madison, WI, May 31, 1989.

## CHAPTER 22

Evidence for Long-Range Transport of Toxaphene to Remote Arctic and Subarctic Waters from Monitoring of Fish Tissues

Derek C. G. Muir, N. P. Grift, C. A. Ford, A. W. Reiger, M. R. Hendzel, and W. L. Lockhart

INTRODUCTION

The presence of toxaphene in fish from the Great Lakes and Siskiwit Lake, a remote lake on Isle Royale in Lake Superior, has been reported in a number of monitoring studies.[1-6] Rapaport and Eisenreich[7] demonstrated that toxaphene deposition in peat bogs in northern Minnesota and northern Ontario reached a maximum in the mid-1970s, corresponding to the maximum production and use of this insecticide in the United States. The results from peat cores and from monitoring fish in the Great Lakes region, and studies of atmospheric transport of toxaphene,[8,9] suggest that remote lakes in northern Ontario and Manitoba could also have received similar inputs of this insecticide from the atmosphere.

There is little information on the geographical extent of toxaphene deposition north of the Great Lakes or on the degree of contamination of fish in this large region. Bidleman et al.[10] reported toxaphene at concentrations ranging from 36 to 44 pg/m^3 in air at the Ice Island in the Arctic Ocean west of Axel Heiberg Island, but there are no reports of the insecticide in lake or river waters in northern Canada. Toxaphene was the major organochlorine detected in benthic marine amphipods at the Ice Island.[10] Monitoring of marine fish in the Canadian Arctic indicated that toxaphene was one of the major organochlorine contaminants in arctic cod (*Boreogadus saida*) collected at Resolute and Arctic Bay and in Pacific herring (*Clupea harengus pallasi*) from the Beaufort Sea.[11]

The objective of this study was to test the hypothesis that toxaphene levels in fish from Canadian Arctic and subarctic watersheds should decline with latitude, reflecting increased distances from North American sources. We assumed that precipitation, vapor exchange, dry deposition, and transformations in the atmosphere should have removed the insecticide as it was carried north from major use areas in the southern United States and that this should be reflected in lower concentrations in fish tissues from this region. To eliminate the possibility of contamination because of past use of toxaphene as an insecticide and piscicide, we selected fish from remote lakes and rivers where atmospheric transport and deposition represented the only likely pathway for introduction of toxaphene to the watershed.

The choice of fishes for the survey was dictated by the need to determine the extent of contamination of species that form important components of the diet of native people in the Canadian Arctic. Arctic char (*Salvelinus alpinus*) are an important food source for residents of coastal communities. A study of dietary contamination at Broughton Island (N.W.T.) indicated that toxaphene and other organochlorines were present in these fish.[12] Burbot (*Lota lota*) liver has been consumed traditionally by native people in the Mackenzie River valley. Burbot are a predacious, bottom-feeding fish inhabiting rivers in arctic regions and deep lakes in southern Canada.[13] Initial studies of organochlorine contaminants in burbot liver samples collected in this region indicated relatively high levels in comparison with other fish tissues.[14]

MATERIALS AND METHODS

Samples

Liver samples were obtained from 77 burbot collected between 1985 and 1988 at nine sites ranging from Lake 625 (49°45′ N), in the Experimental Lakes Area in northwestern Ontario, to Fort McPherson (67°26′ N), on the Peel River in the Northwest Territories (Table 1 and Figure 1). Burbot were collected with baited hooks (Mackenzie River and Lake 625 samples) or by gill netting by Department of Fisheries and Oceans personnel or staff of the Department of Renewable Resources of the Government of the Northwest Territories. Fish were labeled, frozen in plastic bags, and shipped to Winnipeg in insulated coolers. In the laboratory, fish were partially thawed; sex, fork length, and weight were recorded; and otoliths were removed for age determinations. Portions of livers were sampled for organochlorine analysis and frozen (-40°C) until analysis.

Sixty samples of anadromous arctic char were obtained from six coastal communities in the Northwest Territories between 1985 and 1987 by staff of the Department of Fisheries and Oceans and Northwest Territories Department of Renewable Resources (Table 1). Samples were frozen and shipped to Winnipeg. Fish were weighed and their fork length recorded. After a 50–60 g portion of muscle was removed, the remaining tissue was homogenized in a meat grinder to yield a "whole fish" sample.

Table 1. Locations and Characteristics of Burbot and Arctic Char Samples.

Location	Latitude, Longitude	Mean Lipid (%)	Mean Weight (kg)	Mean Age (yr)	Males	Females	Toxaphene (ng/g lipid wt)
Burbot (*Lota lota*)							
L625 ELA[a]	49°45'N, 93°48'W	30.3 ± 10.1	0.76 ± 0.43	9.3 ± 2.3	6	6	1723 ± 541
Lake Winnipeg	50°42'N, 96°34'W	32.0 ± 11.1	0.71 ± 0.61	6.1 ± 1.7	2	6	807 ± 205
Trout Lake	51°15'N, 93°30'W	26.8 ± 11.7	1.23 ± 0.63	13.0 ± 3.1	3	4	2337 ± 770
South Indian Lake	56°47'N, 98°56'W	20.5 ± 14.4	1.25 ± 0.59	9.4 ± 1.2	4	10	1467 ± 324
Slave River	60°15'N, 111°55'W	36.3 ± 5.7	1.42 ± 0.41	9.6 ± 0.9	2	3	1431 ± 689
Ft. Simpson[b]	61°52'N, 122°21'W	41.7 ± 9.7	2.92 ± 1.00	13.2 ± 1.8	3	1	1132 ± 683
Ft. Good Hope[b]	66°15'N, 128°38'W	32.1 ± 13.4	2.61 ± 1.34	13.3 ± 3.1	12	7	1080 ± 491
Arctic Red River[c]	67°26'N, 133°44'W	37.1 ± 6.8	3.20 ± 1.13	13.8 ± 2.8	1	3	1700 ± 1346
Ft. McPherson[c]	67°26'N, 134°53'W	28.8 ± 5.2	3.51 ± 3.43	11.5 ± 2.4	2	2	931 ± 904
Arctic char (*Salvelinus alpinus*)							
Broughton Is.	67°34'N, 63°54'W	9.5 ± 0.7	2.00 ± 0.56	—	4	2	1631 ± 788
Kingnait Fiord	65°52'N, 65°43'W	12.5 ± 0.7	1.30 ± 0.39	—	—	—[d]	758 ± 112
Diana River	62°50'N, 92°24'W	11.5 ± 0.7	1.94 ± 0.36	—	6	4	670 ± 99
Cambridge Bay	69°03'N, 105°10'W	12.5 ± 1.5	2.65 ± 0.39	16.0 ± 2.0	4	6	1293 ± 415
Rat River	67°47'N, 136°19'W	12.4 ± 0.8	0.66 ± 0.35	7.0 ± 1.0	2	8	120 ± 26
Phillips Bay	69°17'N, 138°30'W	13.1 ± 1.1	0.79 ± 0.24	—	—	—[d]	323 ± 71

Note: Geometric means ± standard deviation from log transformed data.
[a]Experimental Lakes Area, northwestern Ontario.
[b]Samples collected from the Mackenzie River.
[c]Samples collected from the Peel River.
[d]10 samples; sex not determined.

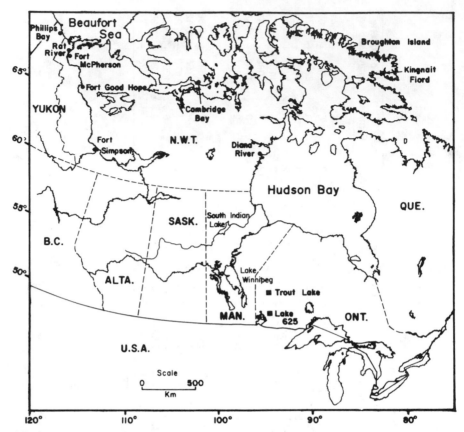

Figure 1. Map of northern Canada showing approximate locations of sampling sites for burbot and arctic char.

Sample Analysis

Toxaphene determinations in fish tissues generally followed the method of Norstrom and Won[15] for organochlorine analyses except for the extraction steps. Char whole fish homogenates (10 g) were mixed with sodium sulfate and extracted with dichloromethane (DCM)–hexane (1:1) using a Polytron (Brinkman Instruments) homogenizer. The extracts were taken to dryness in a tared round bottom flask and lipid levels determined gravimetrically. Sub-samples of burbot liver (2 g) were homogenized by blending with dry ice. The liver tissue was then mixed with sodium sulfate and ball-milled (30 min) with hexane. The extract was allowed to stand for 4 hr, centrifuged, and a portion (1/10) removed for lipid determination. The hexane extract was evaporated to about 1 mL and taken up in DCM–hexane (1:1).

Lipids in fish tissue extracts were separated from toxaphene by gel perme-ation chromatography (GPC) on SX-3 Biobeads using an Autoprep 1001A with DCM–hexane as the eluant. GPC eluates were then fractionated on Florisil (1.2% deactivated with water) into three fractions:[15]

- F1 — hexane
- F2 — hexane–DCM (85:15)
- F3 — hexane–DCM (1:1)

The chromatography on Florisil separated polychlorinated biphenyls (PCBs) in F1 from most toxaphene components, chlordane-related compounds, and p,p'-DDT in F2.

Florisil eluates were analyzed by capillary gas chromatography (GC) with ^{63}Ni–electron capture detection using a 60 m × 0.25 mm DB-5 column with H_2 carrier gas. GC conditions were identical to those used in previous studies.[16] Toxaphene was confirmed by GC/mass spectrometry (GC/MS) on a 30 m × 0.25 mm DB-5 column using a HP 5987 or a HP-5970 mass selective detector (MSD) operated in electron impact mode at 70 eV. GC/MS conditions were identical to those of Norstrom et al.[17] The presence of toxaphene was also confirmed by treatment of samples with HNO_3–H_2SO_4 (1:1) to destroy chlorinated aromatics such as PCBs and DDT-related compounds.[18]

Quantification of toxaphene was accomplished by summing the areas of eight major peaks previously identified by GC/MS and multiplying by a response factor based on the areas of the same peaks in the standard (T1 to T8 in Figure 2). The toxaphene analytical standard was obtained from the EPA standards repository.

Data Analysis

Toxaphene concentrations in burbot liver and char whole fish samples were converted to lipid weight basis and then were $\log_{10}$ transformed because preliminary statistical analysis indicated that at some locations standard deviations were proportional to means. The general linear models procedure of SAS[19] was used with an analysis-of-covariance model to examine the main effects of location and sex, and regressor effects of fish weight and age, on toxaphene concentrations. Comparisons between mean concentrations of toxaphene, fish weight, and age at each location were then made using Tukey's test with the mean square error from the analysis of covariance. Means, standard errors, and correlation coefficients were calculated with $\log_{10}$ transformed data.

RESULTS

Samples

The nine collection sites for burbot formed a northwesterly transect from Lake 625, a small headwater lake near the U.S.-Canada border, to Fort McPherson, N.W.T., on the Peel River about 200 km south of the Beaufort

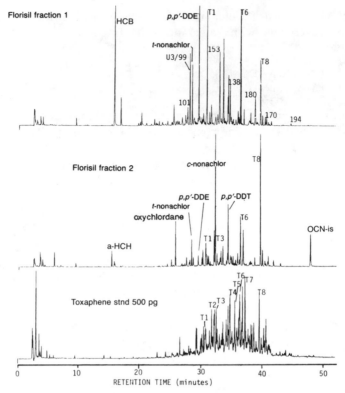

Figure 2. Capillary GC/ECD chromatograms (60 m × 0.25 mm DB-5 columns, H_2 carrier gas) of a burbot liver extract Florisil fractions 1 and 2. Lower chromatogram shows the toxaphene standard, which elutes primarily in fraction 1 under the conditions used. Major PCBs are identified by their IUPAC numbers. U3 is a *t*-nonachlor isomer.[17]

Sea (Figure 1). Burbot ranged from 0.71 to 3.51 kg mean weight and from 6 to 14 years in mean age at the nine sites (Table 1). Burbot from Lake Winnipeg were significantly younger (p < 0.05) than those from all other sites. Burbot collected from riverine habitats in the Mackenzie River valley sites were generally heavier than those collected in Manitoba and Ontario lakes. The migratory behavior of burbot in the Mackenzie River is unknown, but the sampling sites were sufficiently far apart that different populations were thought to have been sampled.[14] Extractable lipid in burbot liver ranged from 7 to 63% and was generally higher in samples in fish from northern locations (Table 1).

Collection sites for anadromous arctic char formed an east-west transect; all sites were between latitude 62° and 69° N, while longitude ranged from 63° to 138° W (Table 1). Char from the western Arctic were of significantly lower weight (p < 0.05) than those from other sites.

Evidence for Toxaphene

The toxaphene pattern in burbot liver in Florisil fraction 2 showed extensive transformation in comparison with a standard (Figure 2). Toxaphene peaks were also observed in Florisil fraction 1, along with PCBs, p,p'-DDE, and *trans*-nonachlor (Figure 2). Following treatment of both fractions with HNO_3-H_2SO_4 (1:1), chlorinated aromatics were removed, leaving most of the toxaphene and chlordane-related compounds intact (Figure 3). Comparison of the toxaphene standard in Figure 2 and 3 shows that toxaphene peak T8 was lost in the nitration procedure.

Arctic char (whole fish) showed a pattern of toxaphene peaks similar to that in burbot liver, although peaks T1 and T8 were not as prominent in Florisil fraction 1 (Figure 4). Full scan electron impact mass spectra of toxaphene peaks T1 and T6, which elute mainly in Florisil fraction 1, were obtained with an extract of a pooled sample of arctic char liver (Figure 5). T1 had an ion

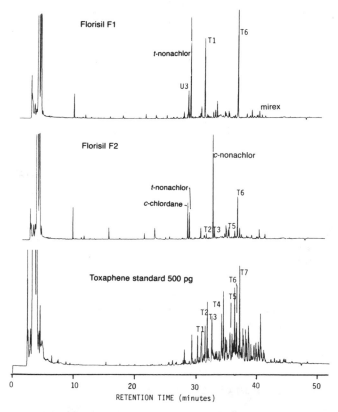

Figure 3. Capillary GC/ECD chromatograms of burbot liver extracts, Florisil fractions 1 and 2 and a toxaphene standard following nitration. Prominent peaks are toxaphene or chlordane-related; PCBs and DDT components have been destroyed. GC conditions are identical to those in Figure 1.

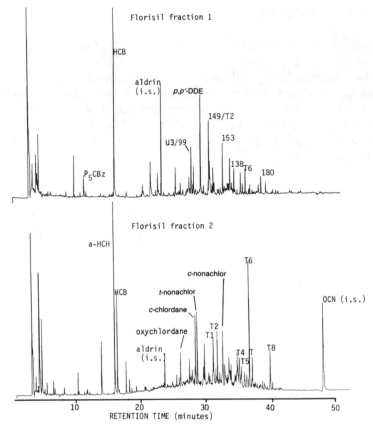

Figure 4. Capillary GC/ECD chromatograms of arctic char (whole fish) extracts, Florisil fraction 1 (upper) and 2 (lower). GC conditions are identical to those in Figure 1.

cluster of low intensity at m/z 377 corresponding to a heptachlorobornane fragment ($C_{10}H_{10}Cl_7$) while T6 had a weak cluster at m/z 412 (Figure 5). The mass spectrum of T1 was almost identical to that of the toxaphene component toxicant B[20].

Further analysis by negative chemical ionization mass spectrometry (NCI-MS), using the techniques described by Bidleman et al.[10], indicated that T1 was an octachlorobornane with an intense M-Cl peak centered at m/z 377. T6 was identified as a nonachlorobornane by NCI-MS based on a characteristic M-Cl fragment at m/z 412.

Geographical Variation of Toxaphene

Toxaphene was found in all burbot liver samples at concentrations ranging from 107 to 5248 ng/g (lipid weight basis). Toxaphene residues in burbot liver consisted of six to eight major peaks, with T1, T6, and T8 (Figure 2) account-

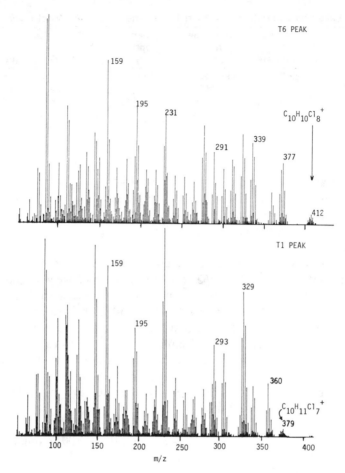

Figure 5. Electron-impact mass spectra of peaks T1 and T6 (see Figures 2 to 4) in a pooled arctic char liver extract. Further analysis by NCI-MS indicated T1 was a octachlorobornane and T6 a nonachloroborane.

ing for about 75% of the peak area in all samples. This pattern was quite consistent between fish, and no trend with latitude or location was evident.

Results are reported as geometric means ($\pm$ 95% confidence limits) on an extracted lipid basis to facilitate comparisons between locations (Table 1). Mean concentrations of toxaphene in liver of male burbot were not significantly different from means for females at each location (t-test at $p < 0.05$). Results for males and females were therefore pooled for presentation in Table 1. Results for Lake Winnipeg burbot were omitted from the analysis of covariance because of significant age and weight differences from other locations.

Burbot liver from Trout Lake had the highest toxaphene concentrations, and those from Lake Winnipeg had the lowest levels (Table 1). Analysis of covariance indicated no significant effects of location but did show a signifi-

Table 2.   Lake Characteristics and Mean Toxaphene Levels in Burbot Liver.

Lake	Surface Area (km²)	Drainage Area (km²)	Mean Depth (m)	Water Residence Time (yr)	Toxaphene[a] (ng/g lipid wt)
Winnipeg	23,750	953,250	18.0	2.9–4.6	807 ± 285
South Indian	2,014	203,000	9.8	0.4–0.8	1467 ± 324
Trout	368	1,300	13.7	22.3	2338 ± 796
ELA 625	1.5	2	b	b	1723 ± 541

[a]Geometric mean ± 95% confidence limits.
[b]Not available.

cant sex-location interaction. Comparisons of means of log transformed data indicated no significant differences in toxaphene levels between any of the northern and southern locations (Tukey's test at $p < 0.05$). Correlations of toxaphene concentrations at each location against latitude were not significant.

Toxaphene was found in all arctic char (whole fish) that were analyzed, ranging in concentration from 53 to 4520 ng/g (lipid weight). Highest toxaphene levels were found at Broughton Island in the eastern Arctic, and the lowest levels at Rat River (near Fort McPherson) in the western Arctic (Table 1). Analysis of covariance revealed that location had a significant effect ($p < 0.01$) on toxaphene levels in char. Comparisons of means of log transformed data showed that toxaphene levels in fish from the two western Arctic sites were significantly lower ($p < 0.05$) than each of the other locations. Regression of toxaphene concentrations against longitude gave the following equation ($r^2 = 0.50$; r significant at $p < 0.01$):

$$\log \text{toxaphene} = (3.738 \pm 0.143) - (0.01 \pm 0.001)(\text{longitude})$$

Toxaphene concentrations were not significantly correlated with age of burbot for the data set as a whole; however, some correlations were observed within each location. Toxaphene levels were positively correlated with age of male burbot from Lake 625 ($r = 0.93$; $N = 6$) but not with a pooled group of males from Mackenzie valley sites ($N = 32$). There were no significant correlations of age of females at any locations. In arctic char, toxaphene levels were not correlated with fish weight (all locations) or age (two sites).

DISCUSSION

Evidence for an Atmospheric Pathway

The presence of toxaphene in burbot liver can be due only to atmospheric transport and deposition for most of the locations that were sampled. Lake 625 and Trout Lake are headwater lakes in northwestern Ontario (Table 2).

Both are uninhabited and not directly accessible by road. South Indian Lake forms part of the Churchill River system, which drains northern Saskatchewan and Manitoba. The population density in this watershed is < 1 person per 10 km², and agricultural activity is negligible; however, there is mining activity in the region. The Mackenzie River and Peel River watersheds also have very low population densities. There is little agricultural activity in the Mackenzie River watershed except in the Peace River region in northern Alberta (Figure 1). There has been some use of toxaphene in the Lake Winnipeg watershed, which drains predominately agricultural land extending from North Dakota, west to the Rocky Mountains and into northwestern Ontario on the east (Table 2). Agriculture Canada statistics indicate that toxaphene use in Manitoba, Saskatchewan, and Alberta ranged from 3600 to 6755 kg/yr from 1976 to 1978,[21] but sales of toxaphene were discontinued by the manufacturer in 1982. Assuming similar amounts were used in North Dakota and northern Minnesota, the total quantities used in the Lake Winnipeg watershed probably represent less than 0.1% of the total toxaphene used in the United States and Canada in the late 1970s.[8,21]

Although atmospheric inputs would seem to be the major source of toxaphene to the lakes, toxaphene residues in burbot liver are not related to watershed drainage area, lake size, or water residence time (Table 2). Lower toxaphene levels in Lake Winnipeg may be due to the significantly younger age of the burbot population that was sampled. For the other three lakes, there is relatively little difference in toxaphene levels in fish despite great differences in limnological characteristics. The results of Johnson et al.[22] for organochlorine loadings to remote lakes in the Great Lakes region also support the view that drainage area and lake size are not important factors. They found that loadings of organochlorines from runoff into the Turkey and Emmett/Bartley lakes in Ontario were less than 5% of total lake loadings of these chemicals, except for hexachlorobenzene (where it was 66%). However, in the Great Lakes, toxaphene levels in lake trout (*Salvelinus namaycush*) of similar age class show a strong relationship with lake surface area and turnover rate, with the highest levels in Lake Superior fish and the lowest in Lake Erie and Lake Ontario.[6] Precipitation has been shown to account for the major proportion of inputs of organochlorines to Lake Superior[23] and Siskiwit Lake.[1] The similar levels of toxaphene in burbot in Trout Lake, Lake 625, and South Indian Lake may be indicative of similar amounts of atmospheric deposition and, in the case of South Indian Lake and Lake Winnipeg, of the negligible contribution of the drainage basin to toxaphene loading.

Geographical Variations of Toxaphene

The lack of a significant decline in levels of toxaphene in burbot with latitude is contrary to observations with organochlorines and other contaminants that originate in the midlatitudes. Other organochlorines — hexachlorocyclohexane (HCH), PCBs, and chlordane- and DDT-related compounds — monitored in the same burbot liver samples declined significantly with increasing latitude.[16] Limited data suggest that some atmospherically borne contaminants, such as ^{137}Cs in lichen[24] and α-HCH in polar bear liver,[17] decline with increasing latitude in the Canadian Arctic.

The toxaphene data from arctic char show a significant increase from west to east. This trend was also reported by Norstrom et al.[17] for chlordane, DDT-related compounds, and dieldrin in polar bear liver. Unlike the data for burbot, the results for char are consistent with the model of movement of toxaphene in air masses passing over the southern United States (as observed by Rice et al.[8]) and moving north and northeast over the Hudson Bay lowlands, northern Québec, and the eastern Canadian Arctic archipelago. Char move into marine waters during the spring ice breakup and return in the autumn to overwinter in freshwater.[25] Thus, toxaphene residues in arctic char may reflect contamination of both freshwater and marine waters in the Arctic from atmospheric deposition. Bidleman et al.[10] reported that toxaphene-related compounds (polychlorinated camphenes) were the major organochlorine contaminants in air, seawater, and amphipods from the Arctic Ocean west of Ellesmere Island. Toxaphene was the major organochlorine contaminant in arctic marine fish surveyed in the mid-1980s[11] and is also detected in beluga (*Delphinapterus leucas*) in the region.[26]

Effect of Fish Age and Habitat on Toxaphene Levels

The lower toxaphene levels in char from the western Arctic may also be a reflection of the smaller size and younger age of these fish (Table 1), although, in general, age and weight were not correlated with residue level at each location. Fish collected at Rat River (mean age: 7 yr) were younger than those at Cambridge Bay (mean age: 16 yr) because heavy exploitation has reduced the numbers of larger char available.[27] Veith et al.[28] showed a positive correlation existed between fish length and DDT content in lake trout from Lake Superior, and Burgermeister et al.[29] reported a similar relationship between weight and PCB content in burbot from a Swiss lake. Swackhamer and Hites[1] found increasing concentrations of DDE, nonachlor, and most hepta- and octachloro-PCBs with fish size for lake trout from Siskiwit Lake but did not observe this relationship for toxaphene. In burbot, toxaphene levels were not positively correlated with age or weight for the data set as a whole.

The lack of variation of toxaphene with age or size of fish seen in this study and also by Swackhamer and Hites[1] is surprising since it is observed for other organochlorines. A possible explanation for this is that toxaphene is excreted and metabolized more rapidly than chlorinated aromatics like PCBs and *p,p'*-DDE. Half-lives for toxaphene in fish range from 63 days in lake trout to 38 days in fathead minnows,[30-32] compared to greater than 200 days for most penta- to nonachlorobiphenyls.[30] The effect of elimination rate on bioaccumulation of toxaphene is best assessed with the aid of a food chain model. We used the model of Thomann and Connolly,[33] which has previously been used to

Table 3. Predicted Toxaphene and PCB Concentrations in Aquatic Food Chains Using the Thomann and Connolly Food Chain Model.

Compartment	BCF	K_d (day^{-1})	α (%)	F (g/g/day)	Concentration (ng/g wet wt) via water	via food
Toxaphene						
Detritus/phytoplankton	—	—	—	—	85	—
Crustacea	32,000	0.04	35	0.105	16	125
Small fish	60,000	0.01	70	0.017	30	168
Large fish	62,500	0.005	80	0.009	31	356
PCB[b]						
Detritus/phytoplankton	—	—	—	—	20	—
Crustacea	75,300	0.01	35	0.105	75	74
Small fish	21,500	0.004	70	0.017	22	443
Large fish	55,000	0.001	80	0.009	55	3340

[a]BCF data from the summary of Sullivan, J. R., and D. E. Armstrong, "Toxaphene Status in the Great Lakes," University of Wisconsin Sea Grant Institute Report (1985). Elimination rate data (k_d) from the summary on Niimi.[30] Water concentration of toxaphene assumed to be 0.5 ng/L (Sullivan and Armstrong). Concentration in detritus/phytoplankton from measurements reported by Rice and Evans.[35]
[b]PCB data from Thomann[34] amd Thomann and Connolly.[33] Water concentration assumed to be 1 ng/L (Arochlor 1254 equivalent).

predict PCB levels in Lake Michigan lake trout. The model has four compartments: organic detritus/phytoplankton as the base of the food chain, filter-feeding crustaceans (*Mysis relicta*), small fish, and predatory fish. The steady-state concentration of toxicant (CF_i) at each trophic level i is given by:

$$CF_i = \frac{k_{ui} \, CW + \alpha \, F_i \, CP_{i-1}}{k_{di}}$$

where k_{ui} = uptake rate constant from water (days^{-1})
    CW = water concentration of toxicant (μg/L)
    α = assimilation efficiency of the toxicant from food
    F_i = feeding rate (g/g fish/day)
    k_{di} = depuration rate (0.693/half-life in days)
    CP_{i-1} = concentration (μg/kg) in prey (lower trophic level)

To determine values of CF_i in each compartment, data for toxaphene concentrations in Lake Superior water and laboratory-derived bioconcentration factors and half-lives were used. A steady state (no growth and single feeding and depuration rates) within each compartment was assumed to simplify calculations. Assimilation efficiency of PCBs and toxaphene were assumed to be the same. Feeding rates for each compartment were taken from Thomann and Connolly[33] and Thomann.[34]

Modeling results (Table 3) indicate that although toxaphene does biomagnify (i.e., concentrations are greater in predacious fish than in fish feeding on crustaceans), food chain transfer of toxaphene is less than for PCBs. Concentrations in large fish are predicted to be 1.9-fold greater than in small fish for toxaphene at steady state versus 7.4-fold greater for PCBs (Table 3). The model also suggests that uptake from water is a more important contributor to body burdens of toxaphene than it is for PCBs at all trophic levels. The relatively (compared to PCB) rapid elimination of toxaphene and higher proportion accumulated from water suggests that residues in fish will respond more rapidly to reductions in toxaphene inputs to lake waters than would be the case for PCBs. The model fails to predict the concentrations of toxaphene (wet weight basis) seen in large predacious fish in Lake Superior,[5] such as lake trout, by three-to eightfold. This may be due to lower elimination rates for large fatty fish than those observed with smaller fish in the laboratory. Concentrations in detritus/phytoplankton may be higher than those used in Table 3; Rice and Evans[35] have reported concentrations in plankton ranging from 85 to 560 ng/g dry weight.

Habitat differences between the northern riverine burbot and lake-resident burbot may also influence the results. High suspended solids loads, and short water residence times, could serve to reduce the bioavailability of toxaphene in riverine habitats compared to the southern lakes. Few differences have been reported between the diets of northern burbot and those inhabiting lakes. Adult burbot in Alaskan rivers feed on bottom-feeding fish[36] as do burbot in Manitoba lakes.[37] Juvenile burbot feed on deep-water invertebrates, mollusks, and crayfish, but relatively few fish.[13] The lower toxaphene levels in the Lake Winnipeg burbot, which were significantly younger and smaller than the fish at other locations, may thus be due to feeding at a lower trophic level.

Comparison with Data for Other Remote Lakes

Toxaphene levels in burbot liver and arctic char whole fish homogenates are lower than levels reported in fish from Siskiwit Lake and Lake Superior, when compared on a lipid weight basis (Table 4).[1-6,38-40] All studies reported in Table 4 used capillary GC for separation of toxaphene components from other organochlorines, but comparison between locations is complicated by different methods of quantification and, in some cases, lack of data on lipid content of the fish. Comparison of the results for *Salvelinus* species (i.e., lake trout and char) in Lake Superior with those in the Arctic suggest that there is a decline in toxaphene levels with increasing latitude, although this is not apparent in burbot. Toxaphene levels in burbot from Trout Lake and Lake 625 were similar to those reported in lake trout muscle (estimated on a lipid basis) from the same region in a survey by Ontario Ministry of Environment.[38] There are relatively few published data for toxaphene or other organochlorines in burbot. Pyysalo and Antervo[41] did not observe toxaphene in burbot from Finnish waters using selected ion monitoring. Schmitt et al.[5] reported toxaphene residues in rainbow trout (*Salmo gairdneri*) and round whitefish (*Prosopium cylindraceum*) from the Kenai river in Alaska at levels similar to those in

Table 4.   Comparison of Toxaphene in Fish from Lakes and Remote Rivers Receiving Atmospheric Inputs.

Species	Tissue[a]	Location/Year	Mean Concentration (ng/g lipid wt)	Reference
Lake trout	WF	Siskiwit L. 1983	11,000	1
Lake trout	WF	Siskiwit L. 1983	~2,900[b]	2
Lake trout	WF	Siskiwit L. 1983	~4,400[b]	3
Lake trout	WF	L. Superior 1979[c]	23,470	5
Lake trout	WF	L. Superior 1979[d]	8,920	5
Lake trout	WF	L. Superior 1980	7,130	6
Lake trout	M	Ramoos Lake, Ont., 1984	~2,480[b]	38
Lake trout	M	Trout Lake, Ont., 1984	~630[b]	38
Lake trout	M	Snook Lake, Ont., 1984	1,000[b]	38
Arctic char	WF	L. Vattern, Sweden, 1978	9,000	39
Arctic char	E	L. Drachensee, Austria, 1978	125	40
Arctic char	WF	Phillips Bay, Yukon, 1986	333	this study
Arctic char	WF	Broughton Island, N.W.T., 1986	1,632	this study
Burbot	WF	Fairbanks, AK., (Chena R.) 1979	<100	5
Burbot	L	Ft. McPherson (Peel R.) 1987	931	this study
Burbot	L	Trout Lake 1987	2,338	this study
Rainbow trout	WF	Soldatna, AK., (Kenai R.) 1978	1,538	5
Whitefish	WF	Soldatna, AK., (Kenai R.) 1978	1,563	5
Pike	E	L. Carrigellen, Ireland, 1979	240	40

[a]WF = whole fish, L = liver, M = Muscle, E = eggs.
[b]Results were converted to lipid weight basis assuming 10% lipid.
[c]Keweenaw Point, MI.
[d]Whitefish Point, MI.

Mackenzie River burbot but did not detect toxaphene (detection limit: 100 ng/g wet weight) in burbot from the Chena River near Fairbanks, Alaska. There are a number of reports on PCBs, DDT, and other organochlorines in burbot (whole fish or muscle) indicating that this species has similar levels to other piscivorous fish,[28,29,42] but these studies did not determine toxaphene residues.

CONCLUSIONS

Toxaphene concentrations in burbot liver were not correlated with latitude, unlike results for other atmospherically derived pollutants in the Canadian Arctic. Factors such as fish weight, age, diet, and differences in diet and habitat of the northern riverine burbot and lake-resident burbot sampled at the southern sites did not account for the lack of a trend with latitude. One possible explanation of the lack of a latitudinal trend for toxaphene in burbot liver could be that additional sources such as Eastern Europe and Asia may contribute to the loading of this contaminant in Canadian Arctic and subarctic waters.

The data for arctic char tend to support the hypothesis of declining loadings of toxaphene from south to north because levels were five- to eightfold lower than other *Salvelinus* species (i.e., lake trout) in Lake Superior. The toxaphene levels in char also declined longitudinally from east to west, which is consistent with flows of air masses from south to north in eastern North America. However, size and age of fish may have influenced residue levels in char because fish sampled in the western Arctic were significantly smaller than those in the east.

Lake and watershed size had no effect on toxaphene levels in burbot liver from the Manitoba and northwestern Ontario lakes in this survey. This is consistent with observations that runoff contributes relatively little loading of organochlorines to remote lakes compared to atmospheric deposition but is contrary to the pattern of toxaphene concentrations in lake trout in the Great Lakes.

Toxaphene residues in burbot were generally not correlated with fish size or age. This lack of age/size effect has been observed for toxaphene and other more water-soluble organochlorines in lake trout in Siskiwit Lake, but not for PCBs and DDE. Use of the food chain model of Thomann and Connolly[33] predicted that the relatively rapid elimination of toxaphene would lead to small differences in residue levels between small and large predatory fish (1.9-fold) compared to those for PCBs (7-fold).

ACKNOWLEDGMENTS

We wish to thank the following personnel of the Department of Fisheries and Oceans, Central and Arctic Region, for sample collections: K. Chang-Kue, K. Mills, and R. Fudge for collection of burbot and W. Bond, J. Reist, P. Bobinski, L. Dalke, and G. Carder for collection of arctic char. We are grateful to M. Simon and R. J. Norstrom (Canadian Wildlife Service, Environment Canada, Ottawa) and to T. F. Bidleman (University of South Carolina, Columbia, SC) for providing mass spectral analysis of toxaphene; to D. A. Metner (Department of Fisheries and Oceans, Winnipeg) for additional burbot sample collection, preparation, and data archiving; and to D. M. Whittle and D. B. Sergeant (Department of Fisheries and Oceans, Burlington, Ont.) for helpful reviews of the manuscript.

REFERENCES

1. Swackhamer, D. L., and R. A. Hites. *Environ. Sci. Technol.* 22:543–48 (1988).
2. Swackhamer, D. L., M. J. Charles, and R. A. Hites. *Anal. Chem.* 59:913–17 (1987).
3. Petty, J. D., T. R. Schwartz, and D. L. Stalling. In *New Approaches to Monitoring Aquatic Ecosystems,* T. P. Boyle, Ed., ASTM STP 940 (Philadelphia, PA: American Society for Testing and Materials, 1987), pp. 165–72.
4. Gooch, J. W., and F. Matsumura. *J. Agric. Food Chem.* 16:349–55 (1987).
5. Schmitt, C. J., M. A. Ribick, J. L. Ludke, and T. W. May. U.S. Dept. of Interior, Fish and Wildlife Service. Resource Publ. 1542, Washington, DC (1983).
6. Whittle, D. M. Dept. of Fisheries and Oceans, Great Lakes Fisheries Research Lab., Burlington, Ont., personal communication (1986).

7. Rapaport, R. A., and S. J. Eisenreich. *Atmos. Environ.* 20:2367-79 (1986).
8. Rice, C. P., Samson, P. J., and G. E. Noguchi. *Environ. Sci. Technol.* 20:1109-16 (1986).
9. Chapter 15.
10. Bidleman, T. F., G. W. Patton, M. D. Walla, B. T. Hargrave, W. P. Vass, P. Erickson, B. Fowler, V. Scott, and D. J. Gregor. *Arctic* 42:307-13 (1989).
11. Muir, D. C. G., R. Wagemann, W. L. Lockhart, N. P. Grift, B. Billeck, and D. Metner. Environmental Studies No. 42, Indian and Northern Affairs Canada, Ottawa (1987).
12. Muir, D. C. G., N. P. Grift, and C. A. Ford. Report to Dept. of Indian Affairs and Northern Development, Ottawa (1986).
13. Scott, W. B., and Crossman, E. J. "Freshwater Fishes of Canada," Bulletin 184, Fish. Res. Board Canada, Ottawa (1973).
14. Lockhart, W. L., D. A. Metner, D. A. J. Murray, R. W. Danell, B. N. Billeck, C. L. Baron, D. C. G. Muir, and K. Chang-Kue. "Studies to Determine Whether the Condition of Fish from the Lower Mackenzie River is Related to Hydrocarbon Exposure," Environ. Studies Report No. 61, Indian and Northern Affairs Canada (1989).
15. Norstrom, R. J., and H. Won. *J. Assoc. Offic. Anal. Chem.* 68:130-35 (1985).
16. Muir, D. C. G., C. A. Ford, N. P. Grift, D. A. Metner, and W. L. Lockhart. *Arch. Environ. Contam. Toxicol.* (in press).
17. Norstrom, R. J., M. Simon, D. C. G. Muir, and R. E. Schweinsburg. *Environ. Sci. Technol.* 22:1063-70 (1988).
18. Musial, C. J., and J. F. Uthe. *Intern. J. Environ. Anal. Chem.* 14:117-26 (1983).
19. *SAS User's Guide: Statistics* (Cary, NC: SAS Institute, 1987).
20. Saleh, M. A. *J. Agric. Food Chem.* 31:748-51 (1983).
21. Agriculture Canada, Pesticide Division, Ottawa, personal communication (1988).
22. Johnson, M. G., J. R. M. Kelso, and S. E. George. *Can. J. Fish. Aquat. Sci.* 45:170-78 (1988).
23. Eisenreich, S. J. In *Sources and Fates of Aquatic Pollutants,* R. A. Hites and S. J. Eisenreich, Eds., Adv. Chem. Ser. 216 (Washington, DC: American Chemical Society, 1987), pp. 393-470.
24. Hutchison-Benson, E., J. Svoboda, and H. W. Taylor. *Can. J. Bot.* 63:784-91 (1985).
25. Grainger, E. H. *J. Fish. Res. Board Can.* 10:326-70 (1953).
26. Muir, D. C. G., C. A. Ford, R. E. A. Stewart, T. G. Smith, R. F. Addison, M. E. Zinck, and P. Béland. In Beluga Monograph, J. Geraci and T. G. Smith, Eds., *Bull. Can. J. Fish. Aquat. Sci.* (in press).
27. Reist, J. D. Department of Fisheries and Oceans, personal communication (1988).
28. Veith, G. D., D. W. Kuehl, F. A. Puglisi, G. E. Glass, and J. G. Eaton. *Arch. Environ. Contam. Toxicol.* 5:487-99 (1977).
29. Burgermeister, G., M. Bedrani, and J. Tarradellas. *Eau de Québec* 16:135-43 (1983).
30. Niimi, A. J. *Rev. Environ. Contam. and Toxicol.* 99:1-46 (1987).
31. Mayer, F. L., Jr., P. M. Mehrle, Jr., and W. P. Dwyer. "Toxaphene Effects on Growth, Reproduction and Mortality of Brook Trout," EPA-600/3-75-013 (1975).
32. Mayer, F. L., Jr., P. M. Mehrle, Jr., and W. P. Dwyer. "Toxaphene Chronic Toxicity to Fathead Minnows and Channel Catfish," EPA-600/3-77-069 (1977).

33. Thomann, R. V., and J. P. Connolly. *Environ. Sci. Technol.* 18:65–71 (1984).
34. Thomann, R. V. *Can. J. Fish. Aquat. Sci.* 38:280–96 (1981).
35. Rice, C. P., and M. S. Evans. In *Toxic Contaminants in the Great Lakes,* J. O. Niriagu and M. S. Simmons, Eds. (New York, NY: John Wiley and Sons., 1984), pp. 163–94.
36. Chen, L.-C. Biol. Papers, University of Alaska 11:53 (1969).
37. Lawler, G. H. *J. Fish. Res. Board Can.* 20:792–97 (1963).
38. Johnson, A. F. Ontario Ministry of the Environment, Toronto, personal communication (1988).
39. Jansson, B., R. Vaz, G. Blomkvist, S. Jensen, and M. Olsson. *Chemosphere* 4:181–90 (1979).
40. Zell, M., and K. Ballschmiter. *Fres. Z. Anal. Chem.* 300:387–402 (1980).
41. Pyysalo, H., and K. Antervo. *Chemosphere* 14:1723–28 (1985).
42. Armstrong, F. A. J., and A. Lutz. *Tech. Rep. Fish. Mar. Serv.* 683 (1977).

## CHAPTER 23

Chlorinated Pesticides and Polychlorinated Biphenyls in the Atmosphere of the Canadian Arctic

T. F. Bidleman, G. W. Patton, D. A. Hinckley, M. D. Walla, W. E. Cotham, and B. T. Hargrave

INTRODUCTION

The presence of DDT residues and polychlorinated biphenyls (PCBs) in fish, marine mammals, and birds of the Arctic has been documented since the early 1970s.[1,2] Recent investigations of organochlorines (OCs) in arctic fauna have expanded the list to include chlorobenzenes, hexachlorocyclohexanes (HCHs), dieldrin, components of technical chlordane and their metabolites, and polychlorinated camphenes (PCCs, e.g., toxaphene).[3-7] These pollutants are transferred through the few species comprising polar food chains and can reach high levels in marine mammals due to their storage of large quantities of fat.

Atmospheric transport and subsequent deposition is a likely source of OCs to the arctic ecosystem. Hoff and Chan[8] found *trans-* and *cis*-chlordane (TC, CC) and *trans-* and *cis*-nonachlor (TN, CN) in the air at Mould Bay, Northwest Territories. Airborne PCBs, hexachlorobenzene (HCB), HCHs, and CC were reported from several stations in the Norwegian Arctic: Spitzbergen, Bear Island, Hope Island, and Jan Mayen Island.[9-12] DDTs and HCHs were carried through the atmosphere over the Bering Sea.[13] Evidence of OC deposition is provided by measurements of PCBs and pesticides in snow from several areas of the Canadian Arctic.[14-16] Addison et al.[7] noted that over the last decade PCBs have declined more rapidly than total DDT compounds in arctic ringed seal (*Phoca hispida*) compared to seal populations from eastern Canada. Furthermore, levels of untransformed *p,p'*-DDT in ringed seal were not significantly different between 1969 and 1981. The authors suggested that DDT input continued for a longer time in the Arctic due to air transport from Asia.

A better understanding of OC contamination in the Arctic is important because of the fragile ecosystem and the native people who rely on fish and marine mammals for their food supply. In 1986, groups from Bedford Institute of Oceanography (Dartmouth, Nova Scotia) and Arctic Laboratories (Sidney, British Columbia) initiated an investigation of OC input to and transfer through arctic food chains.[16] We determined OC pesticides and PCBs in air, snow, and surface water of the high Arctic as part of this project.

METHODS

Location and Dates

The Canadian Polar Continental Shelf Project (PCSP) established a research camp in 1984 on a tabular iceberg that calved off the Ward Hunt Ice Shelf on Ellesmere Island. The island provides a safe platform for oceanographic observations in a region where shipboard work is not practical because of extensive ice cover. The 7 km × 4 km × 45 m thick Ice Island was located off the northwest shore of Axel Heiberg Island at 81° N, 101° W, at the time of our work (Figure 1). Collections were made approximately 2 km from the research camp. Air samples were taken from August 25-September 5, 1986; air, snow, and seawater samples were taken from June 7–15, 1987. The air temperature during these times was -10 to +2°C.

Figure 1.   Location of the PCSP Ice Island during August-September 1986 and June 1987.

Sample Collection

Air volumes of 1400–4000 m³ were pulled through a 20 cm × 25 cm Gelman AE glass-fiber filter (GFF), followed by two 7.8 cm diameter × 7.5 cm thick plugs of polyurethane foam (PUF, density 0.022 g/cm³) at flow rates of 0.6–0.8 m³/min using a Rotron DR313 pump. Duplicate samples were collected on several occasions. Procedures for the cleanup of PUF and GFFs and sample collection are given elsewhere.[17-19] Electrical power was supplied by gasoline generators operated 50–100 m downwind of the pumps. After sampling, PUF plugs were placed into metal cans, which were sealed with masking tape. GFFs were wrapped in solvent-rinsed aluminum foil.

Seawater was collected at a depth of 10 m using a National Bureau of Standards sampler, which allowed a 3.5-L glass bottle to be opened and closed at depth. One surface sample (0.5 m) was taken by manually submerging a bottle. The salinity was 32 parts per thousand, and the water temperature was −1.7°C.

Snow was collected by removing the top surface layer (upper 5 cm) into a stainless steel pot, using a solvent-cleaned metal shovel. Two samples were taken prior to a snow event of June 12, 1987; this "older" snow had accumulated before our arrival on June 5. The snow event sample was taken by scraping the "new" snow off the older snow surface. Snow samples were covered and allowed to melt at room temperature ($\approx 20°C$) in the Ice Island laboratory.

Melted snow and seawater (2–3 L) were pulled at 0.8–1.3 L/hr through 47-mm diameter Gelman AE binderless GFFs followed by two 500-mg cartridges of C_8-bonded silica (J. T. Baker SPE). This step was done on the Ice Island, and the GFFs and cartridges were then shipped in glass jars with polytetrafluoroethylene-lined caps to Columbia, South Carolina, for analysis. Procedures for cleanup of GFFs and C_8 cartridges and details of the preconcentration steps have been described by Hinckley and Bidleman.[20]

Spiked PUF plugs for analyte recovery checks were prepared on the Ice Island by pipetting 1.0 mL of a calibration standard containing 5–25 ng/mL OC pesticides onto clean plugs. Seawater (2.0-L volumes) was spiked with 170 ng each of α-HCH and γ-HCH, and the water was pulled through the collection system on the Ice Island. Blank PUF plugs, GFFs, and C_8 cartridges were carried to and from the Ice Island with samples.

Extraction and Cleanup of Air Samples

PUF plugs were extracted in a Soxhlet apparatus with petroleum ether for 8 hr. GFFs were cut into strips, placed into round-bottomed flasks, and refluxed with dichloromethane for 8 hr. GFF and PUF extracts were concentrated by rotary evaporation and nitrogen blow-down, and the GFF extracts were transferred to hexane during this step. Samples were cleaned up and split into two fractions on a column of 3 g silicic acid containing 5% added water overlaid with 2 g neutral alumina containing 6% added water. This procedure was initially developed to separate aliphatic from aromatic hydrocarbons[21] and was subsequently modified slightly to permit fractionation of OCs as well as hydrocarbons. Fraction 1 contained chlorobenzenes, PCBs, and *p,p'*-DDE; fraction 2 contained TC, CC, TN, CN, *p,p'*-DDT, HCHs, dieldrin, endosulfan, and PCCs. Both fractions were shaken with 18 M sulfuric acid (except for aliquots analyzed for dieldrin and endosulfan), isooctane was added, and the fractions were blown down to a final volume for analysis.

Extraction and Cleanup of Water Samples

The C_8 cartridges were eluted with 3 mL 1:1 ethyl ether-hexane, additional hexane was added to the extracts, and the ethyl ether was removed by a nitrogen stream. GFFs were cut into strips and refluxed with acetone (used because of its water miscibility) for 4 hr, and the extracts were concentrated into hexane by rotary evaporation and nitrogen blow-down. A small volume of water remaining after this step was removed with a Pasteur pipette. Extracts were shaken with 18 M sulfuric acid, and δ-HCH was added as an internal standard before analysis.

Gas Chromatographic Analysis

Gas chromatography with electron capture detection (GC/ECD) was carried out on a 25 m × 0.2 mm i.d. HP-5 bonded phase fused silica column (polydimethylsiloxane, 5% phenyl, 0.33 μm film thickness, Hewlett-Packard Corp.) using Varian 3700 or Carlo Erba 4160 instruments. Samples were injected splitless (Grob technique, 1–2 μL volume, 30 sec split time) with the column oven at 90°C. After a 1-min hold, the oven was programmed at 5–15°C/min to a final temperature of 270°C. Other GC conditions were carrier gas, H_2 at 20–40 cm/sec; injector 240°C; and detector 320°C.

Chromatographic data were collected and processed with a Hewlett-Packard 3390A or Shimadzu Chromatopak CR3A integrator. Air sample components were calculated using peak areas with response factors derived from external standards. PCBs were quantified as Aroclors 1242 and 1254 using the sum of sample and standard peak areas, and also as the sum of PCB congeners. Response factors for the latter were obtained from peak areas of the individual peaks in Aroclor standards and the weight percentages of each congener in Aroclors 1242 and 1254.[22] HCHs in water and snow samples were quantified by peak areas relative to the δ-HCH internal standard. Pesticide and PCB standards were obtained from the Environmental Protection Agency Repository for Pesticides and Industrial Organic Chemicals, Research Triangle Park, North Carolina.

In addition to GC/ECD analysis, air samples were examined for technical chlordane components, PCCs, DDE, and DDT by gas chromatography/mass spectrometry (GC/MS), using a Finnigan 4521C instrument and the same

Table 1. Ions Monitored for GC/MS Confirmation of Organochlorines

Electron Impact Mode	
p,p'-DDE	246, 248, 316, 318
p,p'-DDT	235, 237
Negative Ion Mode	
Chlordanes and nonachlors	300, 302, 334, 408, 410, 444
PCCs	309, 311, 343, 345, 379, 381, 413, 415

column as used for GC/ECD work. The carrier gas was He at 20–40 cm/sec. Samples were injected splitless (Grob technique, 2 μL volume, 60 sec split time) with the column at 90°C. After a 1-min hold, the oven was programmed at 15°C/min to 220°C, then at 5°C to 300°C. Chlordane components and PCCs were quantitatively determined by negative ion mass spectrometry and multiple ion detection (MID). The ion source was maintained at 130°C and 0.1 torr methane. Ions monitored for chlordanes, nonachlors, and PCCs (Table 1) were selected from those used previously for negative ion work.[23,24] OCs were quantified by comparing sample peak areas to those of external standards or to a mirex internal standard. Qualitative analysis of DDE and DDT was done in the electron impact mode at 70 cV using MID. The ion source temperature was 150°C. Ions monitored for DDE and DDT are given in Table 1.

RESULTS AND DISCUSSION

Analytical Recovery and Collection Efficiency

Average percent recoveries of HCB, α-HCH, and γ-HCH from PUF air sampling plugs fortified with OCs on the Ice Island were 68 ± 3, 69 ± 5, and 72 ± 6% (n = 4). Quantities of HCB and HCHs in atmospheric samples were calculated by correcting the values found for the spiked plug yields. Recoveries were ≥ 82% for OCs less volatile than γ-HCH, and no corrections were made in calculating air results.

Pentachlorobenzene (PeCB) and HCB were the only OCs that showed substantial breakthrough from the front to the back PUF plugs. Back-plug quantities of other OCs did not differ from blanks. Quantities of PeCB were approximately equal on the two plugs for all samples, indicating probable breakthrough past the back plug, and no quantitative results for PeCB could be obtained. HCB on back traps varied from 27–99% of front trap quantities, depending on the sampling temperature and air volume. HCB concentrations were calculated by summing the quantities on the two plugs; upper limits are given in Table 2 for samples in which breakthrough to the back plug exceeded 50% of the front trap value.

The mean percent recovery of HCHs from water samples spiked on the Ice Island and preconcentrated onto C_8 cartridges was 78 ± 8% (n = 2). HCH

Table 2. Airborne Organochlorines at the Ice Island.

Collection Date	Sample	Air (m³)	HCB[b]	α-HCH	γ-HCH	TC	CC	TN	CN	p,p'-DDE	p,p'-DDT	PCBs[c] Aroclors Light	Heavy	Total	PCBs Congeners Total	PCCs[d]	Dieldrin	Endosulfan I
1986																		
Aug. 25–27	1	1344	131	591	27	1.7	3.2	2.0	—[e]	0.2	1.4	6.7	13.0	19.7	19.4	25	5.0	9.7
Aug. 27–29	2A	1449	233	513	33	0.7	2.0	1.2	—	0.2	0.5	8.1	6.3	14.4	18.3	53	2.7	6.2
	2B	1655	199	469	29	1.0	3.0	1.6	—	0.1	0.7	6.7	3.8	10.5	12.9	78	3.3	2.7
Aug.29–Sept. 1	3A	1738	≤190	612	37	1.3	3.7	2.0	0.7	0.2	1.3	8.6	7.5	16.1	14.4	47	0.1	8.1
	3B	2080	217	731	36	1.1	3.6	1.9	0.5	>0.1	1.0	9.6	4.3	13.9	14.8	44	0.5	6.3
Sep. 1–5	4A	4208	147	417	24	0.6	1.9	0.9	0.2	<0.1	0.5	7.1	2.1	9.2	9.9	30	0.3	8.0
	4B	4003	209	490	29	1.1	2.4	0.8	0.2	<0.1	0.7	10.0	4.7	14.7	16.0	30	1.4	8.5
X̄			189	546	31	1.1	2.8	1.5	0.4	0.1	0.9	8.1	6.0	14.1	15.1	44	1.9	7.1
1987																		
June 7–9	5	2294	>44[f]	283	41	1.6	2.7	2.1	—	0.5	2.6	9.6	5.4	15.0	9.3	21	0.2	1.8
June 9–11	6A	1738	—	377	45	2.5	4.1	3.2	0.4	—	0.8	—	—	—	—	41	—	—
	6B	1759	155	370	52	2.4	4.9	4.0	0.8	1.0	1.0	16.0	5.5	21.5	17.0	45	—	—
June 11–13	7A	1694	≥119	349	53	2.1	3.8	2.8	0.7	3.5	3.2	12.0	6.5	18.5	26.1	38	—	—
	7B	1544	132	302	41	1.9	3.3	5.1	0.8	5.0	1.6	15.5	12.0	27.5	21.7	35	—	—
June 13–15	8	2131	154	358	36	3.4	5.1	4.9	—	4.6	4.3	12.4	6.0	18.4	13.8	34	1.0	5.0
X̄			147	340	45	2.3	4.0	3.7	0.7	2.9	2.3	13.1	7.1	20.2	17.6	36	0.6	3.4

Concentration[a] (pg/m³)

[a] Concentrations based on GC/ECD, except CN which was analyzed by GC/MS.
[b] HCB breakthrough past the second plug may have occurred for samples designated as ≥.
[c] Light PCBs calculated as Aroclor 1242, heavy PCBs as Aroclor 1254.
[d] PCCs calculated as toxaphene.
[e] Not determined.
[f] Front trap only.

concentrations in samples were corrected for this recovery factor. No break-through of HCHs to the back C_8 cartridge was found for any samples, nor were HCHs found on the GFFs which preceded the cartridges.

Concentrations and Possible Sources of Airborne Organochlorines at the Ice Island

OCs are likely to move into the Arctic along the same pathways as other anthropogenic contaminants. Descriptions of meteorological conditions in the Arctic and their relationship to pollutant transport have been given by Barrie[25,26] and Rahn.[27] In the winter the arctic air mass extends southward over the North American and Eurasian continents to about 40° N. From approximately November through May, influx of pollutants from lower latitudes causes the Arctic to be shrouded in a haze blanket several kilometers thick. The haze is a mixture of accumulation mode aerosols (0.1-1.0 μm diameter), containing sulfate, ammonium, soot carbon, trace elements, and other anthropogenic materials, and coarse particles (> 1-10 μm diameter) composed largely of crustal and sea salt elements. Due to a combination of source strengths and meteorological situations, Europe and Asia contribute the major share of pollutant loadings to the Arctic. Eurasian sources of SO_2 available to the Arctic are more than double those in North America, and these emissions traverse snow-covered land areas where deposition processes are minimal. By contrast, aerosols moving out of North America are carried over the North Atlantic and are more subject to precipitation scavenging. A quantitative study of sulfur flux into the Arctic for the period July 1979-June 1980 indicated that 57% came from Eastern and Western Europe, 42% from the Soviet Union, and only 6% from North America.[26]

The above discussion is not intended to rule out North America's potential contribution of OCs to the Arctic. Usage of DDT and HCHs in the eastern half of the Northern Hemisphere is overwhelming compared to North America, but other pesticides such as chlordane and PCCs have been heavily applied in North America until recently and the relative importance of their eastern vs western sources has not been established. Atmospheric removal processes for trace organic substances are different from those of sulfate aerosols. In tropical and temperate latitudes, most OCs exist in the atmosphere as gases with only small fractions adsorbed to airborne particles. Scavenging ratios for OCs are about 10-100 times lower than those of particulate pollutants, even if one considers washout of the vapor-phase fraction.[28] Thus, the en route loss of OCs along the path from North America to the Arctic via the North Atlantic may be less than for sulfur compounds. In general, the longer residence time of OCs may result in their being more uniformly mixed throughout the troposphere than aerosols.

In the summer the arctic air mass is confined to the polar region, and pollutants from lower latitudes are largely excluded. Barrie[25] presented seasonal variations in atmospheric sulfate at three Canadian Arctic stations, which showed peak aerosol levels between January-April and minimum concentrations in the summer. From these trends, our June samples may have caught the tail of the haze season, whereas the August-September samples were taken when particle concentrations were lowest. GFFs from both trips remained white or showed only faint traces of grey.

Aerial concentrations of OCs from the analysis of PUF traps are given in Table 2. GFFs were analyzed for a few samples (2A, 2B, 7A, and 7B), but none showed detectable quantities of OCs. A comparison of our results to other arctic data is presented in Table 3.

Hexachlorobenzene

Concentrations of HCB at the Ice Island were within the range of others reported in the Arctic (Table 3). HCB is predominantly a by-product of the manufacture of chlorinated benzenes and solvents, polyvinyl chloride, and nitroso rubber. It has also been used as a fungicide and is a contaminant in certain herbicides. The total quantity of HCB released to the environment has been estimated at 1000–2000 metric tons/year.[29,30] The input of HCB to the atmosphere has been derived from its accumulation in peat bog cores near the United States-Canada border.[29] A slow rise in HCB loading occurred from 1915–1945, followed by an increase to peak loading in 1967. HCB input decreased during the 1970s and appears to have been constant over the past decade.

HCB levels in the northern troposphere at widely separated locations are similar, with most falling in the range of 100–200 pg/m^3.[24,31,32] At temperatures near 0°C and above, HCB resides in the atmosphere in the gas phase with only a minor fraction attached to particles.[33] Unlike the HCHs, the Henry's law constant of HCB does not indicate dissolution of the vapor into raindrops.[28,31-33] As a result, HCB is not readily removed from the atmosphere by gas or particle scavenging and may have a longer atmospheric residence time compared to other high molecular weight OCs. This is probably the reason why HCB and other OCs show less winter-summer differences than haze aerosols.

Hexachlorocyclohexanes

HCH levels were similar to those reported by others for the Northern Hemisphere (Table 3). α-HCH was the most abundant OC, averaging 546 pg/m^3 for August–September 1986 and 335 pg/m^3 for June 1987. Mean concentrations of γ-HCH were 31 pg/m^3 and 45 pg/m^3 for the same periods.

The extensive usage of HCH as an insecticide throughout the Northern Hemisphere can be appreciated from average statistics taken from the Food and Agricultural Organization Production Yearbooks.[34] Technical HCH (formerly "benzene hexachloride", BHC), a mixture of isomers consisting of approximately 70% α-HCH, 7% β-HCH, 13% γ-HCH, 5% δ-HCH, and 5%

Table 3. Comparison of Organochlorine Levels in Arctic Air, pg/m^3, Range (Mean)

	HCB	α-HCH	γ-HCH	Total HCH	TC	CC	TN	CN	Total Chlordanes	DDTs	Total PCBs[a]
Ice Island This study, 1986–87	≥119–233 (≥171)	283–731 (451)	24–53 (37)	324–767 (488)	0.6–3.4 (1.6)	1.9–5.1 (3.4)	0.8–5.1 (2.5)	0.2–0.8 (0.5)	3.6–13 (8.0)	0.5–8.9 (2.7)	2.1–13 (6.4)
Ice Island[b]											
May–June 1986	(73)	(425)	(70)	(495)	—	—	—	—	(3.6)[c]	(<1)	(<2)
Aug.–Sept. 1986	(63)	(253)	(17)	(270)	—	—	—	—	(1.9)[c]	(<1)	(<1)
Mould Bay[d] June–July 1984	—	—	—	—	0.5–1.7 (1.0)	1.1–1.8 (1.5)	1.0–1.5 (1.3)	<0.2–0.4 (0.2)	2.1–3.7 (4.0)	—	—
Spitzbergen[e]											
Fall 1982	75–169 (123)	757–1416 (1192)	0.1–9 (2)	—	—	0.6–3.2 (1.2)	—	—	—	—	2–47 (15)
Winter–Spring 1983	112–187 (144)	151–734 (485)	21–102 (66)	—	—	1.1–2.3 (1.7)	—	—	—	—	32–145 (62)
Fall 1983	158–227 (186)	407–695 (548)	34–67 (46)	—	—	—	—	—	—	—	—
Winter–Spring 1984	29–389 (151)	121–787 (273)	12–70 (29)	—	—	0.6–5.1 (2.0)	—	—	—	—	4–51 (16)
Summer 1984	20–201 (154)	260–774 (488)	24–100 (47)	—	—	1.7–5.4 (2.8)	—	—	—	—	0.9–43 (21)

Table 3, continued

	HCB	α-HCH	γ-HCH	Total HCH	TC	CC	TN	CN	Total Chlordanes	DDTs	Total PCBs[a]
Bear Island[e]											
Fall 1982	78–200 (120)	450–1550 (774)	0.1–29 (4)	—	—	0.5–1.7 (0.8)	—	—	—	—	6–38 (17)
Winter–Spring 1983	87–201 (124)	110–469 (282)	23–80 (47)	—	—	1.2–3.1 (1.7)	—	—	—	—	9–66 (23)
Fall 1983	88–180 (134)	277–477 (326)	14–32 (22)	—	—	—	—	—	—	—	—
Summer 1984	42–149 (57)	38–305 (155)	5–41 (18)	—	—	0.6–2.1 (1.3)	—	—	—	—	4–14 (8)
Hope Island, 1982–83[f]	100–250	250–1700	<5–75	—	—	1.0–2.0	—	—	—	—	10–70
Jan Mayen, 1982–83[f]	50–200	400–1600	<5–50	—	—	0.5–2.0	—	—	—	—	50–300
Bering Sea, July 1979[g]	—	—	—	540–1700 (980)	—	—	—	—	—	4–15 (9)	—

[a]Pentachlorobiphenyls or Aroclor 1254.
[b]Hargrave et al.[16]
[c]cis-chlordane and trans-chlordane only.
[d]Hoff and Chan.[8]
[e]Pacyna and Oehme.[12]
[f]Oehme and Ottar.[11]
[g]Tanabe and Tatsukawa.[13]

other isomers,[35] is the main product used in Asia. Large tonnages of technical HCH have been applied in India (23,400 metric tons/year in 1980–85)[34] and in the People's Republic of China, where one plant produced 20,000 metric tons annually.[36] Technical HCH has been applied in Mexico from 1980–85 at 230 metric tons/year,[34] but in the United States technical HCH has not been used since 1978.[37] In Europe pure γ-HCH (lindane) is used almost exclusively, with particularly heavy consumption in Italy (1440 metric tons/year during 1980–84).[34]

The predominance of α-HCH in the northern troposphere has been attributed to the large tonnages of technical HCH applied in eastern countries[36] and to conversion of γ-HCH to α-HCH in the environment.[11,12] Support for the regional source hypothesis is provided by a report of extraordinarily high HCH concentrations, approximately 1000 times above arctic levels, in the atmosphere of Delhi, India, in 1980–82.[38] Also, Atlas and Schauffler[32] identified heptachlorocyclohexanes, which are minor constituents of technical HCH products, in air samples from the North Pacific.

However, use of technical HCH cannot entirely account for the fact that α-HCH:γ-HCH ratios in arctic air are often $\geq$ 10 and at times can exceed 500.[12] By comparison, ratios of average α-HCH:γ-HCH concentrations from three locations in Delhi, India, were only 3.3–4.9.[38] In the Norwegian Arctic the α-HCH:γ-HCH ratios were 6–9 in the winter to spring and 9–580 in the summer to fall.[11,12] Ratios of average HCH isomer concentrations found at the Ice Island were consistent with these trends: 7.6 in late spring (June 1987) and 17.6 in summer (August-September 1986)(Table 2).

Oehme and coworkers[10-12] suggested that the proportion of α-HCH is higher in older air masses because of photochemical transformation of γ-HCH to α-HCH.[11,12] The high proportion of α-HCH in arctic air argues for a selective loss of γ-HCH from the atmosphere but does not prove that photochemical isomerization is responsible. More information is needed about the atmospheric stability of the two isomers and other removal processes such as precipitation scavenging and dry deposition.

Chlordanes

The sum of four chlordane components, TC, CC, TN, and CN, averaged 5.8 pg/m^3 in August-September 1986 and 10.7 pg/m^3 in June 1987 (Table 2). The August-September mean is close to the total of the same four isomers measured by Hoff and Chan[8] at Mould Bay (76.5° N, 118° W) in June 1984 (4.0 pg/m^3), but our June mean was nearly three times higher. Average CC concentrations at the Ice Island for the two expeditions (2.8 and 4.0 pg/m^3, Table 2) were also slightly higher than those found at Spitzbergen and Bear Island (0.8–2.8 pg/m^3).[12] Considering that samples were collected and analyzed by different groups at different times, the chlordane levels in Table 3 agree remarkably well.

In 1986, 1818 metric tons of chlordane and 340 metric tons of the related pesticide heptachlor were distributed in the United States for structural termite control, about 2.3–2.5 times less than the quantities sold in 1980.[39] Usage of chlordane in Japan for the same purpose has amounted to 500 metric tons/year.[40] Highest reported concentrations of chlordane in ambient air have been in the southern United States, and levels over the North Atlantic increase with proximity to the United States.[24]

Ratios of average concentrations of chlordane components (R = CC:TN:TC) at the Ice Island were August-September R = 1.0:0.54:0.39 and June R = 1.0:0.93:0.60. This order of abundance was also found by Hoff and Chan,[8] whose June 1984 R = 1.0:0.89:0.56. The expectation is for TC to be the dominant isomer in air since its concentration and volatility are the highest of the chlordanes and nonachlors in the technical chlordane mixture, but TC appears to be depleted in arctic air.

TC:CC ratios at arctic stations and lower latitudes are shown in Figure 2. The predicted TC:CC ratio (1.7) for technical chlordane vapor in equilibrium with its liquid phase was calculated from the composition of technical chlordane[41] and the liquid-phase saturation vapor pressures of TC and CC,[33] assuming Raoult's law. The observed TC:CC ratio (1.9) for air samples collected in an area where chlordane is routinely applied to building foundations (Colum-

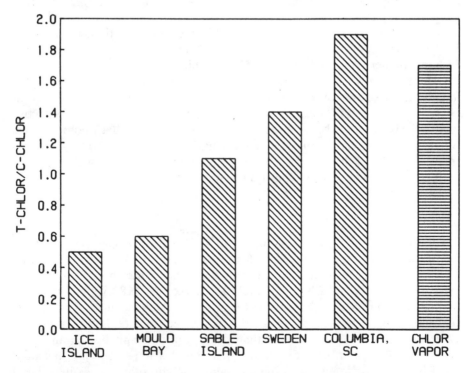

Figure 2.   Ratios of *trans*-chlordane:*cis*-chlordane in air from several locations: Ice Island (this work), Mould Bay,[8] Sable Island,[43] Sweden,[24] and Columbia, SC.[42] For comparison, the calculated ratio in technical chlordane vapor is also shown.

bia, South Carolina)[42] was close to the predicted value. It is interesting to note that the TC:CC ratio was lower in locations removed from source areas: Sweden,[24] Sable Island (44° N, 60° W),[43] Mould Bay,[8] and the Ice Island. Like the HCH isomers, differences in ratios among chlordane components may be an indication of selective transformation/removal processes and, thus, clues to the transit time of OCs in the atmosphere. The isomer composition of foreign technical chlordane mixtures would be helpful in interpreting atmospheric data.

The growing importance of chlordanes as global pollutants was demonstrated in a recent survey of OCs across northern Canada.[3,4] The sum of technical chlordane components was similar to PCB levels in arctic cod (*Boreogadus saida*) ringed seal (*Phoca hispida*), and polar bear (*Ursus maritimus*). Chlordane residues (primarily oxychlordane and nonachlor III) in polar bear adipose tissue increased fourfold between 1969 and 1982-83.[1] In the Bering Sea, chlordane residues were found in surface seawater, zooplankton, and marine animals from several trophic levels.[6]

Polychlorinated Camphenes

Aerial concentrations of PCCs at the Ice Island were similar for both expeditions. Oehme and Stray[9] identified traces of PCCs at Spitzbergen, but they were not quantified. No other information is available about the presence of PCCs in arctic air, although they have been found in ambient air in the continental United States,[18,44-46] over the North Atlantic,[47] and in southern Sweden.[24] Of the OCs in arctic air, PCCs were third in abundance, ranking ahead of total chlordanes, PCBs, and DDTs (Table 2). In a related study we found relatively high PCC concentrations in Arctic Ocean surface water, zooplankton, and benthic amphipods from the Ice Island and in snow from other areas of the Canadian Arctic.[48] Muir et al.[4] reported that PCC levels were the highest of all OCs in arctic cod.

Toxaphene, a PCC product, was heavily used in the southern United States during the mid-1970s, with a peak consumption of nearly 22,000 metric tons in 1976.[34] Between 1980-82, a total of 30,500 metric tons of toxaphene was used.[34] Toxaphene production was halted at the end of 1982, but provisions were made for "existing stocks" to be used through 1986.[49] The extent of such use is unknown. Between 1980-85, toxaphene applications continued in Mexico at an average of 1280 metric tons/year.[34]

Whether the PCCs in the arctic atmosphere arise by volatilization of residues in North American soils or by transport from other source areas is an open question. PCCs are among the major organic contaminants in fish from the Great Lakes.[50-52] They have been found in peat bogs from the Great Lakes region and northeastern Canada, even though the predominant usage of toxaphene was in the cotton-growing regions of the southern United States.[29,53] The input function of toxaphene to the peat bogs closely paralleled the production of the pesticide in the United States.[53] These findings attest to atmospheric deposition of PCCs at least as far north as the United States-Canada boundary.

However, the United States and Mexico may not be the only sources of PCCs. PCCs have been found in birds and aquatic life from the Baltic Sea,[5,54] in rain from southern Sweden,[55] and in human milk from Sweden[56] and Finland,[54] even though PCC products are not used in Scandinavia. Examination of air movements suggested that transport from eastern Europe may have supplied a portion of the PCCs found in air samples from southern Sweden.[24] Use of PCCs during the 1980s at 11–109 metric tons/year has been reported by Czechoslovakia, Poland, and Hungary.[34] PCCs are manufactured in the Soviet Union and used for a variety of agricultural applications,[57] but tonnages have not been disclosed. PCCs were among the major OCs found in rain in the western Mediterranean Sea area,[58] implying a regional source.

DDT and DDE

Relative to the other OCs, the DDT group were only minor constituents of Ice Island air samples (Table 2). Concentrations of p,p'-DDT (DDT), determined by GC/ECD, ranged from 0.5–1.3 pg/m^3 in August-September 1986 and 0.8–4.3 pg/m^3 in June 1987, but these values may have been inflated by coeluting PCC components, as discussed in the GC/MS section. Levels of p,p'-DDE (DDE) were < 0.1–0.2 pg/m^3 in August-September and 0.5–5.0 pg/ m^3 in June. The wide range in concentrations of the DDT compounds may indicate that their atmospheric input to the Arctic is more variable than for the other OCs.

These results are the first indications of DDT compounds in the Canadian Arctic atmosphere. Others have searched without success for DDT and DDE in air at stations in the Norwegian Arctic.[9-12] Tanabe and Tatsukawa[13] reported total DDT residues that averaged 7.6 pg/m^3 over the Bering Sea in 1979. Evidence of atmospheric loading of DDT compounds to the Arctic in the mid-1980s comes from findings of DDT and DDE in snow cores from the Ice Island[16] and the Northwest Territories.[15]

Rapaport et al.[59] found that the major proportion of total residues in peat bog cores from near the United States-Canada border consisted of untransformed p,p'-DDT and suggested that atmospheric transport of DDT from Central America or Mexico might be responsible. Larsson and Okla[60] measured fallout of total DDT compounds (ΣDDT) in Sweden during 1984–85 and found that the flux decreased significantly since an earlier study in 1972–73.[61,62] Moreover, the DDT:DDE ratio decreased from 4.1 in 1972–73 to 2.6 in 1984–85. Thus, the DDT restrictions introduced in Sweden in 1971 have been effective in reducing DDT contamination from within the country. In the 1984–85 study, aerial concentrations and fallout fluxes of ΣDDT were highest in southern Sweden and declined northward, and the proportion of DDE:DDT increased at northern stations. This suggested that long-range transport from southern sources outside of Sweden was continuing to supply DDT to the Swedish environment.

Goldberg[63] predicted that the world production of DDT through 1981 would be 116,000 metric tons/year, virtually the same quantity as in the 1960s. Between 1976 and 1978, the United States exported 35,000 metric tons of DDT, most of which went to tropical countries.[64] In India, 7500 metric tons of DDT were used for public health purposes during 1978-79.[65] Whether these tonnages are being approached today is unknown since many countries do not report their pesticide consumption. DDT usage from average statistics quoted by the FAO[34] falls far short of the above figures, but nevertheless points to India (1660 metric tons/year, 1980-85), Mexico (367 metric tons/year, 1980-85), and Africa (Kenya and Gambia, 270 metric tons in 1983) as recent sources. Levels of total DDT compounds measured in the atmosphere of Delhi, India, in 1980-82 were nearly 7000 times higher than the maximum concentrations found at the Ice Island.[38]

Dieldrin and Endosulfan

These cyclodienes are destroyed by treatment with 18 M sulfuric acid. They were identified by their GC/ECD retention times in samples that had been fractionated by alumina-silicic acid but had not been shaken with acid. Considering that the samples analyzed for dieldrin and endosulfan I received a less vigorous cleanup than for the other OCs, and that we did not attempt GC/MS confirmation, the atmospheric concentrations in Table 2 should probably be considered upper limits.

Hargrave et al.[16] reported that dieldrin in air at the Ice Island in 1986 was < 6 pg/m³, but dieldrin was found in snow, ice cores, and seawater. Gregor and Gummer[15] found dieldrin and endosulfan 1 in snow from several locations in the Northwest Territories and the Ice Island. Dieldrin is a common pollutant in polar bear, ringed seal, and arctic cod from the Canadian Arctic.[3,4]

Production of dieldrin and aldrin (which is converted to dieldrin in the environment) in the United States was stopped in 1974; however, between 1981 and 1985 aldrin was imported at a rate of 450-980 metric tons/year to meet the demands of the termite control industry.[39] Usage of "aldrin and similar insecticides" (which probably includes other cyclodienes) in 1983-85 was reported to be 100 metric tons/year in Mexico, 130-209 metric tons/year in Kenya, and 410-1060 metric tons/year in southeast Asia (Burma and India).[34]

Polychlorinated Biphenyls

PCBs were calculated as Aroclor 1242 (light PCBs) and Aroclor 1254 (heavy PCBs), and also as individual congeners. Total PCBs derived by these two methods agreed well (Table 2). The distribution of PCB congeners in the average Ice Island air sample is shown in Figure 3. Capel et al.[22] developed a multilinear regression technique to interpret PCB congener data in terms of

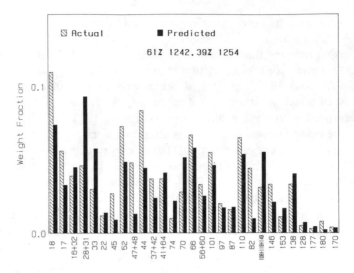

Figure 3. Distribution of PCB congeners (identified by their IUPAC numbers) in the average Ice Island air sample (striped bars). The best fit to the air data was provided by a mixture of 61% Aroclor 1242 and 39% Aroclor 1254 (predicted distribution, black bars).

mixtures of Aroclor fluids. In the case of the PCBs in Ice Island air, the best fit is provided by a mixture of 61% Aroclor 1242 and 39% Aroclor 1254. Levels of heavy PCBs at the Ice Island were considerably below those reported for stations in the Norwegian Arctic (Table 3).

The peak output of PCBs in the United States (38,000 metric tons/year, about one-third of world production) occurred in 1970, at which time about half the PCBs were used as transformer coolants and dielectric fluids in capacitors, and the other half were used for a variety of industrial applications including plasticizers, hydraulic and heat transfer fluids, inks, and adhesives. By 1975, PCB production had dropped to about half the 1970 figure, nearly all of which was used as transformer and capacitor fluids, and the 1978 consumption was only 240 metric tons.[66,67] The cumulative U.S. production of PCBs from 1930–74 has been estimated to be 5.9×10^5 metric tons.[68]

The PCB history in the Organization for Economic Cooperation and Development (OECD) countries has largely paralleled trends in the United States.[66,67] World production has been estimated to be 1.2×10^6 metric tons.[69,70] Of this, approximately 31% has been released to the environment, 4% has been degraded or incinerated, and 65% is still "land-stocked" in use or deposited in dumps and landfills.[69]

Future trends in PCB contamination are uncertain. Bletchly[70] predicted that the disposal of PCBs in old transformers and capacitors would peak in the next decade. In view of this, Tanabe[69] felt that PCB levels in the oceans and remote areas would be unlikely to decline in the near future. PCBs in adipose tissue of polar bears from the Northwest Territories approximately doubled between 1969 and 1983–84.[3] Larsson and Okla[60] found that, unlike DDT, PCB fallout in Sweden did not decline between 1972–73 and 1984–85, despite restrictions against open use since 1971. They suggested that inputs of PCBs to the atmosphere are continuing from inappropriate storage or incomplete burning of PCB waste. On the other hand, Addison et al.[7] found that PCBs in the blubber of arctic ringed seal declined by about 60% between 1972 and 1981 and attributed the decrease to controls on PCB usage. Atmospheric loadings of PCBs to peat bog cores along the United States-Canada border have paralleled PCB production. Peak inputs occurred during the late 1960s and showed a downward trend in the 1970s.[29]

GC/MS Analysis of Organochlorines in Air

Eleven air samples were analyzed for chlordanes and PCCs by GC/MS in the negative ion mode. TC, CC, TN, and CN were clearly identified on front PUF plugs, with only traces on back plugs (Figure 4) and none detectable in blanks. A reconstructed ion chromatogram of PCCs containing 6–9 chlorine atoms (Figure 5) shows that the most volatile constituents of the technical mixture are atmospherically transported to the Arctic. This preference is shown in more detail in chromatograms of the PCC homologs (Figures 6 and 7). No traces of PCCs were found on back PUF sample plugs or blanks.

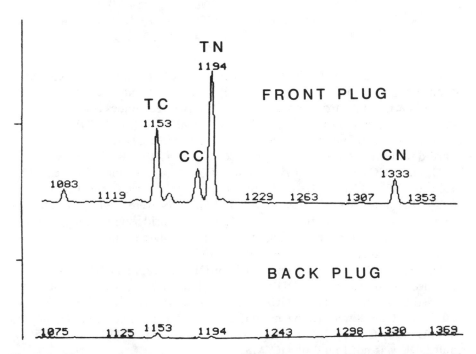

Figure 4. Reconstructed ion chromatograms of chlordane components in air sample 6A. CC appears lower than TC, even though atmospheric levels of CC were higher, because of the different responses of TC and CC at the ions selected.

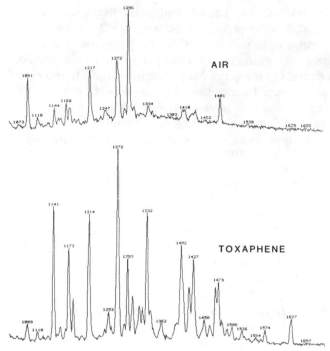

Figure 5. Reconstructed ion chromatogram of PCCs in air sample 6B compared to a toxaphene standard.

Chlordanes and PCCs in the 11 samples were quantified by GC/ECD
against external standards and by GC/MS using either external standards or a
mirex internal standard. A comparison of the two techniques is given in Table
4. Ratios of quantities found by GC/MS and GC/ECD (MS:ECD) were calculated for each compound and sample. Average MS:ECD ratios ($\pm$ one standard deviation) were TC = 0.55 $\pm$ 0.26, CC = 1.10 $\pm$ 0.53, TN = 0.74 $\pm$
0.23, and PCCs = 1.09 $\pm$ 0.57. These ratios are significantly different from
1.00 for TC and TN ($\alpha <$ 0.01), but not different for CC and PCCs ($\alpha >$
0.25). Perhaps GC/ECD results for TC and TN were slightly inflated by
interferences underlying these peaks. However, considering the analytical difficulties at these ultratrace levels, we feel that the agreement between the two
techniques was satisfactory, even for TC and TN.

Samples 5, 7B, and 8 were examined for DDE by capillary GC/MS/MID in
the electron impact mode. DDE was identified in 7B and 8, both of which
showed clear DDE peaks in GC/ECD chromatograms. GC/MS estimates of
DDE in 7B and 8 were 1.5 pg/m³ and 5.2 pg/m³, compared to 5.0 pg/m³ and
4.6 pg/m³ by GC/ECD. Sample 5 showed only 0.5 pg/m³ DDE by GC/ECD,
and DDE was not found by GC/MS.

Samples 5, 7A, 7B, and 8 were checked for DDT. DDT could not be found
by GC/MS in 5 and 7B, even though GC/ECD chromatograms showed a peak

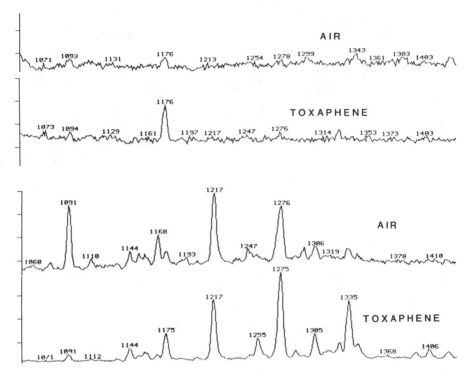

Figure 6. PCC homologs in air sample 6B compared to a toxaphene standard. Top: Cl_6 components, $(M\text{-}Cl)^- = 309 + 311$; bottom: Cl_7 components, $(M\text{-}Cl)^- = 343 + 345$.

at the DDT retention time. Sample 7A contained too much background inter-ference to draw conclusions. The DDT content of sample 8 was 2.4 pg/m³ by GC/MS and 4.3 pg/m³ by GC/ECD. Sample 8 had the second highest air volume of the June set and yielded the most DDT for analysis. Even so, the DDT peak was a small one, and DDT in the other samples may have been below the detection limit by GC/MS. It is also possible that the GC/ECD results for DDT were inflated by interferences, with PCC components being a possibility. Since DDT was confirmed by GC/MS in only one sample, the levels in Table 3 should be considered upper limits.

Hexachlorocyclohexanes in Snow and Seawater

Because of the small water volumes extracted, HCHs were the only OCs detected in melted snow and seawater (Table 5). Concentrations of α-HCH and γ-HCH in freshly fallen snow were six and two times higher than average levels in old snow. The α-HCH:γ-HCH ratios were 2.8 and 0.86 in new and old snow. It is not clear why the fresh snow had higher HCH concentrations and a different proportion of isomers compared to old snow; however, con-centrations of the HCHs and their isomer ratio in the new snow sample were

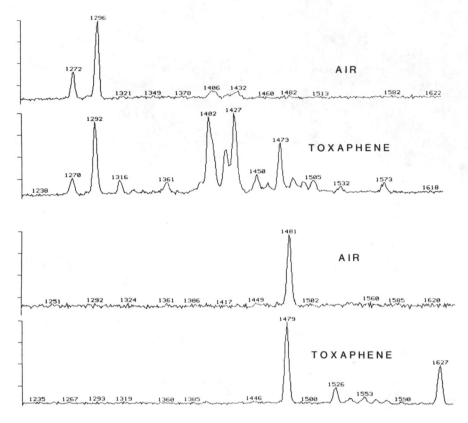

Figure 7.   PCC homologs in air sample 6B compared to a toxaphene standard. Top: Cl_8 components, $(M-Cl)^- = 379 + 381$; bottom: Cl_9 components, $(M-Cl)^- = 413 + 415$.

similar to those found for surface snow (unknown whether old or new) from the Ice Island in 1986 (Table 5). Gregor and Gummer[15] found a large variation in total HCH concentrations, 0.6–11 ng/L, and α-HCH:γ-HCH ratios, 1.1–3.7, in snow cores collected from the Northwest Territories in 1986.

Average concentrations of α-HCH and γ-HCH in surface seawater were 7.1 and 0.81 ng/L. Hargrave et al.[16] found HCH levels about 25–40% lower in 1986 (Table 5). Mean air and water HCH concentrations from the June 1987 trip were used to estimate the state of air-sea equilibrium relative to the air-water partition coefficient (Henry's law constant, H). Mackay and Shiu[71] reviewed the physical properties of γ-HCH and recommended $H_\gamma = 3.1 \times 10^{-6}$ atm m^3/mol for fresh water, which is similar to a recent determination by Fendinger and Glotfelty[72] of 2.0×10^{-6} atm m^3/mol at 23°C. Lindane (γ-HCH) is 23% less soluble in 0.5 M sodium chloride than in distilled water,[73] so the appropriate seawater $H_\gamma = 2.5$–3.8×10^{-6} atm m^3/mol. The Henry's law con-

Table 4.   Comparison of GC/ECD and GC/MS Analysis of Chlordanes and PCC

Sample	Method	Concentration (pg/m³)			
		TC	CC	TN	PCC
1	ECD	1.7	3.2	2.0	25
	NIMS	0.81	2.3	1.1	44
2A	ECD	0.69	2.0	1.2	53
	NIMS	0.76	4.1	1.5	39
2B	ECD	0.97	3.0	1.6	78
	NIMS	0.73	4.1	1.4	45
3A	ECD	1.3	3.7	2.0	47
	NIMS	0.54	7.5	1.8	61
3B	ECD	1.1	3.6	1.9	44
	NIMS	0.29	4.4	1.0	35
4A	ECD	0.57	1.9	0.86	30
	NIMS	0.18	0.93	0.48	14
4B	ECD	1.1	2.4	0.85	30
	NIMS	0.25	1.4	0.45	15
6A	ECD	2.5	4.1	3.2	41
	NIMS	1.1	3.2	2.1	36
6B	ECD	2.4	4.9	3.9	45
	NIMS	1.8	5.3	3.6	99
7A	ECD	2.1	3.8	2.8	38
	NIMS	1.3	3.0	2.3	49
7B	ECD	1.9	3.3	5.1	35
	NIMS	1.3	3.1	2.6	54

stant for α-HCH (H_γ), determined in seawater at 23°C, is 1.1 x 10^{-5} atm m³/mol.[74] We selected H_γ = 1.1×10^{-5} and H_γ = 3.8 x 10^{-6} atm m³/mol.

Henry's law constants of the HCHs have not been determined at different temperatures, so H_α and H_γ were adjusted to the Arctic Ocean surface temperature (-2°C) by assuming the temperature dependence of H for PCB mixtures. Burkhard et al.[75] estimated that H for Aroclors 1221 through 1260 decreased by factors of 8–14 between 25°C and 0°C. We assumed that H of the HCH isomers decreased by a factor of 11 between 23°C and -2°C. The resulting constants at -2°C used to calculate HCH concentrations in arctic surface water in equilibrium with HCH in the atmosphere were H_α = 1.0×10^{-6} and H_γ = 3.5×10^{-7} atm m³/mol.

Observed concentrations of HCHs in Arctic Ocean surface water are compared with equilibrium values in Table 6. α-HCH appears to be close to equilibrium, but γ-HCH is undersaturated. At this point we do not know whether the difference between the two isomers is real or an artifact of uncertainties in the H values. A likely source of error in these calculations is the extrapolation of H_α and H_γ to -2°C, and a better knowledge of these constants

Table 5.   Seawater and Snowmelt Concentrations of HCHs at the Ice Island

| Collection Date | Sample | Depth(m) | Concentration (ng/L) | |
			α-HCH	γ-HCH
Seawater				
This work				
June 9, 1987	1	10	6.4	0.83
June 9, 1987	2A	10	8.9	1.00
	2B	10	6.6	0.72
June 10, 1987	3	10	7.0	0.61
June 11, 1987	4A	10	8.1	0.65
	4B	10	6.0	0.79
June 12, 1987	5	1	7.0	1.10
	$\overline{X}$		7.1 ± 1.0	0.81 ± 0.18
Hargrave et al.[16]				
May–June 1986		10–60	4.4 ± 0.6	0.57 ± 0.02
August–September 1986		10–60	4.5 ± 0.5	0.65 ± 0.06
Snow				
This work				
June 7, 1987	1	old[a]	0.34	0.35
June 9, 1987	2	old	0.29	0.38
June 13, 1987	3	new	1.9	0.67
Hargrave et al.[16]				
May–June 1986			1.5 ± 0.2	0.64 ± 0.09
August–September 1986			1.2 ± 0.3	0.21 ± 0.05
Gregor and Gummer[15]				
May 1986			2.8	1.1

[a]See text.

Table 6.   Measured and Air-Water Equilibrium Concentrations of HCHs in Beaufort Seawater

| | Concentration (ng/L) | |
	Measured	Equilibrium[a]
α-HCH	7.1	7.6
γ-HCH	0.81	2.9

[a]HCH in seawater in equilibrium with average air concentrations in June 1987 (Table 2), calculated using Henry's law constants at −2°C (see text).

as a function of temperature is needed. Also, the stability of the two HCHs in seawater is not known.

CONCLUSIONS

Due to analytical constraints, only a few of the vast number of synthetic organic chemicals have been sought in this study. Our results and those of other investigators cited herein show that contamination of the Arctic is not restricted to the well-publicized DDTs and PCBs, but includes chlorobenzenes and most of the other OC pesticides as well. The presence of OCs in air and snow has established an atmospheric link to the Arctic, and it is likely that other halogenated and nonhalogenated compounds having similar physical properties will follow similar transport and deposition pathways.

A hindrance to interpreting seasonal and temporal variations in OCs in arctic air is the lack of good source information for these compounds. From FAO statistics we have only a vague picture of heavy OC pesticide usage in tropical countries. Needed are tonnages of OCs emitted into the atmosphere (or at least applied) by latitude and longitude. Such gridded data for SO_2 emissions has enabled a quantitative model of sulfur transport to the Arctic to be constructed.[26]

In order to estimate loadings of trace organic substances to the Arctic, atmospheric deposition and resuspension processes need to be better understood. Such processes could include

- adsorption of organic vapors to haze aerosols, followed by precipitation scavenging and dry deposition
- air-sea gas exchange
- revolatilization of deposited material during the melting season

Finally, it is tempting to think of the Arctic as a sink or a "cold trap" for organic chemicals. Although the deposition of trace organic substances is enhanced by cold temperatures, it is unlikely that all such compounds entering the arctic air mass remain there. Barrie[26] pointed out that because of low precipitation, poor vertical mixing, and an aerodynamically smooth surface, the residence time of particulate pollutants in arctic air is relatively long. Fluxes of sulfur during the winter months occur both into and *out of* the Arctic.[25,26] Because of their longer atmospheric residence times, OCs may be even less efficiently deposited than aerosols. Thus, pesticides from the other side of the world may reach the subarctic latitudes of North America by transit "over the top."

ACKNOWLEDGMENTS

This work was supported by the NATO Scientific Affairs Division (Grant No. NATO 04–0667–86), the University of South Carolina Venture Fund, and the Canadian Polar Continental Shelf Project. Thanks to the Department of Fisheries and Oceans and to Arctic Laboratories for extending the invitation for us to participate in this project and for their help with field sampling, and to the PCSP staff on the Ice Island. Contribution 808 of the Belle W. Baruch Institute.

REFERENCES

1. Addison, R. F., and T. G. Smith. *Oikos* 25:335-37 (1974).
2. Bowes, G. W., and C. J. Jonkel. *J. Fish. Res. Bd. Can.* 32:2111-23 (1975).
3. Norstrom, R. J., M. Simon, D. C. G. Muir, and R. E. Schweinsburg. *Environ. Sci. Technol.* 22:1063-71 (1988).
4. Muir, D. C. G., R. J. Norstrom, and M. Simon. *Environ. Sci. Technol.* 22:1071-79 (1988).
5. Andersson, Ö., C.-E. Linder, M. Olsson, L. Reutergårdh, U.-B. Uvemo, and U. Wideqvist. *Arch. Environ. Contam. Toxicol.* 17:755-65 (1988).
6. Kawano, M., T. Inoue, T. Wada, H. Hidaka, and R. Tatsukawa. *Environ. Sci. Technol.* 22:792-97 (1988).
7. Addison, R. F., M. E. Zinck, and T. G. Smith. *Environ. Sci. Technol.* 20:253-55 (1986).
8. Hoff, R. F., and K.-W Chan. *Chemosphere* 15:449-52 (1986).
9. Oehme, M., and H. Stray. *Fres. Z. Anal. Chem.* 311:665-73 (1982).
10. Oehme, M., and S. Manø. *Fres. Z. Anal. Chem.* 319:141-46 (1984).
11. Oehme, M., and B. Ottar. *Geophys. Res. Lett.* 11:1133-36 (1984).
12. Pacyna, J. M., and M. Oehme. *Atmos. Environ.* 22:243-57 (1988).
13. Tanabe, S., and R. Tatsukawa. *J. Oceanogr. Soc. Japan* 36:217-26 (1980).
14. McNeely, R., and W. D. Gummer. *Arctic* 37:210-23 (1984).
15. Gregor, D. J., and W. D. Gummer. *Environ. Sci. Technol.* 23:561-65 (1989).
16. Hargrave, B. T., W. P. Vass, P. E. Erickson, and B. R. Fowler. *Tellus* 40B:480-93 (1988).
17. Billings, W. N., and T. F. Bidleman. *Environ. Sci. Technol.* 14:679-83 (1980).
18. Billings, W. N., and T. F. Bidleman. *Atmos. Environ.* 17:383-91 (1983).
19. Simon, C. G., and T. F. Bidleman. *Anal. Chem.* 51:1110-13 (1979).
20. Hinckley, D. H., and T. F. Bidleman. *Environ. Sci. Technol.* 23:995-1000 (1989).
21. Keller, C. D., and T. F. Bidleman. *Atmos. Environ.* 18:837-45 (1984).
22. Capel, P. D., R. A. Rapaport, S. J. Eisenreich, and B. B. Looney. *Chemosphere* 14:439-50 (1985).
23. Jansson, B., and U. Wideqvist. *Internat. J. Environ. Anal. Chem.* 13:309-21 (1983).
24. Bidleman, T. F., U. Wideqvist, B. Jansson, and R. Söderlund. *Atmos. Environ.* 21:641-54 (1987).
25. Barrie, L. A. *Atmos. Environ.* 20:643-63 (1986).
26. Barrie, L. A. "Arctic Air Pollution: A Case Study of Continent to Ocean to Continent Transport," In: Knap, A. H. (ed.) *The Long-Range Atmospheric Transport of Natural and Contaminant Substances*, NATO ASI Series #297, Kluwer Academic Publishers, Dordrecht, The Netherlands, pp. 137-148 (1990).
27. Rahn, K. A. *Atmos. Environ.* 15:1447-55 (1981).

28. Bidleman, T. F. *Environ. Sci. Technol.* 22:361–67 (1988), correction in *Environ. Sci. Technol.* 22:726–27 (1988).
29. Rapaport, R. A., and S. J. Eisenreich. *Environ. Sci. Technol.* 22:931–41 (1988).
30. Courtney, K. D. *Environ. Res.* 20:225–66 (1979).
31. Atlas, E. L., and C. S. Giam. In *The Role of Air-Sea Exchange in Geochemical Cycling,* P. Buat-Menard, Ed. (Dordrecht, Netherlands: Reidel, 1986), pp. 295–330.
32. Chapter 12.
33. Bidleman, T. F., W. N. Billings, and W. T. Foreman. *Environ. Sci. Technol.* 20:1038–43 (1986).
34. *Production Yearbook, Vols. 31–40* (Rome: Food and Agricultural Organization of the United Nations, 1978–87).
35. Tatsukawa, R., T. Wakimoto, and T. Ogawa. "BHC Residues in the Environment," in *Environmental Toxicology of Pesticides,* F. Matsumura, C. M. Bousch, and T. Misato, Eds. (New York, NY: Academic Press, 1972), pp. 229–38.
36. Tanabe, S., R. Tatsukawa, M. Kawano, and H. Hidaka. *J. Oceanogr. Soc. Japan* 38:137–48 (1982).
37. "Ambient Water Quality Criteria for Hexachlorocyclohexane," Office of Water Regulations and Standards, U.S. EPA-440/5-80-054 (October 1980), p. A-1.
38. Kaushik, C. P., M. K. K. Pillai, A. Raman, H. C. Agarwal. *Water Air Soil Poll.* 32:63–76 (1987).
39. "Chlordane, Heptachlor, Aldrin, and Dieldrin: Technical Support Document", U.S. Environmental Protection Agency, Office of Pesticide Programs, Washington, DC (July 1987), pp. III-2 to III-3.
40. Hirose, C. *J. Pest. Sci.* 2:187–200 (1977).
41. Sovocool, G. W., R. G. Lewis, R. L. Harless, N. K. Wilson, and R. D. Zehr. *Anal. Chem.* 49:734–40 (1977).
42. Foreman, W. T., and T. F. Bidleman. *Environ. Sci. Technol.* 21:869–75 (1987).
43. Bidleman, T. F., and R. F. Addison. University of South Carolina and Bedford Institute of Oceanography, unpublished data (1988).
44. Rice, C. P., P. J. Samson, and G. E. Noguchi. *Environ. Sci. Technol.* 20:1109–16 (1986).
45. Atlas, E. L., and C. S. Giam. *Water Air Soil Poll.* 38:19–36 (1988).
46. Chapter 14.
47. Bidleman, T. F., E. J. Christensen, W. N. Billings, and R. Leonard. *J. Marine Res.* 39:443–64 (1981).
48. Bidleman, T. F., G. W. Patton, M. D. Walla, B. T. Hargrave, W. P. Vass, P. Erickson, B. Fowler, V. Scott, and D. J. Gregor. *Arctic* 42:307–13 (1989).
49. *Federal Register* 47(229):53784–93 (1982).
50. Rice, C. P., and M. S. Evans. In *Toxic Contaminants in the Great Lakes,* J. O. Nraigu and M. S. Simmons, Eds. (Ann Arbor, MI: Ann Arbor Science, 1984), pp. 163–94.
51. Swackhamer, D. L., M. J. Charles, and R. A. Hites. *Anal. Chem.* 59:913–17 (1987).
52. Swackhamer, D. L., and R. A. Hites. *Environ. Sci. Technol.* 22:543–48 (1988).
53. Rapaport, R. A., and S. J. Eisenreich. *Atmos. Environ.* 20:2367–79 (1986).
54. Pyysalo, H., and K. Antervo. *Chemosphere* 14:1723–28 (1985).
55. Sundström, G. "Toxaphene in the Swedish Environment: Transport Via Aerial

Fallout," *Proc. 17th Nordic Symp. Water Res.*, Vol. 1, Nordforsk Environmental Protection Series Publication (1981), pp. 331–36.

56. Vaz, R., and G. Blomqvist. *Chemosphere* 14:223–31 (1985).
57. Izmerov, N. F. "Toxaphene," USSR Commission for the United Nations Environment Programme, International Register of Potentially Toxic Substances (IRPTC), Scientific Reviews of Soviet Literature on Toxicity and Hazards of Chemicals Series, IRTPC-32, Center of International Projects, GKNT, Moscow (1983).
58. Villeneuve, J.-P., and C. Cattini. *Chemosphere* 15:115-20 (1986).
59. Rapaport, R. A., N. R. Urban, P. D. Capel, J. E. Baker, B. B. Looney, S. J. Eisenreich, and E. Gorham. *Chemosphere* 14:1167–73 (1985).
60. Larsson, P., and L. Okla. *Atmos. Environ.* 23:1699–1711 (1989).
61. Södergren, A. *Nature* 236:395–97 (1972).
62. Södergren, A. *Environ. Qual. Safety*, Supp. III, pp.803–10 (1975).
63. Goldberg, E. D. *Proc. Royal Soc. London* 189B:277–89 (1975).
64. Fowler, D. L., and J. N. Mahan. *The Pesticide Review,* U.S. Department of Agriculture, Washington, DC (1977–80).
65. Jalees, K., and R. Vemuri. *Int. J. Environ. Stud.* 15:49–54 (1980).
66. Addison, R. F. *Environ. Sci. Technol.* 17:486A-94A (1983).
67. Addison, R. F. *Canadian Chemical News* 38(2):15–17 (1986).
68. *Polychlorinated Biphenyls* (Washington, DC: National Academy of Sciences, 1979).
69. Tanabe, S. *Environ. Pollut.* 50:5–28 (1988).
70. Bletchly, J. D. In *Proc. of PCB Seminar*, M. C. Barros, H. Koemann, and R. Visser, Eds., Ministry of Housing, Physical Planning, and Environment, The Netherlands (1984), pp. 343–72.
71. Mackay, D., and W.-Y. Shiu. *J. Phys. Chem. Ref. Data* 10:1175–99 (1981).
72. Fendinger, N. J., and D. E. Glotfelty. *Environ. Sci. Technol.* 22:1289–93 (1988).
73. Masterson, M. L., and T. P. Lee. *Environ. Sci. Technol.* 6:919–21 (1972).
74. Atlas, E. L., R. Foster, and C. S. Giam. *Environ. Sci. Technol.* 16:283–86 (1982).
75. Burkhard, L. P., D. E. Armstrong, and A. W. Andren. *Environ. Sci. Technol.* 19:590–96 (1985).

## CHAPTER 24

Deposition and Accumulation of Selected Agricultural Pesticides in Canadian Arctic Snow

Dennis J. Gregor

INTRODUCTION

The source of organochlorine pesticides that were recently found in arctic marine and freshwater biota at unexpectedly high concentrations[1,2] is clearly not local. Conceivably, both the atmosphere and background marine contamination are important links in the system. The ultimate significance of these contaminants in this environment, however, is the impact upon the indigenous natives who use natural foods for a substantial portion of their diet and who may therefore be ingesting greater quantities of these compounds than residents of more southerly latitudes where the substances are produced and used.[3,4]

The history of Canadian observations of arctic aerosol chemistry has been summarized by Whelpdale and Barrie[5] and Hoff and Barrie.[6] Only for a short period in the summer of 1984 was this air sampling network used for the collection of aerosols for the purpose of measuring chlorinated compounds, specifically, chlordane and its oxidation/metabolization products. Chlordane was only observed in the gas phase at the pg/m^3 level.

Similarly, there is only limited published evidence of the deposition and presence of chlorinated pesticides in the Canadian Arctic aquatic environment. The earliest known work on organochlorine compounds in snow samples from the Canadian Arctic was in 1970 from Mount Logan in the Yukon Territory, far to the west and south of the sites sampled here.[7] Nineteen snow samples were taken at depths of 1–15 m; DDT has not quantified in any of these samples at a detection limit of 5 ng/L. McNeely and Gummer[8] reported measurable concentrations of polynuclear aromatic hydrocarbons, lindane (γ-HCH) and its isomer α-HCH, dieldrin, and DDT from a number of snowpack samples from east central Ellesmere Island. The Finnish Huure expedition to the North Pole from the northern tip of Ellesmere Island in 1984 collected 1 L water-equivalent snow samples for analyses of a broad range of chlorinated pesticides.[9] None of the target pesticides were detected above the estimated detection limit of 0.5 ng/L.

Studies at the Canadian Ice Island (see Figure 1), conducted concurrently with the present study, investigated the presence of organochlorine pesticides in the arctic food web, including the snow.[10] However, an extensive spatial and temporal survey of the presence of pesticides and other trace organic substances in the Canadian Arctic snowpack was essential as a first step in understanding the importance of atmospheric transport and deposition to the contaminant burden of Arctic biota. Selected results of a study that commenced in 1986 for this purpose are reported here.

METHODS

Annual snowpack accumulations were sampled at 12 sites from 62°28' N to 80°59' N and from 73°30' W to 118°48' W) (Figure 1) beginning in late April and continuing through to mid-May 1986, thereby providing a time-integrated sample essentially at the termination of the snow accumulation period. This also coincides with the termination of the period identified as most critical for atmospheric transport of anthropogenic pollutants to the Arctic.[5,11] To minimize the effect of contamination on the quantification of ultratrace levels of man-made synthetic organic compounds during collection, melting, and extraction and to ensure sufficient quantity of material for compound identification using gas chromatography (GC), large quantities of snow were collected. The methods used are described in Gregor and Gummer.[12] In general, field methods consisted of sampling up to 200 L of snow in 2 mil thick Teflon bags placed inside cardboard boxes. The bags were sealed airtight in the field using a clip sealer.

The specific sampling sites for the spatial component of the survey were selected to represent a seasonal accumulation with a snow depth of about 30 cm or more. This did not ensure that each sample was representative of the snow accumulation season, but since the snow was windblown in most areas (resulting in accumulation only in sheltered locations), this was a practicable approach.

Site number 9 (Figure 1) on the Agassiz Ice Cap on north central Ellesmere Island was selected as the site for the temporal survey. In 1986 this site was sampled to a depth of 2.2 m, which encompassed five annual snow layers. When this site was revisited in May 1987, a 7-m pit was excavated, and samples from each annual layer, including those layers sampled the previous year, were collected, back to and including the winter of 1970/71. The same procedures were used during both years at this site. In both 1986 and 1987, replicate samples were collected from the 1985/86 winter snow layer. Indentification of annual layers was determined by means of the specific conductance pattern, measured essentially continuously down the pit wall, together with the presence of summer snowmelt layers.

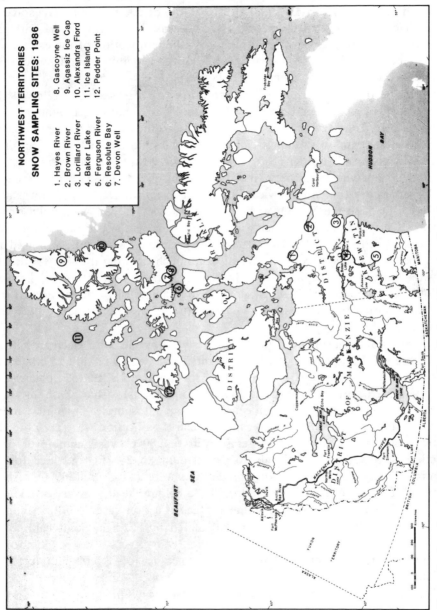

NORTHWEST TERRITORIES
SNOW SAMPLING SITES: 1986

1. Hayes River
2. Brown River
3. Lorillard River
4. Baker Lake
5. Ferguson River
6. Resolute Bay
7. Devon Well
8. Gascoyne Well
9. Agassiz Ice Cap
10. Alexandra Fiord
11. Ice Island
12. Pedder Point

Figure 1.   Location map of 1986 snow survey sites, Northwest Territories, Canada.

The snow was melted at room temperature within the sealed bags. Approximately 40 L of meltwater was passed through a continuous flow, large volume liquid phase extractor (LVX)[13,14] at a flow rate of 500 mL/min. The solvent, pesticide grade methylene chloride (DCM) was thoroughly mixed with the water by an external centrifugal pump. At the termination of the extraction, the separated solvent and a small quantity of the remaining sample water were removed and stored at temperatures below 0°C.

Final extraction, cleanup, and separation of these samples followed the procedures developed in Environment Canada's National Water Quality Laboratory.[15] In summary, these procedures included the transfer of the sample to clean glass separatory funnels (1 L) for final separation of the solvent. The DCM was passed through a 5-cm layer of muffled Na_2SO_4 in an Allihn suction funnel, all of which was prewashed with 50 mL DCM. The H_2O remaining in the separatory funnel was twice rinsed with 25 mL of DCM, which was also passed through the Na_2SO_4 without suction. The Na_2SO_4 was finally rinsed with 20 mL of isooctane, which was drained using gentle suction. This mixture was concentrated to 1 mL by rotary evaporation to 3–5 mL, followed by evaporation under a nitrogen stream. Column chromatography on 3% deactivated silica gel was used to fractionate the extract with 40 mL of hexane, followed by 60 mL of DCM–hexane (1:1 by volume). Isooctane (10 mL) was added to each fraction, which was again concentrated to a final volume of 1 mL.

The organochlorine pesticides were analyzed by GC with electron capture detectors (ECD) on dual capillary columns. The two 30 m × 0.25 mm fused silica columns were SPB1 (polydimethyl siloxane) and SPB5 (5% phenyl and 95% methyl polysiloxane) (listed as Supelco products) with a phase thickness of 0.25 μm. The sample was injected using an autosampler in splitless mode with a time delay of 0.5 minute, after which the injection was split onto the two columns. An initial oven temperature of 80°C was maintained for 2 minutes, increased at 4°C/min to a maximum of 280°C, and held for 16 minutes. Injector and detector temperatures were 250°C and 300°C, respectively. The carrier gas was hydrogen with a column head pressure of 90 kPa, and with argon–methane as the makeup gas at a flow rate of 30 mL/min.

Compound concentrations were calculated by a single point calibration using an external standard mix of the target compounds. Compounds had to be confirmed on both columns with a tolerance of ± 0.02 minute to be quantified.

Extraction efficiencies were assessed in 1986 by spiking known but variable quantities of endrin into the sample bags prior to extraction. In 1987, 100 mL of a spiked methanol solution was added during the extraction procedure at 2 mL/min in addition to the endrin spike. Each 100 mL of this solution contained 20 ng of 1,3,5-tribromobenzene and 1,2,4,5-tetrabromobenzene, 110 ng of δ-HCH and 100 ng of perthane.

RESULTS

Extraction Recovery Efficiencies

The recovery efficiencies of the LVX technique, based on endrin recoveries for 19 samples collected in 1986, averaged 98%. Five of the samples were not spiked, but a number of these samples contained low levels of endrin. The quantity of endrin present in the unspiked samples could amount to between 20 to 40% of the 100 ng spike, suggesting real recoveries of the order of 60 to 80%.

The 1987 samples from the Agassiz Ice Cap showed generally good recoveries for all five spikes considered. Based on 20 samples, average recoveries were 99% for δ-HCH, 61% for endrin, 79% for perthane, 63% for 1,3,5-tribromobenzene, and 78% for 1,2,4,5-tetrabromobenzene.

1986 Spatial Survey

The concentration of targeted hexachlorocyclohexanes (HCHs) and cyclodiene group pesticides frequently detected in the surface snow samples collected in 1986 are listed in Tables 1 and 2, respectively. The most prevalent compounds, usually comprising more than 75% of the total organochlorines measured, are the HCHs, specifically, the pesticide lindane and the isomer α-HCH. These are followed in abundance by pesticides of the cyclodiene group with dieldrin > α-endosulfan > cis- and trans-chlordane > heptachlor epoxide. Heptachlor, a more insecticidal constituent of commercial chlordane and the parent compound of heptachlor epoxide, was detected infrequently.

The results of replicate blank tests for the Teflon bags are also shown in Tables 1 and 2. Only the HCHs were detectable in bag blanks. These were well below the reported sample concentrations and therefore have been considered to be negligible. These pesticides were not detectable in either the water blanks used in the preparation of bag blanks or solvent blanks.

Also reported in Tables 1 and 2 are the detection limits as determined by the laboratory for routine 1-L samples.[15] Since these samples are concentrated from a larger volume, the theoretical detection limit (TDL) for each compound can be calculated by dividing the DL by the sample volume (Table 1). The TDL has been calculated here using 40 L as a representative sample volume. Note that a number of the samples are reported as an average of several splits or replicates.

1987 Temporal Survey

Pesticide residue concentrations of HCHs, heptachlor epoxide, dieldrin, and cis-and trans-chlordane in annual snow layers for each of the five years sampled in 1986 and for each of the 17 years sampled in 1987, including repeat sampling of the five years collected in 1986, are shown in Figures 2 and 3.

Table 1.   Annual Snowpack Concentration and Estimated Winter Flux from the Atmosphere to the Snow of Hexachlorocyclohexanes in the Canadian Arctic, 1986.

Station Name	Sample Volume (L)	α-HCH		Lindane	
		Concentration (pg/L)	Winter Flux (ng/m²)	Concentration (pg/L)	Winter Flux (ng/m²)
Hayes R.[a]	41.0	6800	1877	2380	656
Brown R.	41.5	8120	2962	3050	1113
Ferguson R.[a]	38.3	1715	642	725	271
Lorillard R.[a]	41.5	6420	1972	1740	535
Baker Lake[a]	35.8	875	311	360	128
Resolute[a]	35.2	8070	2524	2425	759
Agassiz L-1[b]	35.6	6660	640	4446	427
Alexandra	37.4	430	65	220	33
Ice Island	21.0	2830	424	1130	170
Gascoyne	40.5	1670	522	850	266
Devon	40.8	810	253	500	156
Pedder	39.2	1320	682	1190	615
Bag blanks[c]	32.0	60		20	
DL[d]		1300		400	
TDL[e]		33		10	

[a]Average volume, concentration, and flux based on a split sample (i.e., two analyses).
[b]Average volume, concentration, and flux based on three replicate samples, of which two were split (i.e., 5 analyses).
[c]Average volume and concentration based on three replicate samples.
[d]DL: Detection limit, defined as the lowest concentration that can be determined in a real sample matrix by means of spiking the analyte into solvent for extraction and being carried through the entire procedure and which is statistically distinct from the background response of a sample carried through the same procedure.
[e]TDL: the theoretical detection limit, determined by dividing the DL by a representative sample volume (here taken to be 40 L).

Figure 3 does not include α-endosulfan because this pesticide is only present at high concentrations in the surface snow and is frequently nondetectable at depth.

For the 1985/86 snow layer, five samples were analyzed in 1986, and three replicates were collected from this layer in 1987. In general there is good agreement among the replicates: coefficients of variations are between 10 and 20% for the five samples of 1986 for the five compounds considered here. The coefficients of variation for the three replicates from the same layer collected in 1987 increased to between 15 and 35% for the same compounds, but this is due to the decrease in absolute concentrations in this layer for all compounds between the 1986 and 1987 sampling seasons. This phenomenon, which also applies to α-endosulfan, will be discussed below. For the other three annual layers with repeat samples, the 1986 values tend to be somewhat higher than the 1987 measurements. This could largely be a statistical phenomenon due to the absence of replicates in these layers or could be the result of minor analytical differences. In any event, the differences shown in Figures 2 and 3 are not

Table 2.   Annual Snowpack Concentration and Estimated Winter Flux from the Atmosphere to the Snow for the Major Cyclodiene Pesticides in the Canadian Arctic, 1986.

Station Name	Heptachlor Epoxide (pg/L)	Cis- and trans- Chlordane (pg/L)	α-Endo sulfan (pg/L)	Dieldrin (pg/L)	Cyclodiene Flux ng/m^2
Hayes R.[a]	155	325	510	630	447
Brown R.	230	450	750	900	850
Ferguson R.[a]	65	130	275	205	253
Lorillard R.[a]	95	185	290	310	270
Baker Lake[a]	25	20	95	130	96
Resolute[a]	225	545	610	1300	838
Agassiz L-1[b]	360	740	1095	1400	346
Alexandra	100	170	240	830	201
Ice Island	160	160	420	740	222
Gascoyne	ND[c]	300	400	450	360
Devon	70	70	100	250	153
Pedder	140	160	270	420	512
Bag blanks[d]	ND	ND	ND	ND	
DL[e]	60	70	50	180	
TDL[f]	5	5	5	5	

[a]Average volume, concentration, and flux based on a split sample (i.e., two analyses).
[b]Average volume, concentration, and flux based on three replicate samples, of which two were split (i.e., 5 analyses).
[c]ND indicates that the compound was not detected.
[d]Average volume and concentration based on three replicate samples.
[e]DL: Detection limit, defined as the lowest concentration that can be determined in a real sample matrix by means of spiking the analyte into solvent for extraction and being carried through the entire procedure and which is statistically distinct from the background response of a sample carried through the same procedure.
[f]TDL: the theoretical detection limit, determined by dividing the DL by a representative sample volume (here taken to be 40 L).

considered significant with the analytical variance that can be anticipated at these concentrations.

Note that "nondetectable" levels are plotted on the abscissa of Figures 2 and 3. Questionable or unavailable analytical results generally have been edited from the figures. Specifically, the concentrations of lindane and α-HCH for the 1971/72 layer are not available as there was insufficient extract for the required dilution and reanalyses for quantification. Chlordane, dieldrin, and heptachlor epoxide for this year were quantifiable on the undiluted extract. The sample for the 1978/79 layer, the only sample in which lindane was not quantified above the detection limit, showed poor spike recoveries. It is probable that lindane was indeed present in this layer, and therefore it has not been shown as "not detectable." Similarly, the α-HCH measurement of 1000 pg/L for this sample is likely an underestimate; nevertheless, it has been included. The same argument applies to the levels of dieldrin and heptachlor epoxide for the 1978/79 layer, which have been excluded from Figure 4. The results for the 1986 sample for the 1983/84 layer are not available, except for α-HCH, due to analytical problems with this sample.

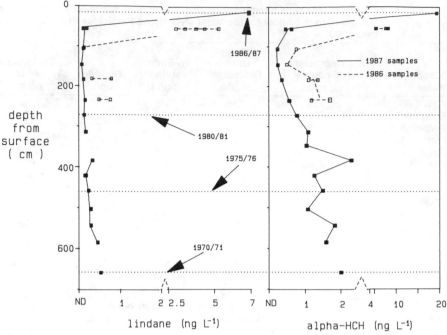

Figure 2. Hexachlorocyclohexane pesticide concentrations in annual snow layers from the Agassiz Ice Cap from the winter of 1970/71 to the winter of 1986/87.

DISCUSSION

Atmospheric deposition of organochlorine pesticides to the Canadian Arctic snowpack is occurring throughout a wide geographic area. Concentrations within snow samples collected in 1986 for lindane and α-HCH range from several hundred pg/L to as much as 8 ng/L. The apparent spatial differences in concentration of these compounds appear to depend to some extent on spatially variable deposition as well as on local dilution, as discussed by Gregor and Gummer.[12]

Representative winter flux calculations for these stations are difficult because the snow mass can not be measured at the time of sampling due to the highly variable distribution of snow. Snowfall data are available for arctic weather stations, but only a few are close to these sampling sites. Moreover, it has been reported[16,17] that actual snow storage in nearby basins was between 130% and 300% more than the snowfall reported at the weather station. These differences were not systematic. In the absence of more representative snow data, it has been necessary to assume that the nearest weather station approximates snow deposition at each sampling site. In this way, the water equivalent fluxes of α-HCH, lindane, and total cyclodiene pesticides have been approximated for the seasonal snow accumulation periods of 1985/86 (Tables 1 and 2).

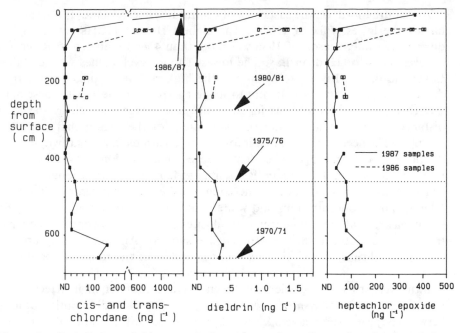

Figure 3. Cyclodiene pesticide concentrations in annual snow layers from the Agassiz Ice Cap from the winter of 1970/71 to the winter of 1986/87.

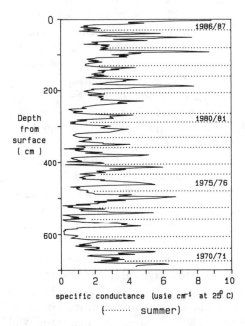

Figure 4. Plot of specific conductance versus depth for the Agassiz Ice Cap from the surface in the spring of 1987 to a depth of 6.9 m.

Alexandra Fiord, on the east coast of central Ellesmere Island, is notable for having the lowest flux of HCHs. The highest flux of HCHs for the snow during the winter season of 1986 was more than 4 $\mu g/m^2$ at Brown River, on the west coast of Hudson Bay. The lowest flux of cyclodienes was at Baker Lake; the Brown River site again had the highest flux of 0.85 $\mu g/m^2$. It is difficult and premature to attempt to suggest a spatial pattern with these data due to the inherent problems in estimating the fluxes, as well as the long distances between stations and the large areas of the region that were not sampled. The need for additional information on atmospheric pathways in the Arctic and local- and regional-scale influences on deposition rates, as well as a more extensive sampling design, is essential to identify a depositional pattern.

The Agassiz Ice Cap, an extremely remote site on north central Ellesmere Island, shows concentrations and winter fluxes of HCHs during the 1985/86 snow season at about the midrange of all sites (1 $\mu g/m^2$). In contrast, concentrations of cyclodienes in the 1986 surface snow of Agassiz Ice Cap are the highest of all sites; the winter flux is in the middle of the range at about 0.35 $\mu g/m^2$.

Intuitively, this cold glacier, on which average summer melt affects only about 3% of the winter snow layer resulting in negligible redistribution of ions between adjacent snow layers,[18] should provide a credible historical record of pesticide deposition. The preservation of the seasonal specific conductance record in the glacial snow is clearly seen in Figure 4. This seasonal profile is consistent with that reported by Barrie et al.[18] As noted by these authors, this profile reflects the trace constituent composition of the atmospheric aerosol, which undergoes a strong seasonal variation with maximum ion concentrations, and therefore specific conductance, occurring during the arctic winter. The dashed horizontal lines in Figure 4 show the approximate depth of the summer layer, determined by the minimum specific conductance measurements and confirmed by the presence of a more dense show lens.

No attempt has been made here to investigate seasonal pesticide deposition. The annual snow accumulation at this latitude and elevation represents, on average, more than 75% of the total annual precipitation[17] and includes the winter season when arctic aerosols are known to peak.[6,11,18]

Annual layers in this study were taken to begin in the fall and to include the subsequent summer melt layer at the top of the annual layer. In this way, any relocation of pesticides within the profile as a result of the melting process would be contained consistently within a single annual layer, thus retaining the integrity of the layer concentrations from this perspective.

As noted above, replicate samples of the 1985/86 surface snow layer were sampled in the spring of both 1986 and 1987. From Figure 2 it is clear that the concentrations of α-HCH and lindane in this snow layer decreased by an order of magnitude over the course of the year. Similarly, the cyclodienes in Figure 3 decreased as well, but generally by lesser amounts. The concentration of α-endosulfan in this layer was 1.1 ng/L in 1986 but was not detectable the following year. Since there is no runoff from this site due to limited melt, and the snow layer below shows consistent low concentrations of all compounds between the two years of sampling, it must be concluded that large quantities of these pesticides disappear from the snow, probably during the summer season. Evidence that this phenomenon occurs annually is provided by the high concentrations in the surface snow layer of 1987 and the similarity between concentrations at depth. Although it is conceivable that this temporal change follows a gross flux trend, it is equally likely that the trend is determined by summer phenomena that regulate the quantity refluxed. The mechanisms have not been investigated, but it is likely that revolatilization and photodegradation of the compounds are important and these will be influenced by annual meteorologic conditions. Clearly, the pesticide residues retained in the glacial snow represent only net deposition of these compounds to the glacier, and further study is required to make any connection between the residue trends and annual gross flux variability and trends.

The importance of this reflux phenomenon is twofold. First, the decreasing trend in the net quantity of pesticide residues in the glacier for the period 1970/71 to 1985/86 (Figures 5 and 6) can not be used to support a decrease in the supply of these organochlorine pesticides to the Canadian Arctic due to global regulation. However, the fact that chlordane and α-endosulfan are less frequently detected in recent years does suggest a decrease in the supply of these two pesticides.

Second, this phenomenon, if it occurs throughout the Arctic, could greatly reduce the quantity of pesticides entering the arctic aquatic environment during snowmelt. For example, the ice cap data indicate that the winter gross flux of lindane and α-HCH, shown in Figure 5, are seven to eight times more than that preserved in the glacial snow. Similarly, the gross flux of chlordane, heptachlor epoxide, and dieldrin is approximately ten, five, and four times more, respectively, than the mass preserved in the glacial snow (Figure 6). Additional sampling on the ice cap and at other stations is essential to confirm and further characterize this reflux from the snow and to better understand annual deposition variability.

Atmospheric transport and deposition of pesticides to snow in the Canadian Arctic is substantial. Loss of the pesticides from the snow during the arctic summer, so that only a relatively small portion remains in the glacial record, is believed to be reported for the first time here. This is very important for estimating the annual supply of toxic trace organic substances to the arctic ecosystem. The quantity of pesticides preserved in the glacial record is therefore a significant underestimate of the total winter flux to the snowpack. Further research is required to understand the mechanisms and pathways of pollutants in the arctic ecosystem, especially with respect to the contribution of the atmosphere to the contaminant burdens of the arctic food chain.

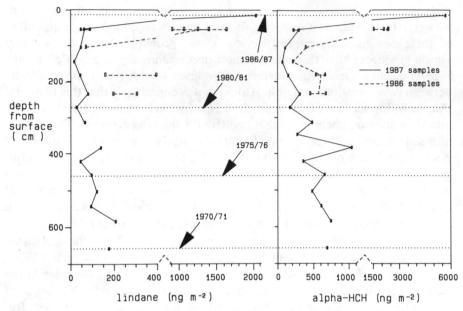

Figure 5. Mass of hexachlorocyclohexane pesticides in annual snow layers from the Agassiz Ice Cap from the winter of 1970/71 to the winter of 1986/87.

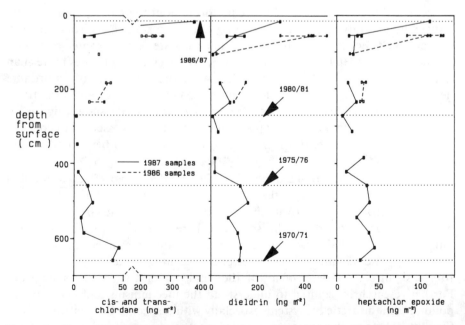

Figure 6. Mass of cyclodiene pesticides in annual snow layers from the Agassiz Ice Cap from the winter of 1970/71 to the winter of 1986/87.

ACKNOWLEDGMENTS

The author wishes to thank the Government of the Northwest Territories, the Canadian Department of Indian and Northern Affairs, and Environment Canada, who supported this project financially. Also, I wish to thank the Polar Continental Shelf Project for logistical support at Resolute. Special thanks are due D. Curtiss, L. James, G. Bangay, and B. Hough of these agencies. R. Koerner was instrumental in the collection of the Agassiz Ice Cap samples and the continued support of W. Gummer and R. L. Thomas is greatly appreciated. The chapter has been greatly improved with the advice of an anonymous reviewer and D. Kurtz.

REFERENCES

1. Muir, D. C. G., R. J. Norstrom, and M. Simon. "Organochlorine Contaminants in Arctic Marine Food Chains: Accumulation of Specific Polychlorinated Biphenyls and Chlordane-Related Compounds," *Environ. Sci. Technol.* 22:1071–79 (1988).
2. Norstrom, R. J., M. Simon, D. C. G. Muir, and R. E. Schweinsburg. "Organochlorine Contaminants in Arctic Marine Food Chains: Identification, Geographical Distribution and Temporal Trends in Polar Bears," *Environ. Sci. Technol.* 22:1063–70 (1988).
3. Wong, M. P. "Chemical Residues in Fish and Wildlife Species Harvested in Northern Canada," Environmental Studies Program, Northern Environment Directorate, Indian and Northern Affairs Canada, Ottawa, Ont. (1985).
4. Kinloch, D. and H. V. Kuhnlein. "Assessment of PCBs in Arctic Foods and Diets—A Sentinel Study in Broughton Island, Northwest Territories," unpublished report, Health and Welfare Canada, Ottawa, Ont. (1986).
5. Whelpdale, D. M. and L. A. Barrie. "Atmospheric Monitoring Network Operations and Results in Canada," *Water Air Soil Poll.* 18:7–23 (1982).
6. Hoff, R. M., and L. A. Barrie. "Air Chemistry Observations in the Canadian Arctic," *Water Sci. Technol.* 18:97–107 (1986).
7. Stengle, T. R., J. J. Lichtenberg, and C. S. Houston. "Sampling of Glacial Snow for Pesticides Analysis on the Higher Plateau Glacier of Mount Logan," *Arctic* 26:335–36 (1973).
8. McNeely, R. and W. D. Gummer. "A Reconnaissance Survey of the Environmental Chemistry in East-Central Ellesmere Island, N. W. T.," *Arctic* 37:210–23 (1984).
9. Paasivirta, J., M. Knuutila, and R. Paukku. "Study of Organochlorine Pollutants in Snow at North Pole and Comparison to the Snow at North, Central and South Finland," *Chemosphere* 14:1741–48 (1985).
10. Hargrave, B. T., W. P. Vass, P. E. Erickson, and B. R. Fowler. "Supply of Atmospheric Organochlorines to Food-Webs in the Arctic Ocean," *Tellus* 40B:480–93 (1988).
11. Rahn, K. A., and G. E. Shaw. "Sources and Transport of Arctic Pollution Aerosol: A Chronicle of Six Years of ONR Research," *Naval Research Reviews* (1982), pp. 3–26.
12. Gregor, D. J., and W. Gummer. "Evidence of Atmospheric Transport and Deposi-

tion of Organochlorine Pesticides in Canadian Arctic Snow," *Environ. Sci. Technol.* 23(5):561–65 (1989).

13. Goulden, P. D., and D. H. J. Anthony. "Design of a Large Sample Extractor for the Determination of Organics in Water," unpublished report, National Water Research Institute, Environment Canada, Canada Centre for Inland Waters, Burlington, Ont. Contribution No. 85-121 (1985).

14. Neilson, M. A., R. J. J. Stevens, J. Biberhofer, P. D. Goulden, D. H. J. Anthony. "A Large Sample Extractor for Determining Organic Contaminants in the Great Lakes," Inland Waters Directorate, Ontario Region, Water Quality Branch, Environment Canada, Burlington, Ont. Technical Bulletin No. 157 (1988).

15. "Analytical Protocol for Monitoring Ambient Water Quality at the the Niagara-on-the Lake and Fort Erie Stations," unpublished report, National Water Quality Laboratory, Environment Canada, Canada Centre for Inland Waters, Burlington, Ont. (1987).

16. Marsh, P., and Woo, M.-K. "Analysis of Error in the Determination of Snow Storage for Small High Arctic Basins," *J. Applied Meteorology* 17:1537–41 (1978).

17. Woo, M.-K., R. Heron, P. Marsh, and P. Steer. "Comparison of Weather Station Snowfall with Winter Snow Accumulation in High Arctic Basins," *Atmos. Ocean* 21:312–25 (1983).

18. Barrie, L. A., D. Fisher, and R. M. Koerner. "Twentieth Century Trends in Arctic Air Pollution Revealed by Conductivity and Acidity Observations in Snow and Ice in the Canadian High Arctic," *Atmos. Environ.* 19:2055–63 (1985).

## CHAPTER 25

Estimates of Human Exposure to Pesticides Through Drinking Water: A Preliminary Risk Assessment

R. Peter Richards and David B. Baker

INTRODUCTION

Concern about the impact of pesticides on drinking water quality has increased in the last several years. Concern over possible ground water contamination has been particularly strong. Water quality guidelines* for the chlorinated hydrocarbon pesticides such as DDT, endrin, lindane, and toxaphene were initially proposed nearly 20 years ago, but concentrations in excess of these guidelines in raw or finished drinking water have rarely been reported, and reported concentrations are generally several orders of magnitude below the guidelines.[1,2] Furthermore, many of these compounds have been banned or restricted and are no longer in general use in the major agricultural areas of the United States. By contrast, the current generation of pesticides are, with some exceptions, more soluble in water and can be found in both raw and finished drinking water drawn from sources that drain agricultural lands. Until very recently, no health guidelines have existed against which ambient concentrations could be compared. Thus, it has not been possible to evaluate the importance of these concentrations of pesticides in drinking water and to compare the risks they pose with those due to other sources.

The extensive water quality monitoring programs run by the Water Quality Laboratory of Heidelberg College (WQL) have been providing data on pesticide concentrations in potential drinking water sources and in finished drinking water in Ohio, primarily in the northwest and north central part of the state, since 1982. This database is sufficiently detailed to provide good estimates of average concentrations for most of the pesticides commonly used by

---

* We use the term "guidelines" in a general sense, to mean concentration levels associated with an estimated risk or other effect. By doing so, we seek to avoid implying anything about legal significance. Although legislated water quality standards exist for some of the older pesticides, such standards have not yet been promulgated for current-generation pesticides in raw or finished drinking water.

Ohio farmers. In this chapter we develop appropriate average concentrations for the four pesticides alachlor, atrazine, cyanazine, and metolachlor, which together account for about 70% of Ohio herbicide use.[3] We also present parallel data for nitrate. We consider concentrations in rivers, pumped storage reservoirs, rainfall, groundwater, and nearshore Lake Erie. The sampling locations are indicated in Figure 1. The average concentrations are compared with guidelines recently proposed by U.S. EPA's Office of Drinking Water, and risks associated with drinking water containing these compounds are compared with other risks drawn from the literature.

ANALYTICAL METHODOLOGY

Pesticides were analyzed using a multiresidue gas chromatography technique that employs dual capillary columns with nitrogen–phosphorus detectors. This technique was developed by our lab for efficient analysis of the pesticides used in Ohio and allows quantification of 19 pesticides that constitute about 80% of pesticide use in Ohio by weight. Details of the procedure are available elsewhere.[4] EPA draft method 507 is very similar.[5]

Analysis of nitrate is by cadmium reduction using EPA method 353.3.[6] Since this method reduces nitrate to nitrite, and nitrite is not separately analyzed, the results are actually for nitrate plus nitrite. However, test analyses have repeatedly shown that nitrite is present in very low concentrations in our samples, compared to nitrate.

DERIVATION OF AVERAGE PESTICIDE CONCENTRATIONS FOR RISK EVALUATION

A health guideline is generally expressed as a concentration that is associated with a specified risk. The risk is based on continuous exposure to that concentration for an extended period of time, usually 70 years.* Thus, evaluation of risk due to ambient concentrations requires that the concentrations be expressed as averages calculated in a manner that reflects exposure patterns comparable to those assumed by the risk model. Actual ambient concentrations often vary widely over time. For example, pesticide concentrations in rainfall follow an annual cycle.[7,8] Pesticides in rivers follow a similar annual cycle, but with strong fluctuations imposed by runoff from storm events. Peak concentrations in rivers are associated with near-peak flows during runoff events, and concentrations at low flow are much lower.[9] Nitrate concentrations in rivers have similar behavior, but concentrations peak later during runoff events because much of the nitrate reaches the rivers through tile drainage. As a consequence of the linkage between concentration and flow, an average

---

* For many compounds, there are also health guidelines and corresponding risk estimates for shorter term exposures. However, for the compounds examined in this chapter, the short-term guidelines are rarely, if ever, exceeded. Thus, we concentrate on lifetime advisory levels.

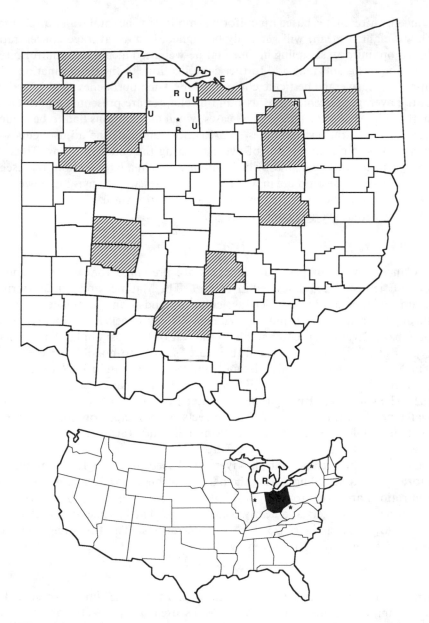

Figure 1. Map showing sampling localities for data used in this chapter. R: river stations,
*: rainfall stations, E: the Lake Erie station, U: pumped storage reservoirs. Shaded
counties within Ohio provided rural well nitrate averages.

concentration that is based on uniform sampling of the total water discharge (a flow-weighted mean) will generally be higher than an average concentration based on uniform sampling in time (a time-weighted mean). Nonuniformity in sampling programs may add further complications to the determination of an appropriate average pesticide concentration. The approaches used in arriving at the average concentrations used in this chapter are presented and discussed in the following paragraphs. In all cases, when assumptions had to be made to adjust concentrations for times or areas not sampled, the adjustments were made so as to err on the side of overestimating the concentrations. Thus, the comparisons with guidelines, which will be made in a later section, represent a kind of worst case estimate in which the guidelines are, if anything, lower than necessary, and the estimated average concentrations are, if anything, higher than the actual average concentrations.

Rivers

Concentrations of pesticides in rivers are presented for five tributaries to Lake Erie, listed and described in Table 1. The Maumee and Sandusky rivers drain predominantly clay-rich agricultural lands, and drainage is mostly through surface runoff and tile systems. Honey Creek is a tributary of the Sandusky River and is comparable to the rest of the basin in land use and soil type, but of course smaller and, therefore, subject to possible scale effects.[10] The River Raisin is of similar size and land use to the Sandusky River but has a coarser soil texture, which permits more infiltration and reduces surface runoff. The Cuyahoga River drains a greater percentage of urban and forested watershed and thus allows land use effects on pesticide concentrations to be evaluated. Of these, the Sandusky, Maumee, and Raisin rivers are currently used as direct sources for potable water.

Average concentrations in the rivers are based on a program that employs more frequent sampling during the periods of high flow, when pesticide concentrations are likely to be high. However, water treatment plants withdraw raw water from rivers at a nearly constant rate. Thus a time-weighted average is employed to characterize these concentrations. The time-weighted mean concentration (TWMC) is given by

$$\text{TWMC} = \frac{\Sigma c_i t_i}{\Sigma t_i}$$

where c_i are the observed concentrations and t_i are the times represented by each sample. Details of the procedure, as used by the WQL, are presented elsewhere.[11] The resulting average concentrations are listed in Table 2.

Reservoirs

The reservoirs used in this study are pumped storage reservoirs that receive raw water from rivers (see Table 1 for details). The water resides in the reservoir for some time before use, allowing breakdown of the pesticides and biological sequestering of the nitrate. In addition, pumping can often be

Table 1.   Sources of Data on Pesticide Concentrations in Water Sources of Potential Use for Drinking Water in the Midwestern United States.

Rivers	Sampling Location	Dates Sampled	% Cropland	Soils	Basin Area (km²)
Maumee	Waterville, Ohio	1983–1986	75.6	clay rich	16,395
Sandusky	Fremont, Ohio	1983–1986	79.9	clay rich	3,240
Honey Cr.	Melmore, Ohio	1983–1986	82.6	clay rich	386
Raisin	Monroe, Michigan	1983–1986	67.1	silty-sandy	2,699
Cuyahoga	Independence, Ohio	1983–1986	4.2	silty	1,831

Pumped Storage Reservoirs	Dates Sampled	Water Source	Tributary To	Residence Time[a]
Attica	summer 1987	Honey Creek	Sandusky River	130 days
Bellevue	summer 1987	Frink Run	Huron River	720 days
Clyde	summer 1987	Beaver Creek	Maumee River	>360 days
Fostoria	summer 1987	East Branch	Portage River	490 days

Rainfall	Location	Dates Sampled	Surrounding Land Use
Indiana	W. Lafayette	summer 1985	cropland
Ohio	Tiffin	summer 1985	cropland
West Virginia	Parsons	summer 1985	forest
New York	Potsdam	summer 1985	forest, dairy

Lake Erie	Location	Dates Sampled
Central Basin	City of Sandusky water intake, about 1 km offshore at 6 m depth, 1.6 km south of mouth of Sandusky Bay	summer 1987

Groundwater	Location	Dates Sampled
Ohio (pesticides)	rural wells throughout the state	1987
Ohio (nitrate)	rural well data for 13 counties	1986–1987
Iowa (pesticides)	203 public water supply wells	1982–1985

[a]Residence time for pumped storage reservoirs was calculated as reservoir volume divided by average demand, using figures from "Northwest Ohio Water Plan," Ohio Department of Natural Resources, Division of Water, Columbus, OH (1986).

avoided at the times of peak concentrations in the rivers. Thus, it would be expected that the concentrations in pumped storage reservoirs would be lower than those in the rivers from which the water is drawn. The data are based on weekly samples of raw and finished water taken during summer 1987; each sample is a composite of daily aliquots drawn from the raw water stream entering the water treatment plant or the finished water leaving it. No systematic differences in concentration between raw and finished water were observed; thus the two sets of data were grouped for averaging. The averages listed in Table 2 were calculated as simple means of the measured values. The averages for the reservoir at Attica were divided by two because of the short residence time of water in that reservoir and because concentrations of pesticides in the winter are generally below detection limit and those of nitrate are much lower than during the summer. This adjustment of the data to a whole-year basis probably tends to produce somewhat high estimates, since summer concentrations are assumed to hold for half the year. The residence times of

Table 2.   Mean Pesticide Concentrations in Various Potential Drinking Water Sources

Source and Location	No. of Samples	Mean Concentration in µg/L				
		Alachlor	Atrazine	Cyanazine	Metolachlor	Nitrate (mg/L)
Rivers						
Maumee	291	0.91	2.37	0.57	1.40	3.93
Sandusky	302	1.27	3.16	0.70	2.84	3.46
Honey Creek	424	1.69	4.39	0.59	3.51	4.41
Raisin	124	0.67	1.30	0.35	0.49	2.55
Cuyahoga	86	0.14	0.55	0.15	0.30	2.54
Pumped Storage Reservoirs						
Attica	17	0.34	1.78	0.43	0.94	1.35
Bellevue	19	0.20	1.04	0.24	0.44	1.46
Clyde	20	0.22	1.94	0.44	0.52	0.32
Fostoria	20	0.32	2.94	0.66	1.06	0.52
Rainfall						
W. Lafayette, IN	14	0.32	0.13	0.01	0.08	
Tiffin, OH	24	0.42	0.13	0.02	0.12	0.80[a]
Parsons, WV	20	0.02	0.03	0.02	0	
Potsdam, NY	21	0.01	0.04	0.01	0	
Lake Erie						
City of Sandusky	19	0.03	0.07	0.05	0.03	0.23
Groundwater						
Ohio rural wells	280	0.03	0.04	0.004	0.01	0.18–3.45[b]
Iowa public water supply wells[c]	203	0.08	0.36	0.06	0.03	

Note: For approaches and assumptions, see text.
[a]Value based on four stations in Ohio. See text.
[b]Range of mean values for 13 Ohio counties, each mean based on 80–500 samples.
[c]Based on data from Kelley, R. D., in *Pesticides and Groundwater: A Health Concern for the Midwest* (Navarre, MN: Freshwater Foundation, 1987), pp. 115–35. His means adjusted for concentrations below detection limit by assuming concentrations of zero for those samples.

the rest of these reservoirs are a year or longer; thus the water present at any given time is aggregated over the annual cycle, and no adjustment for seasonality of concentration was made.

Lake Erie

Lake Erie was represented by the city of Sandusky's raw and finished water. The sampling program and approach to calculating the average value are identical to those just described for the pumped storage reservoirs. No adjustment for seasonality of concentration was made, although seasonal concentration fluctuations have been documented in the nearshore zone of Lake Erie.[12]

Rainfall

Rainfall would be a potential water supply for those who have cisterns. In this case, all precipitation would be collected and eventually used. The data for pesticides consist of integrated samples over entire rainfall events (summer months only) and thus represent samples analogous to what a cistern would receive. Samples were collected in four states (see Table 1), two in predominantly agricultural areas and two in areas where other land uses predominate. The observed concentrations were weighted by rainfall amount prior to calculating the mean concentration, and the result was halved to account for near-zero concentrations in the fall and winter. This adjustment of the data to a whole-year basis produces somewhat high estimates because the rainfall monitored was less than half the usual annual rainfall. No attempt was made to account for possible breakdown during holding time in the cistern.

The rainfall sampling program did not involve nitrate analysis. The National Atmospheric Deposition Program maintains four atmospheric sampling stations in Ohio. They show a range of annual volume-weighted mean concentrations of 0.5 to 0.8 mg/L as N for the sum of nitrate plus nitrite and ammonia, based on approximately 40 station-years of observation. We have used their high figure of 0.8 mg/L for comparison with our data. The ammonia concentrations were small compared to those for nitrate plus nitrite and were included on the assumption that ammonia would be converted to nitrate in a cistern.

Groundwater

Groundwater is represented by data from rural private wells around the state of Ohio and by literature values from public wells in Iowa. The Ohio pesticide data are based on a program in which roughly 70% of the samples were taken from wells where nitrate was present at a concentration of 0.3 mg/L or greater (about 25% of all wells tested). Since the presence of nitrate is suggestive of contamination, these wells seemed more likely to contain pesticides also. The remaining 30% of the pesticide samples were drawn from the 75% of wells where nitrate was absent or present in concentrations less than 0.3 mg/L. Each well was sampled only once. Values reported are simple means of the observed concentrations. Given the approach used to choose the wells to be sampled, the mean values should be higher than those for Ohio rural private wells as a whole.

There is an important difference between the wellwater results and the other results. Pesticides were not detected in the great majority of wells studied, but a few wells had concentrations in the range of the guidelines. Average concentrations in surface water estimate the exposure of identifiable groups of people using those surface water sources, and high and low concentration inputs to those systems average out over time. Each person using a given surface water source receives the same exposure on a per volume basis. On the other hand, "average ground water" is probably not consumed by anyone. Each well is a separate source that serves a separate population and has a concentration history that, for a number of reasons, may or may not be similar to other nearby wells. Thus, the groundwater average is one degree more abstract than the other averages and could perhaps be compared most directly to an overall average for all the pumped storage reservoirs listed or for all the rivers. Also, the groundwater data distribution is strongly right-skewed. Thus, the average values used here are overestimates of the concentrations to which most of the rural population are exposed, but underestimates for those few with contaminated wells.

Nitrate data for ground water are presented somewhat differently. Here the overall sample is much larger, and it is possible to represent nitrates in private rural wells as averages for selected counties in Ohio. The sampling program involved voluntary participation by the well owner, and no intentional bias toward wells likely to be contaminated was imposed by the WQL in running the program. However, the counties chosen for inclusion in this chapter include most of those with relatively high percentages of wells showing some nitrate contamination; thus the values used here are higher than those to which most of the rural population are exposed. Although the concentrations reported for nitrate represent countywide rather than statewide averages, it still remains true that individual wells show a considerable range of concentrations around these values.

Given the intensity of concern about groundwater pollution, it is important to emphasize that the data presented here represents *wellwater*, and most cases of contamination probably do not represent general *groundwater* (aquifer) contamination. Most contaminated wells are shallow dug wells, often with old casings or none at all, and contamination is likely to be local in nature. Although we have no direct evidence about aquifer contamination in Ohio, the geographic distribution of contaminated wells cannot be explained on the basis of aquifer contamination alone.

FORMS AND LEVELS OF RISK

Three Forms of Risk Assessment

Estimates of risk can be divided for convenience into three categories:

1. lifetime chance of death
2. lifetime chance of contracting cancer
3. lifetime chance of other health effects

Each of these categories involves somewhat different approaches to risk estimation, different levels of uncertainty, and different implications for those affected. Representative risk estimates for the first two categories are presented in Table 3. No-effect levels for the third category are presented in Table 4.

Lifetime chance of death usually involves relatively common hazards, for which both the level of risk and the length of our experience are sufficient that the risk can be stated on the basis of direct experience (actuarial statistics, mortality studies, etc.). These risks are generally higher than those in the other two categories and are subject to less uncertainty. For the risk levels used in

Table 3.   Representative Risk Estimates

Category and Cause	Level	Source
Lifetime Chance of Death (50 Years)		
Fire fighter	32,000 in 1,000,000	Crouch and Wilson[a]
Motor vehicle accidents	17,000 in 1,000,000	Crouch and Wilson
Police officer	8,800 in 1,000,000	Crouch and Wilson
Drowning	2,500 in 1,000,000	Crouch and Wilson
Fire	2,000 in 1,000,000	Crouch and Wilson
Hunting	1,500 in 1,000,000	Crouch and Wilson
Electrocution	370 in 1,000,000	Crouch and Wilson
Tornado	42 in 1,000,000	Crouch and Wilson
Lightning	35 in 1,000,000	Crouch and Wilson
Lifetime Chance of Contracting Cancer (70 Years)		
Cigarette smoking	to 80,000 in 1,000,000	Crouch and Wilson
1/4 lb mixed Lake Michigan fish per week	5,800 in 1,000,000	Clark et al.[b]
Natural Background radiation	1,400 in 1,000,000	Crouch and Wilson
Air pollution	1,000 in 1,000,000	Crouch and Wilson
1/4 lb mixed Lake Michigan fish per year	110 in 1,000,000	Clark et al.
1/2 gal whole milk per week	100 in 1,000,000	Crouch and Wilson
2 oz of peanut butter per week	80 in 1,000,000	Crouch and Wilson
Water with 1.5 μg/L alachlor, 2 L per day	3.4 in 1,000,000	EPA[13]

[a]Crouch, E. A. C, and R. Wilson. In *Assessment and Management of Chemical Risks*, J. Rodricks and R. Tardiff, Eds. (Washington, DC: American Chemical Society, 1984), pp. 97–112.
[b]Clark, J. M., L. Fink, and D. DeVault. *J. Great Lakes Res.* 13(3):367–74 (1987).

this chapter, a "lifetime" is taken to be 50 years. The risks presented here are either aggregate risks associated with a particular profession or activity, or risks associated with different kinds of accidents. Such risks range from high values of 32,000 in 1,000,000 for a fire fighter and 14,000 in 1,000,000 from motor vehicle accidents to low values of 42 in 1,000,000 for death from a tornado and 35 in 1,000,000 for death from lightning. These are average risks for the exposed population over a lifetime; an individual's instantaneous risk may differ greatly depending on factors such as age, experience, and frequency and pattern of exposure to risk.[13]

Lifetime chance of contracting cancer is almost always estimated by extrap-

Table 4.   Proposed Lifetime Health Advisory Levels for Drinking Water for Nitrate, Nitrite, and Some Herbicides.

Substance	Level (μg/L)
Atrazine	3
Cyanazine	8.8
Metolachlor	10
Nitrite	1,000
Nitrate	10,000

Source: Health advisories for a number of pesticides and other compounds, released at various times during 1987 by EPA, Office of Drinking Water, Washington, DC.
Note: These levels are for health effects other than cancer and are estimates of no-effect levels.

olation from studies on laboratory animals. A "lifetime" is generally taken to be 70 years. Detailed description of the methodology is beyond the scope of this chapter. Such risks are generally relatively low compared to risks of death described above. They are also subject to considerable uncertainty for many reasons, including possible differences in response of humans and laboratory animals, the extent of extrapolation required, and the possibility that the linear extrapolation approach itself is not valid. A considerable body of evidence suggests that there may be threshold effects for certain types of carcinogens, and even that low-level exposures may be beneficial in some cases (e.g., background radiation, chloroform; see Calabrese and Gilbert[14] for further examples). Exposure guidelines related to lifetime cancer risks are set using the upper 95% confidence level, often for an extrapolated risk of 1 in 1,000,000, and thus they involve a considerable safety factor. Finally, it is worth emphasizing that they are risks of *contracting* cancer, not *dying* from it. Clearly, the risk of death from cancer will depend also on the mortality statistics for the type(s) of cancer involved.

Lifetime risks of other health effects, like cancer risks, are generally evaluated by extrapolation from laboratory animal studies and are based on a 70-year lifetime. However, the current practice of the U.S. EPA for the compounds involved in this chapter is to set guidelines by extrapolation to a no-effect level, with a safety factor of 100 or 1000. Since the extrapolated exposure level is in effect a zero-risk level, no actual risk is associated with it, and risk assessment is limited to comparing actual concentrations to the exposure level. Exposure rates that exceed the exposure level involve some risk of health effects, but the level of risk is not quantifiable in a manner parallel to that for the other two categories.

Comparison of Estimated Average Concentrations with Corresponding Guidelines

Of the four pesticides reported in this paper, alachlor is considered a Group B2 chemical by the EPA at the present time (a demonstrated carcinogen in laboratory animals and a suspected, but not demonstrated, human carcinogen). Based on the most recent estimate of the q_1^* for alachlor (the upper 95% limit on the "oncogenic potency," estimated from the multistage linear model), the upper bound estimate of the risk of cancer from drinking water containing 1.5 $\mu g/L$ alachlor is 3.4 in 1,000,000. This risk is a lifetime (70-year) risk and assumes a 70-kg adult consuming 2 L of water per day at this concentration. The estimate is conservative, and the actual risk is probably lower: by the design of the estimation model, there is only one chance in 20 that the actual risk exceeds this estimate.

Of the remaining pesticides, atrazine and metolachlor are classed in Group C (limited evidence of carcinogenicity in laboratory animals). They are considered possible human carcinogens. Cyanazine is considered a Group D chemical (insufficient evidence to determine carcinogenicity in animals or in humans).

For these pesticides and for nitrate, the EPA has recently suggested lifetime health advisory levels for health effects other than cancer. The lifetime health advisory level for nitrate is based on a known health effect* on infants consuming water with concentrations around 10 mg/L and higher. These proposed no-effect levels are listed in Table 4. Although the data presented in Table 2 represent ambient water from river sources and unfinished water from other sources, little removal of these pesticides or nitrate accompanies the water treatment process, unless carbon filtration is used to reduce pesticide concentrations. Thus, at the level of resolution needed for this chapter, raw water concentrations adequately reflect likely drinking water concentrations. Use of raw water concentrations may in some cases lead to overestimation of the exposure concentrations, but probably not by a factor exceeding two.

Comparison of the estimated exposure levels of Table 2 with the guidelines listed in Table 3 is facilitated by the graphs in Figure 2. In each case the guideline concentration is shown as a vertical bar, and average concentrations listed in Table 2 are plotted by category with concentration increasing to the right. When viewing these graphs, it should be kept in mind that the guidelines are set conservatively to account for uncertainty in the extrapolation procedure and that uncertainties in the averaging of concentrations have been resolved in all cases in favor of higher concentrations. Thus, the graphs present a kind of "worst reasonable case" estimate or risk.

It is clear from these graphs that in most cases, estimated exposures are below the guidelines. The few exceptions are for river water. Two of the three cases where ambient concentrations exceed the guidelines involve Honey Creek, which is not actually used as a direct water source at present. Concentrations in rivers are generally substantially higher than those from other sources, reflecting a lack of time for degradation and mixing with other water sources.

At the same time, average concentrations reported in this chapter are closer to the guidelines, in many cases by several orders of magnitude, than any known drinking water concentrations for the older chlorinated hydrocarbon insecticides. This suggests that careful monitoring of drinking water concentrations and accurate assessment of risks will be especially important for these compounds.

It is also important to keep these risks in context. With two exceptions, concentrations of atrazine are below the current estimated no-effect level. Cyanazine, metolachlor, and nitrate concentrations are all below the estimated no-effect level, cyanazine always by more than an order of magnitude. Alachlor is present in the worst case in concentrations about equal to the reference concentration of 1.5 μg/L, corresponding to an upper bound estimated lifetime risk of cancer of 3.4 in 1,000,000. By comparison, the risk of death from

---

* This effect, methemoglobinemia or "blue baby syndrome," results from conversion of nitrate to nitrite by bacteria that may be present in the digestive system during the first few months. The nitrite reacts with hemoglobin to block its oxygen-carrying capacity. After the age of about six months, increased secretion of hydrochloric acid kills these bacteria, and the risk is eliminated.

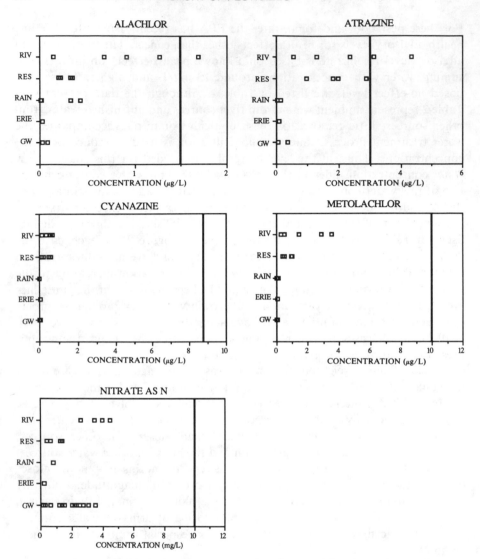

Figure 2. Pesticide and nitrate mean concentrations in potential and actual raw water supplies compared with health guidelines proposed by the U.S. EPA. Guidelines are for lifetime exposure (70 years at 2 L of water per day). Mean concentrations are probably overestimates in most cases. RIV: rivers, RES: pumped storage reservoirs, RAIN: rainfall, ERIE: Lake Erie, GW: groundwater.

lightning is about an order of magnitude higher. A more complete comparison of risks from various sources is presented graphically in Figure 3.

DISCUSSION

True Risk versus Perceived Risk

The public perception of risk often bears only a vague resemblance to the level of risk determined by experts. A common example is the person who drives everywhere because he is afraid he will be killed in a plane crash if he flies. Slovic[15] showed clearly that the perceived risk is strongly influenced by factors such as the magnitude, immediacy, irreversibility, and controllability

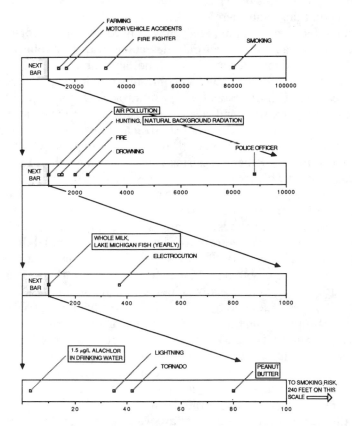

Figure 3. Graphic display of the risks listed in Table 3. Depicted are excess risks of death or of contracting cancer, out of a population of 1,000,000. Risks enclosed in boxes are upper bound estimates of cancer risk extrapolated from the multistage linear model and are presumably higher than the true risks. Risks not enclosed in boxes are based on actuarial data, and more accurately reflect the actual risks to the appropriate populations. The lowest 10% of each risk scale is expanded tenfold to create the next risk scale down the page. If the lowermost bar were extended to cover the same concentration range as the uppermost bar, it would be about 300 feet long.

of the risk (see also Wilson and Crouch[16] and Ames et al.[17] in the same issue). Things that are beyond our control are perceived as being more hazardous than those within our control. Risks of death that play themselves out in bunched fatalities (plane crashes) are perceived as greater than those in which the fatalities are more isolated (automobile accidents). Newly identified risks are often perceived as greater than those we have grown accustomed to.

This difference between risk and perception of risk may be playing itself out at present in our concerns about pesticides in drinking water. The risks that many perceive to exist can be portrayed by statements like the following: Pesticides are synthetic chemicals meant to kill things; therefore they are dangerous, regardless of their concentration. Nitrate is meant to make things grow and is (in part at least) a natural compound; therefore it is not dangerous. And, of course, peanut butter is for kids and cars are for making life easier; they are good things and therefore not dangerous.

Groundwater, once contaminated, may stay contaminated for a long time, and the contaminants are often only detectable with technology that is not available to the average person. Love Canal and subsequent Superfund sites have planted the fear of groundwater contamination in the public mind, and it transfers easily to fear of pesticides in groundwater. On the other hand, rain is perceived to be pure, soft, and clean, and people tend to trust it. Similarly, if the groundwater or the tapwater is feared, many turn to bottled springwater, which has received a positive image through advertising. Rarely is the attempt made to determine whether the springwater is healthier than the source it is replacing, or even whether it actually represents a different source.

Because public perceptions of risk are driven by fear as much as by information, the political process is often driven in directions that may not represent the wisest use of limited funds for matters of environmental health. The data presented here on the concentrations of chemicals in potential drinking water supplies should be extrapolated only with great care beyond the geographic areas where they were obtained. However, to the extent that it is at all representative of other areas, our analysis indicates that contamination of drinking water by these pesticides and by nitrate is a relatively minor problem. It bears watching for future trends, but expensive remediation measures are not presently necessary in most cases. Communities that rely on rivers for their potable water and that decide treatment is warranted will probably find that treatment during a small part of the year can have a major impact on average concentrations, given the typical concentration patterns in rivers.[10,18]

Our data indicate that river water draining agricultural areas with heavy soils is the least desirable source of raw water among those we have considered. Concentrations of pesticides and nitrate are lower in the Cuyahoga River, which is less dominated by agriculture, and in the River Raisin, which has relatively coarse soils. Pumped storage reservoirs drawing water from these same rivers have substantially better water quality. The quality of the river from which the water is pumped and the residence time of the reservoir appear to be the important factors in differentiating between such reservoirs. Rainfall, often perceived as the purest water, appears to have higher concentrations of alachlor than the pumped storage reservoirs, but lower concentrations of the other compounds. Lake Erie, often thought to be highly polluted, and rural wells, whose owners often fear contamination from unknown sources, are the best water sources, based on the data presented here.

These conclusions apply to water quality only in the narrow sense of contamination by the five compounds examined in this chapter. There are obvious limitations to this analysis. Rivers draining largely residential and urban areas will not carry many agricultural compounds because there is not much agriculture, but they may carry a load of industrial and urban pollutants that are not covered here, and some of which may be more hazardous. Similarly, the important pollutants in nearshore areas of Lake Erie, particularly near urban centers, and in groundwater affected by landfill leachate are likely to be other than the ones dealt with in this chapter. However, for the often undercharacterized pollutants of rural diffuse origin, the work presented here not only provides estimates of exposures but indicates that the quality of various sources is different from what might be expected and, in most cases, better.

The Question of Long-Range Transport

In a symposium that documents global transport of older-generation pesticides, it is worth asking whether atmospheric transport of current-generation compounds is also likely to be a problem. Richards et al.[7] and Glotfelty et al.[19] have documented the presence of these pesticides in precipitation and in fog, respectively, and data in the former paper are suggestive of regional transport, though they do not demonstrate it. Glotfelty[8] attributes early spring peaks of atrazine in Maryland to sweet corn production in Florida. The informal consensus of those participating in the symposium seemed to be that global transport of these compounds is unlikely, given their relatively short half-lives. Atrazine is the most likely candidate for long-range transport. To date it is not clear that anyone has looked for it in arctic snow; this should be done.

Whether or not traces of atrazine or other modern pesticides can be found in the Arctic or remote regions, such compounds are unlikely to represent an ecosystem or human health threat outside the areas where they are used. In contrast, the transport of toxaphene and other persistent, bioaccumulating compounds has produced dangerous concentration levels in arctic fish and marine mammals. These compounds have never been used within 1000 miles of the Arctic and have not been extensively used in the United States in the last decade. It is ironic that they may represent more of a threat to arctic Native American populations (through dietary intake) than drinking water, with its burden of widely used modern pesticides, does to those living in the corn belt.

Level of Risk versus Risk Load

A distinction can be drawn between the levels of risk presented in this chapter and what might be called the risk load due to these compounds. The

Table 5. Populations Served by Each Class of Water Supply in Northwest Ohio.

Source	Population	Percent of Total
Lake Erie	554,300	42.5
Rivers	124,150	9.5
Pumped storage reservoirs	187,300	14.3
Rainwater (cisterns)	unknown, but very small	
Public wells[a]	90,780	7.0
Private wells (by difference)[b]	348,200	26.7
Other (quarry)	600	0.0
Total	1,305,330	100.0

[a]Public water supply figures aggregated from data in "Northwest Ohio Water Plan," Ohio Department of Natural Resources, Division of Water, Columbus, OH (1986).
[b]Private well population is assumed to be the difference between the total population and the population served by public water supplies.

implications of a certain level of pollution for human health are a function not only of the quality of the water, but also of the number of people consuming it. The populations served by different water supplies in northwest Ohio are listed in Table 5. Developing a quantitative analysis of risk load is beyond the scope of this chapter and quite possibly not appropriate, given the degree of uncertainty involved in risk assessment and the degree of aggregation and estimation involved in arriving at the concentration figures. Nonetheless, it seems clear that risk load from pesticides in northwest Ohio is lower than the risk levels might indicate because the better water sources serve a disproportionately large percent of the populace.

CONCLUSIONS

Average concentrations of the pesticides alachlor, atrazine, cyanazine, and metolachlor, and of nitrate were estimated for rivers, pumped storage reservoirs, rainfall, Lake Erie, and rural wellwater in Ohio. These averages are estimates of human exposure through drinking water drawn from these sources.

The average concentrations are generally lower than health guidelines for these compounds. Only atrazine and alachlor exceed their respective guidelines in any source, and then only slightly. The averages are designed to be high estimates, and the true averages might not exceed the guidelines. The risks associated with even the highest averages developed in this chapter are small compared to those from other sources.

For the sources studied, the order of overall quality appears to be, from best to worst:

GROUND WATER ≈ LAKE ERIE > RAINFALL ≈ PUMPED STORAGE RESERVOIRS > RIVERS

Atmospheric transport of the pesticides is not likely to be broader than regional in scale, with the possible exception of atrazine, which has a longer half life. Atrazine might possibly reach remote areas such as the Arctic in low concentrations, but since it does not bioaccumulate, there is no danger of adverse consequences of the sort associated with the long-range atmospheric transport of PCBs, DDT metabolites, and toxaphene.

REFERENCES

1. *Federal Register* 48(194):45502-21 (1983).
2. *Drinking Water and Health* (Washington, DC: National Research Council, 1977).
3. Waldron, A. C. "Pesticide Use on Major Field Crops in Ohio — 1982," Ohio Cooperative Extension Service Bulletin 715, Ohio State University (1984).
4. Kramer, J. W., and D. B. Baker. In *Quality Assurance for Environmental Measurements, ASTM STP 867,* J. K. Taylor and T. W. Stanley, Eds. (Philadelphia, PA: American Society for Testing Materials, 1985), pp. 116-32.
5. "Draft Method 507 — For Use in WS023 Only," U.S. EPA, EMSL-Cincinnati, U.S. EPA (1988).
6. "Methods of Chemical Analysis of Water and Wastes," EMSL-Cincinnati, U.S. EPA Report-600/4-79-020 (1979).
7. Richards, R. P., J. W. Kramer, D. B. Baker, and K. A. Krieger. *Nature (London)* 27:129-31 (1987).
8. Chapter 14.
9. Baker, D. B. *J. Soil Water Cons.* 40(1):125-32 (1985).
10. Chapter 17.
11. Baker, D. B. "Sediment, Nutrient and Pesticide Transport in Selected Lower Great Lakes Tributaries," U.S. EPA Report-905/4-88-001 (1988).
12. Rathke, D. E., Ed. "Lake Erie Intensive Study, 1978-1979," GLNPO-Chicago, U.S. EPA Report-905/4-84-001 (1984).
13. Morrall, J. F. *Regulation* (Nov.-Dec. 1986), pp. 25-26.
14. Calabrese, E. J., and C. E. Gilbert. In *Proceedings of the Third National Water Conference* (Philadelphia, PA: Academy of Natural Sciences, 1987), pp. 21-46.
15. Slovic, P. *Science* 236:280-85 (1987).
16. Wilson, R., and E. A. C. Crouch. *Science* 236:267-70 (1987).
17. Ames, B. N., R. Magaw, and L. S. Gold. *Science* 236:271-80 (1987).
18. Miltner, R. J., D. B. Baker, T. F. Speth, and C. A. Fronk. *Journal AWWA* 81:43-52 (1989).
19. Glotfelty, D. E., J. N. Seiber, and L. A. Liljedahl. *Nature (London)* 325:602-5 (1987).

## CHAPTER 26

An Integrated Model of Atmospheric and Aquatic Chemical Fate Useful for Guiding Regulatory Decisions: A Proposal

Paul W. Rodgers, David W. Dilks, and P. Samson

INTRODUCTION

The management of toxic substances is gaining increasing importance in maintaining human health and the quality of our environment. Threats are posed both by the vast quantities of persistent toxic chemicals already released to the environment as well as by the constant generation and release of toxic substances associated with an industrialized world. The role of toxics management is to remedy problems caused by toxic pollutants already existing in the environment and to limit the future release of toxics to levels that will not pose an unacceptable risk to environmental or human health.

Mathematical models play an important role in the management of toxic substances. First, they can be used to integrate knowledge to help better understand ecological cause and effect mechanisms. Scientific studies of a toxic pollutant can be conducted, for example, to describe partition coefficients, volatilization and degradation rates, etc.; mathematical models then combine this information to describe the relative importance of all processes on the pollutant's fate and transport. Knowledge of which processes control a pollutant's behavior helps focus future research and management control programs. Second, models can be used to evaluate remedial actions for existing toxic problems. Models predict not only the expected response of the environment to various remediation alternatives, but also detail system response time. Note that a remediation strategy that results in acceptable concentrations in one year may be more ecologically desirable than one that will take tens of years for the system to respond. The third benefit of models to toxics management is the ability to establish current loading rates that will protect the environment so that acceptable levels are not exceeded. A fourth advantage of models is that they can be used to guide monitoring activities by providing information on which parameters, locations, and frequency of sampling will provide the most useful and cost effective data to support regulatory activities.

Despite the many benefits that models of toxic fate and transport can provide, several challenges exist that potentially limit their usefulness. This chapter focuses upon two of these challenges, one specific to toxics modeling and one to modeling in general:

1. to explicitly link air quality and water quality toxic models for those cases where interdependence between the media are substantial
2. to design models to fully serve the use and interpretation needs of the resource management community

The nature of these problems is first discussed with specific research examples, followed by a proposal for a management-oriented air/water quality model for toxics. The advantages and disadvantages of a linked air/water model are discussed, and a case study site is suggested.

NATURE OF PROBLEM

Two issues directly impact the utility of models for toxics management: (1) interdependence between air quality and water quality and (2) ease of model interpretation.

Interdependence between Air and Water Quality

Models of environmental fate and transport have been historically media-specific, i.e., water quality models for water and air quality models for air. Although interdependence between the media is known to occur for many substances, interdependence has typically been handled through the use of forcing functions. A forcing function is a model input selected by the user that represents environmental conditions that affect model results but are unaffected by model behavior. A good example dates back to the classic Streeter-Phelps[1] dissolved oxygen model, where atmospheric oxygen content is a forcing function in determining dissolved oxygen concentrations in the water. This was a proper use of an atmospheric forcing function, since the oxygen content in the air above the water is reasonably constant and independent of water quality. As models of conventional pollutants (such as dissolved oxygen) have developed, the impact of bottom sediments (e.g., sediment oxygen demand, phosphorus release) on water quality were also included because they explained model behavior better. These impacts were initially treated as forcing functions, even though sediment processes were often directly related to water quality. Conventional pollutant models eventually began including mechanisms to describe sediment quality, and sediment processes in models were allowed to change to reflect changes in water quality.

Water quality models of toxics have evolved along similar pathways. The simplest models concentrated on simulating water column processes only. Although the impact of atmospheric and/or sediment processes are now recognized, toxic concentrations and/or loads outside of the water column are often treated as forcing functions or simplistically represented. The interdependence between water and sediments has been recognized for many toxics; models now often simulate sediment quality as a function of overlying water quality, instead of treating sediment impacts as a forcing function.[2-4]

Although both conventional and toxic pollutant water quality models have expanded to incorporate simulation of the bottom sediments, air quality impacts are still typically addressed with forcing functions. For example, atmospheric loads or concentrations are often simply stipulated as an estimated value. The remainder of this section will describe, both conceptually and through applied research, the interdependence between air quality and water quality models.

Figure 1 demonstrates the conceptual interdependencies that exist between air and water quality. Dry deposition of a particulate pollutant to the lake surface is determined by the concentration of particulate pollutant in the air and atmospheric mixing rates (i.e., air affecting water). Atmospheric mixing rates and wind speed are often determined, however, by the temperature gradient between water and air (i.e., water affecting air). This effect is most pronounced when air masses contact large water bodies. Wind speed affects the rate of vapor diffusion between air and water; the net pollutant flux direction and magnitude is determined by the relative concentration in the air and water (air and water affecting each other). Additionally, large lakes can also affect precipitation patterns and, therefore, the rate of wet pollutant deposition.

The real world ramifications of the interdependence between air quality and water quality can be demonstrated through the example of a PCB model for Lake Ontario.[2,3] A conceptual schematic of the model is shown in Figure 2. Here air quality is treated as a forcing function that impacts water quality in

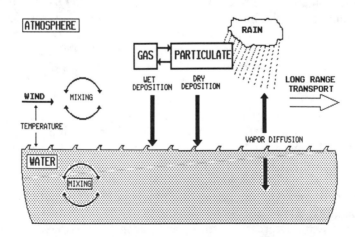

WATER / ATMOSPHERIC INTERFACE

Figure 1.   Water/atmospheric interacting physical and chemical processes.

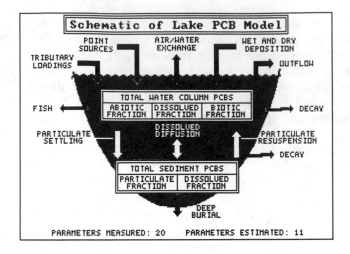

Figure 2.   Toxic model processes. *Source*: Rodgers et al.[1].

two ways: (1) deposition loading of pollutant and (2) the direction and magnitude of vapor diffusion (volatilization).

The model was run retrospectively for the period 1930–1980 to determine which processes had controlled PCB concentrations in the water and sediments. Figure 3 is a component analysis of the sources and losses that impacted the water column concentration during the 50-year simulation period. Note that tributary loading and atmospheric deposition have historically been the dominant sources of PCBs to Lake Ontario, while volatilization was the dominant loss mechanism. The air/water interface has played an important role in the historic fate of PCBs in Lake Ontario, since both the dominant loading and loss mechanisms to the lake occurred across this interface. This impact is especially noteworthy since PCBs are only moderately volatile, and this observation should have substantial influence on resource management activities.

The model also served as a projection tool to predict the response of Lake Ontario and its sediments to different pollution control strategies. During these projection simulations, the sensitivity of Lake Ontario PCB concentrations to air quality was investigated.[2,3] This was done by varying the atmospheric deposition rate and the air concentration of PCBs by ±50%. The results, shown in Figure 4, indicate that assumptions made regarding future air quality had a marked impact on the predicted water quality. The "baseline" represents expected future conditions if 1980 PCB loads were held constant in all future years. Decreasing atmospheric deposition by 50% led to an almost immediate 30% drop from baseline conditions in the predicted water concentration of PCBs. Decreasing the air concentration by 50% led to a 4% drop in water column concentrations by increasing the concentration gradient across

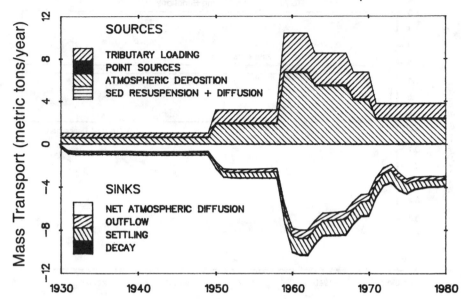

Figure 3.   Component analysis for historic simulation of PCBs in Lake Ontario. *Source*: Rodgers et al.[1].

the air-water interface, and ultimately the net volatilization rate. Presently, the air concentration is only used to compute the concentration gradient across the air-water interface. If a more explicit representation of atmospheric processes were included, as is proposed herein, a decrease in air concentration would also decrease the dry and wet deposition of pollutant, thereby having an even more profound effect on the water and sediment column. Concentration of PCBs in lake sediments also varied dramatically with changing assumptions about air quality, although sediment concentrations took significantly more time to respond (Figure 4).

These model results indicate that any projection of future lake water quality is *highly* dependent upon the assumptions regarding air quality. Water quality model predictions made in the absence of information regarding future air quality are unduly uncertain, and related toxics management decisions are correspondingly of limited value. Similarly, an air transport model investigating transboundary pollution in the Lake Ontario vicinity would benefit if the sources and losses originating from the water body were explicitly represented.

Model Interpretation

Mathematical models of air and water quality have historically been written by scientists for use by scientists, resulting in computer programs that are often difficult to apply and that generate reams of difficult to understand

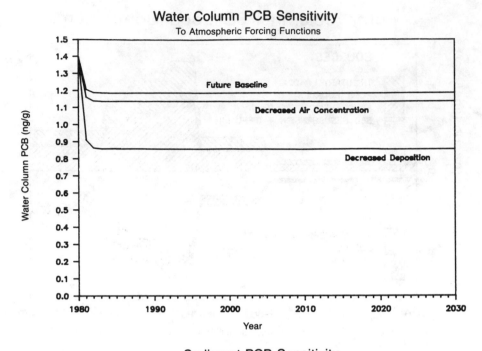

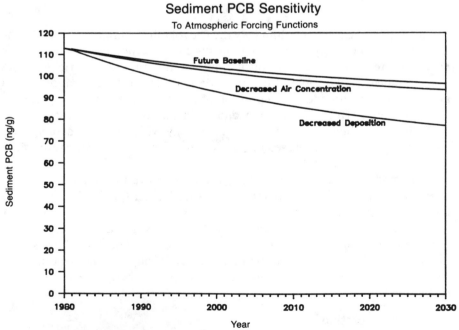

Figure 4.   Sensitivity analysis of changes in atmospheric conditions. Top: water response; bottom: sediment response.

numeric output. These difficulties have tended to insulate environmental managers from the modeling process, requiring them to depend on scientific and computer experts to evaluate and interpret the environmental response of alternative management scenarios, even after the relevant scientific representation of the problem has been developed. The net result has been that fewer sites benefit from more sophisticated assessments and less knowledgeable management decisions are often made.

Figure 5 shows a portion of an input file for a typical water quality model. The entire model input may continue in this style for several pages. This input format poses several problems for management use:

1. The meaning of the different numbers are not readily identified and can only be learned through reference to a user's manual.
2. The majority of inputs to the model are determined during the calibration or parameterization process and are of no direct interest to the environmental manager during the application phase.
3. The model has such rigid format requirements that an input value must be located in *exactly* the required columns or the program will function improperly.

Model results, consisting of pages of tabulated numbers, are often as cryptic as input files. Many of these numbers may be of little direct interest to the manager, and more insightful numbers arc often missing in the output. This type of output requires an interpretive effort from support staff to gain meaningful information from model results. Environmental management could

```
Example simulation of hydrophylic chemical              0 - TITLE
SEGS  SYS  SOL ALFA     START     STOP     STEP    INTRVL  1 - MODEL OPTIONS
   5   02    1  0.0      0.00    365.00    10.0     50.0
   4                                                        2 - DISPERSION
   1    2     48800.   42700.    80.00
   2    3     50200.   42700.    60.00
   3    4     58100.   79200.    70.00
   4    5    100000.   79200.    90.00
              SED TYPE Vol(cu m)  DEP                       3 - PHYSIOGRAPHY
   1    0      0 3.744E+08       4.18
   2    0      0 38.79E+08       7.28
   3    0      0 192.5E+08       7.17
   4    0      0 46.37E+08       6.01
   5    0      0 4.977E+08       4.11
   0                                                        4 - FLOW
   7                                                        5 - KINETICS
   MOLWT       269.8     KOC    1500.0    KOCS    150.0
    FCA         0.04     FCB       0.35    FCS      0.04
  DYNSOL        0.
   1      SOLIDS                                            6 - INITIAL COND
   .  .
   .  .
   .  .
   .  .
```

Figure 5. Example conventional model input.

function far more efficiently if managers were provided user-friendly, interactive access to models that have anticipated information needs of the decision-making process. This would provide immediate feedback about environmental response to management alternatives and would allow for more alternatives to be tested at a fraction of the time and expense currently required.

PROPOSED MANAGEMENT-ORIENTED AIR-WATER TOXICS MODEL

Environmental management of toxic pollutants would be greatly enhanced by a management-oriented model linking air-water quality. This section describes the requirements for such a model and discusses how enhanced model use and interpretation can increase its management utility. Finally, application of the model to Green Bay, Lake Michigan, is suggested as one possible case study site.

Model Requirements

Air quality and water quality models can be divided into distinct categories; only certain types of air and water models are currently amenable to linkage. From an air quality model standpoint, regional-scale transport models (Figure 6a) are not amenable to direct linkage with water quality models, primarily because the impact of water on air quality is limited over such a large terrestrial area. A linked air-water model is instead applicable to meso-scale air transport models and water quality models that interact with atmospheric processes (Figure 6b). Additionally, the air model should also be Eulerian in nature, where particle behavior is tracked deterministically, as opposed to Lagrangian, where transport is simulated based upon probabilities.

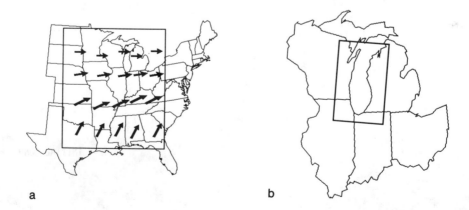

a   b

Figure 6. Geographic representations of atmospheric transport models: (a) regional-scale and (b) meso-scale.

From a water quality standpoint, the linked model should be applied to a system/pollutant where processes occurring at the air-water interface are significant. The suitability of a water quality model to linkage with an air model can be tested in the manner described previously for the Lake Ontario PCB model, i.e., an "unlinked" model can be evaluated to determine the sensitivity of water quality predictions to alternate assumptions on atmospheric processes. Sites where water quality predictions are sensitive to air quality assumptions are prime candidates for linkage to an air quality model.

Management Utility

The utility of the linked air-water quality model to guide regulatory decisions can be maximized through use of an improved model-user interface. Management oriented pre-and postprocessors, applied to a calibrated model, provide inherent communication value and are a significant aid to the decisionmaking process. The first step in this process is to develop a scientifically valid model using an interdisciplinary approach. Concurrently, model development should include a preprocessor that will allow easy access to model parameters of interest to the manager/user (e.g., loading rates and variable environmental conditions), but does not require access to inputs that may not be changed across management scenarios (e.g., rate process coefficients and physical dimensions). Well-conceived preprocessors do not preclude access to model inputs but facilitate user-specific access. Figure 7 shows an example preprocessor input screen from the Lake Ontario PCB model. The preprocessor should clearly describe each model input and allow it to be entered in free-format. This model input screen contains specific loading and environmental inputs for the historic simulation of the Lake Ontario PCB model.[2,3]

The second half of maximizing management utility is to provide a postpro-

Time-Variable Inputs		
SYr, Start of time-variable simulation; yr		1930

PCB Loadings			
Wt,	PCB tributary loadings;	g/yr	1390000
Wp,	PCB point source loadings;	g/yr	60000
Wf,	Atmospheric PCB deposition;	g/yr	2300000
Ca,	Atmospheric PCB concentration; ng/m^3		3.5

Solids Loadings			
Sa,	Abiotic SS concentration;	g/m^3	0.5
Sb,	Biotic SS concentration;	g/m^3	0.15

Hydrology and Temperature			
Q,	Outflow from lake;	m^3/yr	2.126E+011
Temp,	Water surface temperature;	°C	12

	Values
1969	2540000
1970	2540000
1971	1390000
1972	1390000
1973	1390000
1974	1390000
1975	1390000
1976	1390000
1977	1390000
1978	1390000
1979	1390000
1980	1390000

Figure 7. Example preprocessor input screen designed to aid model use and interpretation. Both steady-state and user-selected time variable inputs are easily input.

cessor to summarize and interpret model output in an easily understood format. Microcomputer graphics provide a direct means to condense large quantities of numeric output into a highly descriptive graph. Postprocessor output should not be limited to only plots of concentration over time (e.g., Figure 8), but should also include graphs designed to answer specific questions relevant to management or scientific inquiry. For instance, after reviewing the chemical fate information supplied in Figure 8, reference to the component source/loss mechanisms seen previously in Figure 3 readily distinguishes those processes important to a chemical's fate from less influential processes. In addition, the sensitivity analyses shown in Figure 4 depicted the change in model predictions caused by a unit change in model inputs. The Rodgers et al.[2] model makes available a wide range of graphical output designed to readily answer both scientific and management-oriented questions of concern.

Potential Case Study

U.S. EPA is currently sponsoring a coordinated water quality modeling study on Green Bay. Multiple researchers are working interdependently to simulate pollutant loadings, water circulation, inorganic solids concentrations, organic solids, and toxics. Many federal, state, and institutional resources are being expended in this landmark effort to develop both a regional database and a state-of-the-art water quality model. Since it is anticipated that atmospheric interaction will play an important role in this aquatic environment, atmospheric pollutant levels will be quantified and special studies have been proposed. However, the proposal advanced here for a linked air-water quality model is not a part of the planned activities. Limited resources are already allocated, but the absence of a linked air-water model may equally reflect the

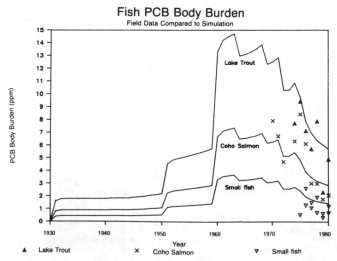

Figure 8. Example toxics model output comparing model output to available field data. *Source*: Rodgers et al.[2]

absence of recognition of its promised utility to management. Perhaps other regional, or otherwise applicable, situations will make an effort to explicitly link these two environmental compartments that so profoundly affect each other's response to toxic management alternatives.

REFERENCES

1. Streeter, H. W., and E. B. Phelps. "A Study of the Pollution and Natural Purification of the Ohio River; III. Factors Concerned in the Phenomena of Oxidation and Reaeration," U.S. Public Health Service, No. 14b (February 1925).
2. Rodgers, P. W., T. A. D. Slawecki, J. V. DePinto, and W. Booty. "LTI Toxics Model Application: PCBs in Lake Ontario—An Exploratory Application," Report to the International Joint Commission, Toxic Substances Committee (February 1987).
3. "Report on Modeling the Loading Concentration Relationship for Critical Pollutants in the Great Lakes," International Joint Commission (October 1988).
4. Thomann, R. V., and D. M. DiToro. "Physicochemical Model of Toxic Substances in the Great Lakes," *Journal of Great Lakes Research* 9(4):474–96 (1983).

## CHAPTER 27

Beyond Long-Range Transport: A Model of a Global Gas Chromatographic System

R. W. Risebrough

INTRODUCTION

This symposium on "Long-Range Transport of Pesticides and Other Toxics" in Toronto followed by 20 years to the week a symposium on "Chemical Fallout" on the other side of Lake Ontario, in Rochester, New York.[1] The Rochester meeting marked a major shift in our thinking about the movements into and through the environment of DDT, dieldrin, related biocides, and the PCBs. Largely because of highly publicized fish kills, rivers had previously been considered the most important transport medium of the chlorinated hydrocarbon biocides. Papers presented in Rochester, and the discussions that accompanied them, clearly established the importance of atmospheric transport in the global movements of these synthetic compounds. In my own contribution to the Rochester symposium,[2] I reported results of a study of the distribution of DDT compounds in coastal and marine fish in California. Levels turned out to be lower in fish from San Francisco Bay, which receives drainage waters from the vast agricultural areas in the Sacramento and San Joaquin valleys, than in fish of the same species from coastal waters. River transport was clearly of minor importance in contributing to the levels of DDT compounds in the local coastal and marine ecosystems.

Reports in the mid-1960s of DDT compounds in antarctic wildlife had provided the first dramatic example of long-range transport.[3,4] In our own work at that time, we detected DDT compounds in dust known to have crossed the Atlantic and had made a crude estimate of input into the Atlantic between the West African coast and the Caribbean.[5] During the intervening 20 years, many additional examples have appeared in the literature, including several contributions to this symposium, specifically devoted to the theme of long-range transport. Although many of these compounds have low vapor pressures, the magnitude of fluxes into the atmosphere at sites of application and use is sufficiently large to produce global distributions.

In this chapter I suggest that focusing on the transport, through the atmosphere, of a synthetic organic compound from point A to distant point B is missing a broader picture, since from point B the compound may move on to points C, D, or E, and, given the appropriate conditions, even back to point A.

Thus, toxaphene has been shown to move from the southern United States, where it was extensively used in the past, to the Great Lakes, where fisheries are now contaminated.[6] Moreover, toxaphene has become a global contaminant; it is present in the arctic atmosphere,[7] with the HCH compounds it is the most abundant chlorinated hydrocarbon in arctic fish,[8,9] and with the α-isomer of HCH it is the most abundant chlorinated hydrocarbon in surface waters of the eastern Pacific.[10,11] Although use of toxaphene in North America north of Mexico has ended, use continues elsewhere in the world. Prediction of the future levels of toxaphene in Great Lakes fish will therefore require a model that goes beyond a description of transport from a point source A, such as the soils of cotton fields in the southern United States, to a distant site B, such as the waters of the Great Lakes.

The earlier concept of "fallout" of organochlorine compounds had been directly derived from the observations of "fallout" of radioactive isotopes produced by thermonuclear explosions, although these invariably passed through the stratosphere in the process of global dispersal and the movements of the organochlorines were considered to be confined to the troposphere. Transport was considered to be essentially a one-way process. This was true enough for the radioactive isotopes, since fluxes back into the atmosphere were negligible. A model based on a one-way transport process would clearly not describe the present global distribution of the chlorinated hydrocarbons and other nonpolar organic compounds of medium molecular weight.

A theoretical basis for a more comprehensive model is well established. Over recent years we have been enlightened by a series of papers by Professor Mackay and his students at the University of Toronto.[12-22] Transfer, or the passage of these compounds from one phase to another, is a two-way process; the net flux between water and air, soil and air, and vegetation and air is determined by the differences in chemical potential, or fugacity. Chemical potentials are evidently highest at source points, with the direction of net flux usually into the atmosphere, but locally the direction of the net flux could be either into or out of the atmosphere, depending on the differences in fugacities between two phases.

The model that emerges is that of a global gas chromatographic system. The obvious analogy is with the gas chromatographic column, in which the molecules of a compound such as p,p'-DDE move many times between gas and liquid phases during passage through the column in maintaining the equilibrium in chemical potential between the two phases. The environmental compounds we are considering, many of which are organochlorines of medium molecular weight, move initially into the atmosphere from source points, and thereafter move in and out of waters, soils, and vegetation, with the directions of the net fluxes being determined by the local relative magnitude of fugacities.

Because of its simplicity, the basic features of the model require no equations for its description; its basic premise is well supported by theory and the direction of fluxes can be predicted. The velocity of transfer, however, would be determined by resistance factors that would depend on many variables. Consideration of these variables is beyond the scope of this chapter, although they would be the focus of continuing research. I shall briefly review some of the literature on movements and distributions between phases, including papers presented at this symposium, looking for data sets that might appear to be inconsistent with the model—inconsistencies that might derive from a significant resistance to transfer from one phase to another in spite of a large difference in fugacity. Under such a circumstance, transfer between phases, i.e., atmospheric transport to marine systems, could essentially be a one-way process.

The compounds considered here are principally the isomers of hexachlorocyclohexane, hexachlorobenzene (HCB), the DDT and chlordane compounds, heptachlor and heptachlor epoxide, dieldrin, endrin, mirex, the chlorinated biphenyls and terphenyls, the chlorinated dibenzodioxins, and the chlorinated dibenzofurans. Other relatively nonpolar compounds of medium molecular weight, i.e., some of the polynuclear aromatics and phthalates, would be expected to behave similarly.

TRANSFER PROCESSES

Not included in this brief review are transfer processes to and from aquatic biota since these are considered as components of the water systems. Similarly, terrestrial food webs are not included, in spite of their importance in the transfer of residues from vegetation to soils, water systems, and the atmosphere. Sediments are considered as a component of the water systems. Except when or where deposition is sufficiently rapid to bury residues at depths where incorporation into food webs or transfer back to the water system is no longer possible, sediments appear to be more appropriately considered as reservoirs rather than sinks. Residues are accumulated from sediments by benthic organisms, including fish,[23-28] and thereby reenter food webs. In the Southern California Bight, input of DDT compounds into the water column from sediments is sufficient to account for continuing high levels of p,p'-DDE in the area, and the failure of residues to decline since the mid-1970s.[29] Nevertheless, in the present version of the model, the sediments are not considered a separate environmental component, but rather a part of the water system.

Air to Water

High levels of several organochlorines in marine wildlife in the 1960s and 1970s prompted concern about contamination of the world's oceans, which lead to symposia such as "Marine Pollutant Transfer"[30] and a report of the National Academy of Sciences, "The Tropospheric Transport of Pollutants and Other Substances to the Oceans."[31] Atmospheric transport of organochlorines to the world's oceans was convincingly demonstrated in a series of papers.[32-37] Experimentally, dry deposition from the atmosphere was demonstrated to both dry and wet surfaces,[38-40] and measured concentrations in rain indicated significant transfer in precipitation.[35,41,42]

Does, however, the measured deposition, wet or dry, to a surface, wet or dry—a determination of a one-way flux—provide an acceptable estimate of net flux in most circumstances? Clearly, this would be the case only if the flux back to the atmosphere were negligible.

Water to Air

The transfer of the organochlorines we are considering here across an air-water interface has been addressed in a number of theoretical papers.[12,13,22,43-48] Reported measurements of PCBs in the atmosphere above Lake Michigan and of PCBs in the water led Murphy et al.[49,50] to the conclusion that the atmosphere rather than the sediments was the current sink for PCBs in this system. In the model developed by Mackay[22] for Lake Ontario, the net flux is from the water to the air by 320 kg/year.

Experimentally, transfer of PCBs from sediment to water and from the water to air has been demonstrated in a model ecosystem.[51,52] The rate of transfer, however, was slow; more than 90% of the PCBs originally applied to the sediment were still present in the sediments after 1.5 years.

Both theory and the few available data, therefore, are consistent with a global chromatographic model of air-water exchange of organochlorines; needed are measurements in an appropriate setting that would determine the kinetics of exchange. One such setting is the western Mediterranean, where the mistral winds from Europe are frequently followed by the sirocco winds from the Sahara. How fast does the content and composition of the organochlorines in surface waters change in response to changes in the levels of the organochlorines in the air mass?

Since temperature is an important factor in the kinetics equations, do the polar regions serve as a "cold-trap" for organochlorines and similar compounds that enter the environment in the temperate and tropical areas?

Air to Soil and Soil to Air

The transfer of organochlorines through dry deposition and in precipitation from air to soil is well documented. There is also an abundant literature on the volatilization of organochlorine pesticides from soils to the atmosphere.[53-60] Both the data of these papers and theory[61] are consistent with the chromatographic model.

Air to Vegetation

Pine needles from the vicinity of Stockholm were among the first environmental samples in which PCBs were detected.[62] Partitioning between the vapor phase and vegetation, particularly leaves with waxy cuticles and leaves or bark with a high resin content, might be expected to result in the sequestration of significant quantities of the chlorinated hydrocarbons present in ambient air. Until recently, relatively few measurements have been made. Monitoring of leafy vegetables, fruit, and dairy products to protect the food supply has consistently detected low levels of the organochlorines,[63] but generally these have been attributed to input from local agricultural soils rather than ambient air.

Levels of several organochlorines, including α-and γ-HCH, p,p'-DDE, p,p'-DDD, o,p'-DDT, p,p'-DDT, HCB, PCBs, and toxaphene, have been reported in pine needles and other foliage in Italy,[64,65] lichens in the south of France,[66] leaves of mango trees in five countries of West Africa,[67] lichens and mosses of the Antarctic Peninsula,[68] lichens in northern Sweden,[69] lichens and mosses in Norway,[70] tree bark,[71] in foliage of a variety of species,[72] in needles of spruce trees, *Picea abies*,[73,74] and in *Sphagnum* sp. of peat bogs in North America.[75] Chlorinated dioxins and dibenzofurans have also been detected in spruce needles.[76] Organochlorines were found to penetrate the spruce needles, not remaining in the outer cuticular parts.[73]

Rapaport and Eisenreich[75] concluded that precipitation was the dominant deposition pathway for the organochlorines accumulated over depth in the peat bogs, in part because of observations by other researchers that deposition is dominated by scavenging of aerosols and gases by rainfall, rather than by dry deposition from either the vapor or aerosol phases.[35,41,42,45] As discussed above, however, determinations of fluxes from the atmosphere do not provide estimates of the net fluxes; correlations between levels recorded in plants and in the atmosphere,[66,77] and consistent differences among species[72] indicate that an equilibrium is reached such that over longer periods of time the levels recorded in plants would be independent of precipitation inputs.

The extent to which chlorinated hydrocarbons reach foliage through the roots and stems is minor in relation to uptake from the ambient atmosphere.[72,78-80]

Vegetation to Air

Organochlorine pesticides applied to leaf surfaces are rapidly lost through volatilization,[58,59] but the data suggesting that an equilibrium prevails between levels of an organochlorine in the ambient atmosphere and in a plant[66,77] would also indicate that levels in the plant would decrease with a decrease in atmospheric levels. Travis and Hattemer-Frey[77] have reviewed available data in terms of relationships among the bioconcentration factors between air and plant, octanol-water partition coefficients, and Henry's law constants. For the very few data so far available, a linear relationship was found between the log of the air-to-plant bioconcentration factor and the log of the octanol-water partition coefficient. A positive linear relationship was also found between the logarithm of the product of the Henry's law constant and the bioconcentration factor with the logarithm of the octanol-water coefficient. Buckley[72] found consistent differences in accumulation of PCBs from one species to another, most likely a function of wax content and other chemical parameters. Such relationships support a model in which residues are lost or acquired by the plants in response to changes in ambient levels in the atmosphere.

It is likely, however, that organochlorines in inner bark layers are not exchangeable with the atmosphere; these layers may be considered a temporary sink for these compounds. Media such as the subsurface layers in peat bogs are a more permanent sink and provide a history of ambient levels over time.[75]

DISCUSSION

I have not been able to find any data, either those presented at this symposium or in the literature, that are not compatible with a gas chromatographic model of global circulation of the organochlorines. Some of the transfer processes, however, are evidently slow. In their model ecosystems, in which PCBs had been added to the sediments, Larsson and Sodergren[52] estimated that only a small fraction of the PCBs added to the system had entered the atmosphere over the 1.5 years of the experiment, although measured levels of PCBs in the air over the mesocosms were up to 2 orders of magnitude higher than background levels. Such processes are evidently highly dependent on temperature, wind speed, turbulence, etc., but the small amount of transfer would suggest that the water-to-air process can be slow under certain conditions.

The experiment of Larsson and Sodergren[52] demonstrated that the existence of another reservoir in the mesocosm, the lipids of fish, increased the flux from the sediments to the water, without, however, increasing the levels in the water over the control. Similarly, the presence of a major reservoir in local vegetation could affect the net fluxes from either soil or water to the atmosphere.

Several papers presented at this symposium reported high levels of the α-isomer of hexachlorocyclohexane in precipitation of southern Canada,[81] and in arctic air, snow, and seawater.[7,82] In many samples it was the most abundant chlorinated hydrocarbon. Since the HCH compounds are no longer used in North America, current sources are in other areas of the world, particularly Asia. α-HCH would appear to be an ideal model compound in continuing investigations.

Is it then justified to describe movements through the environment in terms of "residence times" since this concept assumes a one-way transfer between phases? Only, it would seem, if the chemical potential in one component of a local ecosystem is much larger than in other phases, i.e., after application of an organochlorine pesticide to vegetation or soil. How then are variations in measured concentrations best explained? A number of calculations of residence times in the atmosphere have been based on the approach developed by Junge,[44,83,84] who used variations in the measured concentrations in the atmosphere to estimate residence times in that phase. A detailed review of the existing data sets, and collection of additional data, would appear desirable to determine how much of the variation can be attributed to recent inputs, and how much to local partitioning into other phases.

At this time the modeling efforts are much further advanced than are the measurements of the organochlorines in water systems. These are still nontrivial, with problems of both contamination and recovery.[10] The creation of data sets that include measurements in both air and water of the compounds we are considering would appear to be the highest priority in our immediate efforts. Studies of the kinetics of transfer processes, and determination of the principal factors that affect rates of transfer would appear to be at the next level of priority in increasing our understanding of the global movements and distributions of these ubiquitous compounds.

ACKNOWLEDGMENTS

The Bodega Bay Institute provided the support for the preparation of this chapter.

REFERENCES

1. Miller, M. W., and G. G. Berg, Eds. *Chemical Fallout* (Springfield, IL: Charles C. Thomas, 1969).
2. Risebrough, R. W. In *Chemical Fallout,* M. W. Miller and G. G. Berg, Eds. (Springfield, IL: Charles C. Thomas, 1969), pp. 5–23.
3. Sladen, W. L. J., C. M. Menzie, and W. L. Reichel. *Nature* 210:670–73 (1966).
4. George, J. L., and D. E. H. Frear. *J. Appl. Ecol.* 3(suppl.):155–67 (1966).
5. Risebrough, R. W., R. J. Huggett, J. J. Griffin, and E. D. Goldberg. *Science* 159:1233–36 (1968).
6. Bidleman, T. F., M. T. Zaranski, and M. D. Walla. In *Chronic Effects of Toxic Contaminants in Large Lakes,* N. W. Schmidtke, Ed. (Chelsea, MI: Lewis Publishers, Inc., 1988), pp. 257–84.
7. Chapter 23.
8. Muir, D. C. G., R. J. Norstrom, and M. Simon. *Environ. Sci. Technol.* 22:1071–79 (1988).
9. Chapter 22.
10. de Lappe, B. W., R. W. Risebrough, and W. Walker II. *Can. J. Fish. Aquat. Sci.* 40(suppl. 2):322–36 (1983).
11. Risebrough, R. W., and B. W. de Lappe. Unpublished observations.
12. Mackay, D., and A. W. Wolkoff. *Environ. Sci. Technol.* 7:611–14 (1973).
13. Mackay, D., and P. J. Leinonen. *Environ. Sci. Technol.* 9:1178–80 (1975).

14. Cohen, Y., W. Cocchio, and D. Mackay. *Environ. Sci. Technol.* 12:553–58 (1978).
15. Mackay, D. *Environ. Sci. Technol.* 13:1218–23 (1979).
16. Mackay, D. In *The Handbook of Environmental Chemistry,* O. Hutzinger, Ed., Vol. II, Part A (New York, NY: Springer-Verlag, 1980), pp. 31–45.
17. Mackay, D., A. Bobra, W. Y. Shiu, and S. M. Yalkowsky. *Chemosphere* 9:701–11 (1980).
18. Mackay, D., and S. Paterson. *Environ. Sci. Technol.* 15:1006–14 (1981).
19. Mackay, D., and A. T. K. Yeun. *Environ. Sci. Technol.* 17:211–17 (1983).
20. Mackay, D., W. Y. Shiu, and E. Chau. *Can. J. Fish. Aquat. Sci.* 40(suppl. 2):295–303 (1983).
21. Mackay, D., S. Paterson, and W. H. Schroeder. *Environ. Sci. Technol.* 20:810–16 (1986).
22. Chapter 21.
23. Young, D.R., and D. McDermott-Ehrlich. In *Coastal Water Research Project Annual Report, 1976,* (El Segundo, CA: Southern California Coastal Water Research Project, 1976), pp. 49–55.
24. Elder, D. L., S. W. Fowler, and G. G. Polikarpov. *Bull. Environ. Contam. Toxicol.* 21:448–52 (1979).
25. Connor, M. S. *Environ. Sci. Technol.* 18:31–35 (1984).
26. Karickhoff, S. W., and K. R. Morris. *Environ. Sci. Technol.* 19:51–56 (1985).
27. Varanasi, U., W. L. Reichert, J. E. Stein, D. W. Brown, and H. R. Sanborn. *Environ. Sci. Technol.* 19:836–41 (1985).
28. Officer, C. B., and D. R. Lynch. *Estuar. Coast. Shelf Sci.* 28:1–12 (1989).
29. Risebrough, R. W. *Distribution of Organic Contaminants in Coastal Areas of Los Angeles and the Southern California Bight* (Berkeley, CA: Bodega Bay Institute, 1987).
30. Windom, H. L., and R. A. Duce, Eds. *Marine Pollutant Transfer* (Lexington, MA: D.C. Heath and Company, 1976).
31. *The Tropospheric Transport of Pollutants and Other Substances to the Oceans,* Ocean Sciences Board (Washington, DC: National Academy of Sciences, 1978).
32. Bidleman, T. F., and C. E. Olney. *Science* 183:516–18 (1974).
33. Bidleman, T. F., and C. E. Olney. *Nature* 257:475–77 (1975).
34. Bidleman, T. F., C. P. Rice, and C. E. Olney. In *Marine Pollutant Transfer,* H. L. Windom and R. A. Duce, Eds. (Lexington, MA: D.C. Heath and Company, 1976), pp. 323–52.
35. Bidleman, T. F., and E. J. Christensen. *J. Geophys Res.* 84:7857–62 (1979).
36. Bidleman, T. F., E. J. Christensen, W. N. Billings, and R. J. Leonard. *Mar. Res.* 39:443–64 (1981).
37. Bidleman, T. F., and R. Leonard. *Atmos. Environ.* 16:1099–1107 (1982).
38. Young, D. R., D. J. McDermott, and T. C. Heesen. *Bull. Environ. Contam. Toxicol.* 16:604–11 (1976).
39. Christensen, E. J., C. E. Olney, and T. F. Bidleman. *Bull. Environ. Contam. Toxicol.* 23:196–202 (1979).
40. Villeneuve, J.-P, and C. Cattini. *Chemosphere* 15:115–20 (1986).
41. Murphy, T. J., and C. P. Rzeszutko. *J. Great Lakes Res.* 3:305–12 (1977).
42. Harder, H. W, E. C. Christensen, J. R. Matthews, and T. F. Bidleman. *Estuaries* 3:142–47 (1980).
43. Liss, P. S., and P. G. Slater. *Nature* 247:181–84 (1974).
44. Junge, C. E. In *Fate of Pollutants in the Air and Water Environments,* I. H.

Suffet, Ed., Advances in Environmental Science and Technology Series (New York, NY: Wiley-Interscience, 1977), pp. 7-25.

45. Eisenreich, S. J., B. B. Looney, and J. D. Thornton. *Environ. Sci. Technol.* 15:30-38 (1981).

46. Doskey, P. V., and A. W. Andren. *Environ. Sci. Technol.* 15:705-11 (1981).

47. Bopp, R. F. *J. Geophys. Res.* 88:2521-29 (1983).

48. Smith, J. H., D. Mackay, and C. W. K. Ng. *Residue Reviews* 85:73-88 (1983).

49. Murphy, T. J., L. J. Formanski, B. Bronawell, and J. A. Myers. *Environ. Sci. Technol.* 19:942-46 (1985).

50. Murphy, T. J., M. D. Mullin, and J. A. Meyer. *Environ. Sci. Technol.* 21:155-62 (1987).

51. Larsson, P. *Nature* 317:347-49 (1985).

52. Larsson, P., and S. Sodergren. *Water Air Soil Poll.* 36:33-46 (1987).

53. Caro, J. H., and A. W. Taylor. *J. Agric. Food Chem.* 19:379-84 (1971).

54. Spencer, W. F., and M. M. Cliath. *J. Agric. Food Chem.* 20:645-49 (1972).

55. Spencer, W. F., M. M. Cliath, W. J. Farmer, and R. A. Shepherd. *J. Environ. Quality* 3:126-29 (1974).

56. Spencer, W. F. *Residue Reviews* 59:91-117 (1975).

57. Spencer, W. F., and M. M. Cliath. *Residue Reviews* 85:57-71 (1983).

58. Seiber, J. N., S. C. Madden, M. M. McChesney, and W. L. Winterlin. *J. Agric. Food Chem.* 27:284-90 (1979).

59. Willis, G. H., L. L. McDowell, L. A. Harper, L. M. Southwick, and S. Smith. *J. Environ. Qual.* 12:80-85 (1983).

60. Sleicher, C. A., and J. Hopcraft. *Environ. Sci. Technol.* 18:514-18 (1984).

61. Chapter 20.

62. Jensen, S. *Ambio* 1:123-31 (1972).

63. Davies, K. In *Chronic Effects of Contaminants in Large Lakes,* N. W. Schmidtke, Ed., Vol. I (Chelsea, MI: Lewis Publishers, Inc., 1988), pp. 195-226.

64. Gaggi, C., and E. Bacci. *Chemosphere* 14:451-56 (1985).

65. Gaggi, C., E. Bacci, D. Calamari, and R. Fanelli. *Chemosphere* 14:1673-86 (1985).

66. Villeneuve, J.-P., E. Fogelqvist, and C. Cattini. *Chemosphere* 17:399-403 (1988).

67. Bacci, E., D. Calamari, C. Gaggi, C. Biney, S. Focardi, and M. Morosoni. *Chemosphere* 17:693-702 (1988).

68. Bacci, E., D. Calamari, C. Gaggi, R. Fanelli, S. Focardi, and M. Morosini. *Chemosphere* 15:747-54 (1986).

69. Villeneuve, J.-P., E. Holm, and C. Cattini. *Chemosphere* 14:1651-58 (1985).

70. Carlberg, G. E., E. B. Ofstad, H. Drangsholt, and E. Steinnes. *Chemosphere* 12:341-56 (1983).

71. Meredith, M. L., and R. A. Hites. *Environ. Sci. Technol.* 21:709-12 (1987).

72. Buckley, E. H. *Science* 216:520-22 (1982).

73. Reischl, A., M. Reissinger, and O. Hutzinger. *Chemosphere* 16:2647-52 (1987).

74. Schramm, K.-W., A. Reischl, and O. Hutzinger. *Chemosphere* 16:2653-63 (1987).

75. Rapaport, R. A., and S. J. Eisenreich. *Atmos. Environ.* 20:2367-79 (1986).

76. Reischl, A., H. Thoma, M. Reissinger, and O. Hutzinger. *Naturwiss.* 74:88 (1987).

77. Travis, C. C., and H. A. Hattemer-Frey. *Chemosphere* 17:277-83 (1988).

78. Nash, R. G., and M. L. Beall. *Science* 168:1109-11 (1970).

79. Bacci, E., and C. Gaggi. *Bull. Environ. Contam. Toxicol.* 35:673-81 (1985).

80. Bacci, E., and C. Gaggi. *Bull. Environ. Contam. Toxicol.* 37:850–87 (1986).
81. Strachan, W. M. J. "Atmospheric Deposition of Selected Organochlorine Com-
    pounds in Canada," in D. A. Kurtz, Ed., *Long Range Transport of Pesticides*
    (Chelsea, MI: Lewis Publishers, Inc., 1990).
82. Gregor, D. J. "Deposition and Accumulation of Selected Agricultural Pesticides in
    Canadian Arctic Snow," in D. A. Kurtz, Ed., *Long Range Transport of Pesticides*
    (Chelsea, MI: Lewis Publishers, Inc., 1990).
83. Junge, C. E. *Tellus* 26:477–88 (1974).
84. Junge, C. E. *J. Pure Appl. Chem.* 42:95–104 (1975).

List of Contributors

Roger Atkinson, Statewide Air Pollution Research Center, University of California, Riverside, CA 92521

Elliot L. Atlas, NCAR, P.O. Box 3000, Boulder, CO 80307-3000

David B. Baker, Water Quality Laboratory, Heidelberg College, Tiffin, OH 44883

T. F. Bidleman, Department of Chemistry, Marine Science Program, Belle W. Baruch Institute for Marine Biology and Coastal Research, University of South Carolina, Columbia, SC 29208

Alan C. Buchanan, Missouri Department of Conservation, Columbia, MO 65201

Jane Bush, Missouri Department of Conservation, Columbia, MO 65201

L. D. Clendening, Chevron Oil Field Research Company, P.O. Box 446, La Habra, CA 90633-0446

T. E. Clevenger, Environmental Trace Substances Research Center, University of Missouri, Columbia, MO 65203

M. M. Cliath, USDA-Agricultural Research Service, Department of Soil and Environmental Sciences, University of California, Riverside, CA 92521

R. J. Cooper, Department of Plant and Soil Sciences, University of Massachusetts, Amherst, MA 01003

W. E. Cotham, Department of Chemistry, University of South Carolina, Columbia, SC 29208

A. S. Curtis, Massachusetts Pesticide Analysis Laboratory, Department of Entomology, University of Massachusetts, Amherst, MA 01003

James Czarnezki, Missouri Department of Conservation, Columbia, MO 65201

David W. Dilks, LTI, Limno-Tech, Inc., 2395 Huron Parkway, Ann Arbor, MI 48104

Steven J. Eisenreich, Department of Civil and Mineral Engineering, University of Minnesota, 103 Experimental Engineering Building, Minneapolis, MN 55455

F. F. Ernst, Department of Soil and Environmental Sciences, University of California, Riverside, CA 92521

C. A. Ford, Department of Fisheries and Oceans, Freshwater Institute, Winnipeg, MB R3T 2N6

H. P. Freeman, USDA-ARS, Building 001, Room 111, BARC-West, Beltsville, MD 20705

D. E. Glotfelty (Deceased; formerly, USDA-ARS, Building 001, Room 111, BARC-West, Beltsville, MD 20705)

Dennis J. Gregor, National Water Research Institute, Environment Canada, P.O. Box 5050, Burlington, Ont. L7R 4A6

N. P. Grift, Department of Fisheries and Oceans, Freshwater Institute, Winnipeg, MB R3T 2N6

B. T. Hargrave, Department of Fisheries and Oceans, Bedford Institute of Oceanography, Dartmouth, N.S., B2Y 4A2

M. R. Hendzel, Department of Fisheries and Oceans, Freshwater Institute, Winnipeg, MB R3T 2N6

Bernard D. Hill, Research Station, Agriculture Canada Research Branch, Lethbridge, Alberta T1J 4B1

D. A. Hinckley, Ebasco Services, Inc., 10900 N.E. Eighth Street, Bellevue, WA 98004

J. J. Jenkins, Massachusetts Pesticide Analysis Laboratory, Department of Entomology, University of Massachusetts, Amherst, MA 01003

N. Douglas Johnson, ORTECH International, Sheridan Park Research Community, 2395 Speakman Drive, Mississauga, Ont., L5K 1B3

W. A. Jury, Department of Soil and Environmental Sciences, University of California, Riverside, CA 92521

Narayanan Kannan, Department of Environment Conservation, Ehime University, Tarumi 3-5-7, Matsuyama 790, Japan

S. Kapila, Environmental Trace Substances Research Center, University of Missouri, Columbia, MO 65203

Masahide Kawano, Department of Environment Conservation, Ehime University, Tarumi 3-5-7, Matsuyama 790, Japan

David A. Kurtz, Pesticide Research Laboratory, Department of Entomology, Pennsylvania State University, University Park, PA 16802

Douglas A. Lane, Atmospheric Environment Service, Downsview, Ont., M3H 5T4

M. M. Leech, USDA-ARS, Building 001, Room 111, BARC-West, Beltsville, MD 20705

Hiram Levy II, Geophysical Fluid Dynamics Laboratory/NOAA, P.O. Box 308, Princeton University, Princeton, NJ 08542

W. L. Lockhart, Department of Fisheries and Oceans, Freshwater Institute, Winnipeg, MB R3T 2N6

Donald Mackay, Department of Chemical Engineering and Applied Chemistry, University of Toronto, Toronto, Canada M5S 1A4

Michael M. McChesney, Department of Environmental Toxicology, University of California, Davis, CA 95616

Kathleen E. McGrath, Missouri Department of Conservation, Columbia, MO 65201

R. C. Montone, Instituto Oceanografico da USP, 05508 São Paulo, Brazil

Derek C. G. Muir, Department of Fisheries and Oceans, Freshwater Institute, 501 University Crescent, Winnipeg, MB R3T 2N6

Ralph G. Nash, Bio-Analytical Services, Inc., Box 1708, Decatur, IL 62525

Carl E. Orazio, Environmental Trace Substances Research Center, University of Missouri, Columbia, MO 65203

James F. Pankow, Department of Environmental Science and Engineering, Oregon Graduate Institute, 19600 N.W. Von Neumann Drive, Beaverton, OR 97006

G. W. Patton, Battelle Pacific Northwest Laboratories, MSK613 Battelle Blvd., P.O. Box 999, Richland, WA 99352

Ravi K. Puri, Environmental Trace Substances Research Center, University of Missouri, Columbia, MO 65203

A. W. Reiger, Department of Fisheries and Oceans, Freshwater Institute, Winnipeg, MB R3T 2N6

R. Peter Richards, Water Quality Laboratory, Heidelberg College, Tiffin, OH 44883

R. W. Risebrough, Institute of Marine Sciences, University of California, Santa Cruz, CA 95064

Paul W. Rodgers, LTI, Limno-Tech, Inc., 2395 Huron Parkway, Ann Arbor, MI 48104

P. Samson, Department of Atmospheric and Oceanic Research, 2213 Space Research, University of Michigan, Ann Arbor, MI 48109

S. Schauffler, Department of Oceanography, Texas A&M University, College Station, TX 77843

William H. Schroeder, Environment Canada, Atmospheric Environment Service, 4905 Dufferin Street, Downsview, Ont. M3H 5T4

James N. Seiber, Department of Environmental Toxicology, University of California, Davis, CA 95616

W. F. Spencer, USDA-Agricultural Research Service, Department of Soil and Environmental Sciences, University of California, Riverside, CA 92521

Warren Stiver, Department of Chemical Engineering and Applied Chemistry, University of Toronto, Toronto, Canada M5S 1A4

William M. J. Strachan, National Water Research Institute, Canada Centre for Inland Waters, P.O. Box 5050, Burlington, Ont. L7R 4A6

Shinsuke Tanabe, Department of Environment Conservation, Ehime University, Tarumi 3-5-7, Matsuyama 790, Japan

Ryo Tatsukawa, Department of Environment Conservation, Ehime University, Tarumi 3-5-7, Matsuyama 790, Japan

E. C. Voldner, Environment Canada, Atmospheric Environment Service, 4905 Dufferin Street, Downsview, Ont. M3H 5T4

M. D. Walla, Department of Chemistry, University of South Carolina, Columbia, SC 29208

R. R. Weber, Instituto Oceanografico da USP, 05508 São Paulo, Brazil

G. H. Williams, USDA-ARS, Building 001, Room 111, BARC-West, Beltsville, MD 20705

Arthur M. Winer, Environmental Science and Engineering Program, Department of Environmental Health Sciences, School of Public Health, University of California, Los Angeles, CA 90024

James E. Woodrow, Department of Environmental Toxicology, University of California, Davis, CA 95616

Yukihiko Yamaguchi, Department of Environment Conservation, Ehime University, Tarumi 3-5-7, Matsuyama 790, Japan

A. F. Yanders, Environmental Trace Substances Research Center, University of Missouri, Columbia, MO 65203

Index